Progress in Probability
Volume 27

Diffusion Processes and Related Problems in Analysis, Volume II

Stochastic Flows

Mark A. Pinsky
Volker Wihstutz
Editors

Springer Science+Business Media, LLC

Mark A. Pinsky
Department of Mathematics
Northwestern University
Evanston, IL 60208

Volker Wihstutz
Department of Mathematics
University of North Carolina
Charlotte, N.C. 28223

Library of Congress Cataloging-in-Publication Data

Diffusion processes and related problems in analysis.

(Progress in probability ; 22, 27)
Papers presented at an international conference held October 23-27, 1989 and sponsored by the Mathematics Department of Northwestern University.
Vol. 2- edited by Mark Pinsky, Volker Wihstutz.
Includes bibliographical references.
Contents: v. 1. Diffusions in analysis and geometry -- v. 2. Stochastic flows.
1. Diffusion processes--Congresses. 2. Stochastic analysis--Congresses. I. Pinsky, Mark A., 1940-
II. Wihstutz, V. (Volker), 1940- . III. Northwestern University (Evanston, Ill.). Dept. of Mathematics.
IV. Series: Progress in probability ; 22, etc.
QA274.75D54 1991 519.2 90-47081
ISBN 978-0-8176-3543-5 ISBN 978-1-4612-0389-6 (eBook) CIP
DOI 10.1007/978-1-4612-0389-6

Printed on acid-free paper.

ISBN 978-0-8176-3543-5

Camera-ready text prepared by the Authors.

9 8 7 6 5 4 3 2 1

CONTENTS

Part IV: Invariant Measures in Real and White Noise-Driven Systems

Part V: Iterated Function Systems

PREFACE

During the weekend of March 16–18, 1990 the University of North Carolina at Charlotte hosted a conference on the subject of stochastic flows, as part of a Special Activity Month in the Department of Mathematics. This conference was supported jointly by a National Science Foundation grant and by the University of North Carolina at Charlotte.

Originally conceived as a regional conference for researchers in the Southeastern United States, the conference eventually drew participation from both coasts of the U.S. and from abroad. This broad-based participation reflects a growing interest in the viewpoint of stochastic flows, particularly in probability theory and more generally in mathematics as a whole.

While the theory of deterministic flows can be considered classical, the stochastic counterpart has only been developed in the past decade, through the efforts of Harris, Kunita, Elworthy, Baxendale and others. Much of this work was done in close connection with the theory of diffusion processes, where dynamical systems implicitly enter probability theory by means of stochastic differential equations. In this regard, the Charlotte conference served as a natural outgrowth of the Conference on Diffusion Processes, held at Northwestern University, Evanston Illinois in October 1989, the proceedings of which has now been published as Volume I of the current series. Due to this natural flow of ideas, and with the assistance and support of the Editorial Board, it was decided to organize the present two-volume effort.

A stochastic flow is a stochastic process $C(t,\omega)$ taking values in the group of diffeomorphisms of a differentiable manifold E, whether that be a Euclidean space, a non-flat Riemannian manifold or even a more general state space. In order to capture the essential features of the dynamics, it is crucial to study the associated flow F_t on the underlying probability space. These come together in the *cocycle property* or *semigroup property*, which is written in the form $C(t+s,\omega) = C(t, F_s\omega) \circ C(s,\omega)$ and through the *skew product flow* $(C(t,\omega)(\cdot), F_t)$ on $E \times \Omega$. In particular cases the stochastic flow could be generated by solving a stochastic differential equation driven by Brownian motion or a more general local martingale (generically called *white noise-driven stochastic flows*); in other cases it may be generated by a random family of ordinary differential equations driven by the trajectories of a stationary stochastic process (generically called *real noise-driven stochastic flows*). The latter model is closely related to systems of ordinary differential equations which contain deterministic control functions, familiar in the theory of control. In another direction we may have a stochastic flow in *discrete time*, where the process is generated by a sequence of random functions coming from independent and identically distributed random variables or affine mappings; in the latter case we are naturally led to the notion of *iterated function systems*.

A key tool for investigating stochastic flows is the concept of the *Lya-*

punov exponent, or Furstenberg constant. Using Lyapunov exponents, and suitable linearizations, one can give a spectral description of the cocycle flow associated with the flow. This is the content of the now-famous Osceledec multiplicative ergodic theorem, which is basic for the entire subject of stochastic flows.

The present volume contains 15 papers, which reflect the above-mentioned aspects of stochastic flows. Following the "state-of-the-art", most of these articles discuss stochastic flows which are generated by stochastic differential equations and diffusion processes. Since we intend here to give a broader presentation of the field to the scientific public, we have included some expository articles which, while including new technical results, serve principally as a guide to this field, indicating its rich structure, unifying capability and broad domain of applications.

It is a great pleasure to thank both the contributors of the papers to this volume and the many referees who reviewed the papers. We are also indebted to the staff of of Birkhauser Boston for their continuing patience and cooperation over a period of many months, during the preparation of Volumes I and II. The coordination of Volumes I and II was done with the assistance of the Editorial Board of the Series *Progress in Probability*. We are grateful to the staff of the Mathematics Department of UNC Charlotte for their assistance in both administering the conference and in various stages of manuscript preparation.

Mark A. Pinsky
Department of Mathematics
Northwestern University
Evanston, IL 60208

Volker Wihstutz
Department of Mathematics
University of North Carolina
Charlotte, NC 28223

Conference Participants
Charlotte, North Carolina, March 16-18, 1990

Robert F. Anderson (Charlotte)
Ludwig Arnold (Bremen, FRG)
David Baldwin (Chapel Hill)
Peter Baxendale (Los Angeles)
Petra Boxler (Bremen, FRG)
Kamesh Casukhela (Auburn)
Erhan Cinlar (Princeton)
Richard Darling (Tampa)
John H. Elton (Atlanta)
Steve Gibert (Carrolton)
Kurt Helmes (Lexington)
K. Jayakshmi (Auburn)
Olav Kallenberg (Auburn)
John Kao (Princeton)
Wolfgang H. Kliemann (Ames)
Thomas G. Kurtz (Madison)
Alan Lambert (Charlotte)
J.K. Lin (Boca Raton)
Salah Mohammed (Carbondale)
Arunava Mukherjea (Tampa)
Richard Neidinger (Davidson)
Daniel Ocone (New Brunswick)
Ken Palmer (Coral Gables)
Mark Pinsky (Evanston)
Joe Quinn (Charlotte)
Kameswar Rao (Auburn)
James A. Reneke (Clemson)
Franz Rothe (Charlotte)
Andrzej Russek (Chapel Hill)
Boris Rozovsky (Charlotte)
Douglas Shafer (Charlotte)
Harpreet Singh (Charlotte)
Renming Song (Gainesville)
Erik Van Vleck (Atlanta)
Zoran Vondraček (Gainesville)
Volker Wihstutz (Charlotte)
Jie Xiong (Chapel Hill)

PART I

Diffusion Processes and General Stochastic Flows on Manifolds

STABILITY AND EQUILIBRIUM PROPERTIES OF STOCHASTIC FLOWS OF DIFFEOMORPHISMS

by

PETER H. BAXENDALE

1 Introduction

Let M be a smooth paracompact d-dimensional manifold, and consider the (Stratonovich) stochastic differential equation on M

$$dx_t = V_0(x_t)dt + \sum_{\alpha=1}^{r} V_\alpha(x_t) \circ dW_t^\alpha \tag{1.1}$$

where $V_0, V_1, \ldots, V_r$ are smooth vector fields on M and $\{(W_t^1, \ldots, W_t^r) : t \geq 0\}$ is a standard $\mathbf{R}^r$-valued Brownian motion on some probability space $(\Omega, \mathcal{F}, \mathbf{P})$. If $V_1 = V_2 = \cdots = V_r = 0$ then (1.1) reduces to the ordinary differential equation

$$dx_t = V_0(x_t)dt \tag{1.2}$$

on M; it is well known that the solutions of (1.2) for different initial positions $x \in M$ can be pieced together to give a flow of diffeomorphisms of M, and the theory of dynamical systems describes the behavior of the resulting flow. See for example Ruelle [Ru2].

Let us return to the full equation (1.1). It may be regarded as a randomly perturbed version of (1.2), in which the fixed vector field V_0 is replaced by the (time-dependent) random vector field $V_0 + \sum_{\alpha=1}^{r} \dot{W}_t^\alpha V_\alpha$. Here the random coefficients $\dot{W}_t^1, \ldots, \dot{W}_t^r$ denote the white noise process defined formally as the time derivative of the Brownian motions $W_t^1, \ldots, W_t^r$. As a typical Brownian path has everywhere unbounded variation, the white noise process exists only in a distributional sense, and equation (1.1) cannot be interpreted as a time dependent ordinary differential equation; instead its proper interpretation is as a stochastic differential equation, using the theory of stochastic integration.

Assuming suitable growth conditions on the vector fields $V_0, V_1, \ldots, V_r$ the equation (1.1) has a strong solution $\{x_t : t \geq 0\}$ for each initial position $x_0 = x \in M$. Moreover the strong solutions of (1.1) for different initial positions may be pieced together to give a random process $\{\xi_t : t \geq 0\}$ with values in Diff(M), the space of smooth diffeomorphisms (equipped with the compact-open C^∞ topology). More precisely, the process $\{\xi_t : t \geq 0\}$ defined on $(\Omega, \mathcal{F}, \mathbf{P})$ with values in Diff(M) is characterized by the fact that for each $x \in M$ the one-point motion $\{\xi_t(x) : t \geq 0\}$ is the (unique) strong solution to (1.1) with initial condition $\xi_0(x) = x$. The process $\{\xi_t : t \geq 0\}$ is called the ***stochastic flow of diffeomorphisms*** associated with (1.1). In this paper we survey results on stability properties of $\{\xi_t : t \geq 0\}$ and on the nature of the equilibrium states associated with $\{\xi_t : t \geq 0\}$. That is, we consider $\{\xi_t : t \geq 0\}$ from the point of view of stochastic dynamical systems.

For results on the existence of the stochastic flow of diffeomorphisms and on its regularity properties see for example Elworthy [El1], Kunita [Ku2,Ku3], Ikeda and Watanabe [IW1], or Rogers and Williams [RW]. In this paper we assume throughout that the vector fields $V_0, V_1, \ldots, V_r$ are C^∞, and as a consequence the resulting stochastic flow consists of C^∞ diffeomorphisms. There are corresponding results for C^k vector fields; we do not discuss these here. Notice that in equation (1.1) we have used the Stratonovich stochastic differential $\circ dW_t^\alpha$. If $M = \mathbf{R}^d$ the equation (1.1) can be written in terms of the Itô stochastic differential as

$$dx_t = \overline{V}_0(x_t)dt + \sum_{\alpha=1}^{r} V_\alpha(x_t)dW_t^\alpha \tag{1.3}$$

where

$$\overline{V}_0(x) = V_0(x) + \frac{1}{2}\sum_{\alpha=1}^{r} DV_\alpha(x)(V_\alpha(x)).$$

For a general manifold M there is a similar way of writing (1.1) as an Itô stochastic differential equation within any coordinate patch; an advantage of using the Stratonovich form is that it provides a coordinate-free way to write the equation.

The following gives an alternative way of constructing, and thinking about, the stochastic flow of diffeomorphisms $\{\xi_t : t \geq 0\}$. Define $Z_t = tV_0 + \sum_{\alpha=1}^{r} W_t^\alpha V_\alpha$. Then $\{Z_t : t \geq 0\}$ is a Gaussian random process with values in the space $C^\infty(TM)$ of smooth vector fields on M. It may be regarded as a Brownian motion with drift in

$C^\infty(TM)$. The equation (1.1) may now be written more concisely as

$$dx_t = \circ dZ_t(x_t). \tag{1.4}$$

In the terminology of Kunita [Ku3] $\{Z_t : t \geq 0\}$ is the random forward infinitesimal generator of the Brownian flow $\{\xi_t : t \geq 0\}$. Equation (1.4) expresses the fact that the increments in $\{\xi_t : t \geq 0\}$ (with values in Diff(M)) are determined by the increments in $\{Z_t : t \geq 0\}$ (with values in $C^\infty(TM)$). Roughly speaking, for small h the left increment $\xi_{t+h}\xi_t^{-1}$ in the stochastic flow of diffeomorphisms is (approximately) given by the time 1 flow along the random vector field $Z_{t+h} - Z_t = hV_0 + \sum_{\alpha=1}^{r}(W_{t+h}^\alpha - W_t^\alpha)V_\alpha$. A precise statement, asserting the convergence of a sequence of approximations based on this observation, is given in Ikeda and Watanabe [IW1]. The fact that the Stratonovich stochastic differential appears in the equation for the limit of this sequence of approximations goes back to Wong and Zakai [WZ]. In $\mathbf{R}^d$, if the time 1 flow along $Z_{t+h} - Z_t$ is replaced in the sequence of approximations by the increment $Z_{t+h} - Z_t$ itself then the limit corresponds to a version of (1.1) in which the Stratonovich stochastic differential $\circ dW_t^\alpha$ is replaced by the Itô stochastic differential dW_t^α, see Ikeda and Watanabe [IW1].

It should be clear from the previous paragraph that the stochastic flow of diffeomorphisms $\{\xi_t : t \geq 0\}$ should inherit certain properties from $\{Z_t : t \geq 0\}$. In fact $\{\xi_t : t \geq 0\}$ has the following properties:

(1.5) independent left increments (i.e. if $0 \leq t_0 < t_1 < \cdots < t_n$ then $\xi_{t_i}\xi_{t_{i-1}}^{-1}, 1 \leq i \leq n$, are independent).

(1.6) time homogeneous (i.e. if $t > s$ the distribution of $\xi_t\xi_s^{-1}$ depends only on $t - s$)

(1.7) continuous sample paths with probability 1 (w.p.1).

(1.8) $\xi_0 = Id$.

Notice that (1.5) and (1.6) together replace the semigroup property $\phi_s \circ \phi_t = \phi_{s+t}$ for the deterministic flow $\{\phi_t : t \geq 0\}$ of an ordinary differential equation. Notice also that the continuity statement (1.7) cannot be improved to differentiability since a typical sample path $t \to \xi_t(x)$ will be no better behaved than a typical Brownian path in M. Properties (1.5), (1.6) and (1.8) are the obvious extension to continuous time of properties enjoyed by the products $\eta^{(n)} = \eta_n \circ \eta_{n-1} \circ \cdots \circ \eta_1$ along a sequence

$\{\eta_i\}$ of independent identically distributed random diffeomorphisms. Much of what we will say about the behavior of $\{\xi_t : t \geq 0\}$ will correspond to similar results for the products of independent identically distributed random diffeomorphisms. In many situations the discrete time results were proved first, and the continuous time case appears as a corollary. However in other situations the calculus associated with stochastic differential equations allows us to obtain sharper results. We will not consider here solutions of differential equations driven by real (or colored) noise; neither will we consider the discrete time counterpart, namely products along a stationary ergodic sequence of random diffeomorphisms. In each of these cases property (1.5) is lost, although property (1.6) remains valid and much of the associated ergodic theory still works. For this more general setting see the results of Crauel [Cr1,Cr2] on random dynamical systems.

In this paper we shall concentrate on stability, instability, and equilibrium properties of the stochastic flow of diffeomorphisms $\{\xi_t : t \geq 0\}$ associated with equation (1.1). In particular we will concentrate on properties of the stochastic flow $\{\xi_t : t \geq 0\}$ in Diff(M) rather than on properties of its one-point motion $\{\xi_t(x) : t \geq 0\}$ on M. For the behavior of the diffusion process $\{\xi_t(x) : t \geq 0\}$ see for example Kushner [Kus], Khas'minskii [Kh2], and Arnold and Kliemann [AK1].

At various points in this paper we shall impose certain conditions on the stochastic flow of diffeomorphisms $\{\xi_t : t \geq 0\}$. The main assumption is (5.1), asserting among other things that the one-point motion is recurrent. This assumption is essential. If the one-point motion is transient, then all the technology based on the multiplicative ergodic theorem and Lyapunov exponents collapses. If M is compact then (5.1) reduces to a definition of the notation used. The other assumptions are of a different nature; they assert that various finite dimensional diffusion processes associated with $\{\xi_t : t \geq 0\}$ are non-degenerate in certain precise sense. They are of the form 'assume there is enough noise in the system.' Since there is no fixed relation between the dimension d of M and the dimension r of the noise in (1.1) then these conditions will be satisfied in some generic sense. The results here are at the opposite end of the randomness spectrum from the case of deterministic dynamical systems where there is no noise at all. The reader should be aware that in between these two extremes the problems become much harder, as there is much more subtle dependence upon initial configurations. Techniques of geometric control theory, see [AK1], become important for such problems.

2 The law of a stochastic flow of diffeomorphisms

Let $\{\xi_t : t \geq 0\}$ be the stochastic flow of diffeomorphisms corresponding to equation (1.1). The *one-point motion* is a diffusion process in M with generator

$$L = \frac{1}{2}\sum_{\alpha=1}^{r} V_\alpha^2 + V_0. \tag{2.1}$$

The generator L determines the law of the one-point motion, but it does not uniquely determine the law of the stochastic flow $\{\xi_t : t \geq 0\}$. This is because L does not contain the information which determines the law for the k-point motion $\{\xi_t(x_1), \ldots, \xi_t(x_k)) : t \geq 0\}$ in M^k for $k \geq 2$. The k-point motion is obtained by running simultaneously k copies of (1.1) with different initial points but with the same noise $\{(W_t^1, \ldots, W_t^r) : t \geq 0\}$. Equation (1.1) carries information which allows us to talk about random vector fields; in contrast L allows us only to consider random vectors at a single point of M. The law of $\{\xi_t : t \geq 0\}$ is determined by the law of $\{Z_t : t \geq 0\}$, which in turn is determined by the mean V_0 and covariance kernel

$$\begin{aligned} b(x,y) &= \operatorname{cov}(Z_1(x) \otimes Z_1(y)) \\ &= \sum_{\alpha=1}^{r} V_\alpha(x) \otimes V_\alpha(y) \\ &\in T_xM \otimes T_yM \quad \cong \quad L(T_x^*M, T_yM) \end{aligned}$$

for $x, y \in M$. In terms of V_0 and b the generators $L^{(k)}$ for the k-point motions are given as follows. For $f \in C^2(M)$, $L^{(1)} = L$ is given by

$$(Lf)(x) = \frac{1}{2} tr\, [(d_x(b(x,y)(df(x)))\,|_{y=x}] + (V_0.f)(x) \tag{2.2}$$

where d_x denotes the differential with respect to the variable x, and tr denotes the trace of a linear transformation on T_xM. For $k \geq 2$, $f \in C^2(M^k)$ and $\vec{x} = (x_1, \ldots, x_k) \in M^k$,

$$(L^{(k)}f)(\vec{x}) = \sum_i (L_i f)(\vec{x}) + \frac{1}{2}\sum_{i\neq j} b(x_i, x_j)(d_i f(\vec{x}), d_j f(\vec{x})) \tag{2.3}$$

where L_i and d_i denote the action of L and d acting on f as a function of x_i only. (Equivalent formulae for the Itô version of (1.1) in $\mathbf{R}^d$ are given by Kunita [Ku3].) In [LW] Le Jan and Watanabe call the pair (L, b) the local characteristic system of the stochastic flow $\{\xi_t : t \geq 0\}$. Equations (2.2) and (2.3) imply that this is an equivalent way of presenting the information contained in the pair (V_0, b). It is clear from (2.2) and (2.3) that different pairs (V_0, b) give rise to different $L^{(k)}$ and hence to different laws for $\{\xi_t : t \geq 0\}$. Examples of stochastic flows with distinct laws but the same L have been given by Ikeda and Watanabe [IW2] and Baxendale [Ba3]. However the law of the two-point motion $\{(\xi_t(x), \xi_t(y)) : t \geq 0\}$ in M^2 does determine the law of $\{\xi_t : t \geq 0\}$; this can be seen from (2.3). The fact that the law of $\{\xi_t : t \geq 0\}$ is determined by the law of its two-point motion is essentially due to the Gaussian nature of $\{Z_t : t \geq 0\}$.

The covariance $b(x, y)$ is a positive-semi-definite kernel for the tangent bundle TM. That is, it is symmetric $(b(x, y)^* = b(y, x))$ and it has the property that $\sum_{i,j=1}^n b(x_i, x_j)(\zeta_i, \zeta_j) \geq 0$ whenever $n \geq 1$ and $x_i \in M$, $\zeta_i \in T^*_{x_i}M$ for $1 \leq i \leq n$. It can be shown that, modulo appropriate growth conditions, to every pair of a $V_0 \in C^\infty(TM)$ and a smooth positive-semi-definite kernel b there exists a corresponding stochastic flow of diffeomorphisms. We sketch the construction briefly. Any smooth positive-semi-definite kernel b is the reproducing kernel for some Hilbert space $H \subset C^\infty(TM)$. Assuming that the inclusion $i : H \hookrightarrow C^\infty(TM)$ is continuous, then the triple $(i, H, C^\infty(TM))$ is an abstract Wiener space (see Kuo [Kuo]). The abstract Wiener space determines a Gaussian measure on $C^\infty(TM)$, and scaled versions of this Gaussian measure can be used to determine transition probabilities for a $C^\infty(TM)$ valued Wiener process $\{W_t : t \geq 0\}$. (Recall how the one-dimension Wiener process can be constructed using $N(0, t)$ Gaussian measures on $\mathbf{R}$.) In particular $\{W_t : t \geq 0\}$ is a mean-zero Gaussian process with $\mathbf{E}(W_s(x) \otimes W_t(y)) = \min(s, t)b(x, y)$. Define $Z_t = tV_0 + W_t$ and let $\{\xi_t : t \geq 0\}$ be the stochastic flow of diffeomorphisms given by equation (1.4). The existence of $\{\xi_t : t \geq 0\}$ is given in Kunita [Ku3, Thm 4.2.5 and 4.7.7], see also Baxendale [Ba1] and Le Jan and Watanabe [LW]. A direct calculation now shows that the k-point motions of $\{\xi_t : t \geq 0\}$ have generators $L^{(k)}$ given by (2.2) and (2.3).

We may rewrite our new (1.4) in more familiar form by choosing a complete orthonormal basis $\{V_\alpha : \alpha \geq 1\}$ of H. Then we may realize $W_t = \sum_{\alpha \geq 1} W_t^\alpha V_\alpha$ where $\{W_t^\alpha : t \geq 0\}$, $\alpha \geq 1$, are independent one-dimensional Brownian motions. In this way we regain equation (1.1) except now we allow the possibility of $r = \infty$, i.e. infinite-dimensional noise is allowed. The case $r = \infty$ should not seem unreasonable when equation (1.1) is used to generate $\{\xi_t : t \geq 0\}$, a process taking values in an infinite-dimensional space Diff(M). Moreover it has been shown that if $\{\xi_t : t \geq 0\}$ is any process in Diff(M) satisfying (1.5) to (1.8) and if either M is compact (Baxendale [Ba1]) or the k-point motions satisfy certain regularity properties (Le Jan and Watanabe [LW]) then $\{\xi_t : t \geq 0\}$ arises (in law) as the solution of a stochastic differential equation of the form (1.1) with possibly $r = \infty$. The case of a process satisfying (1.5), (1.6) and (1.8) but allowing jumps is treated by Fujiwara and Kunita [FK].

Henceforth we shall restrict to the case of finite-dimensional noise, although most of the results given will remain valid in the case $r = \infty$. This is because they mostly rely on calculations done for finite-dimensional diffusion processes induced by $\{\xi_t : t \geq 0\}$. For example the k-point motion (with values in M^k), or the action of the stochastic flow on a tangent vector (with values in the tangent space TM).

3 Special stochastic flows of diffeomorphisms

Associated to any Riemannian manifold M there is a natural choice $L = \frac{1}{2}\Delta$, where Δ denotes the Laplace-Beltrami operator on M. The corresponding diffusion process is the Brownian motion on M. In general there appears to be no natural choice for the law of a stochastic flow of diffeomorphisms of M, since any choice of a covariance b would involve a pairing of T_xM and T_yM. While there is no canonical choice of a stochastic flow of diffeomorphisms of M there are nevertheless certain special flows which deserve mention.

1) For any Riemannian manifold M there is a *canonical stochastic flow of diffeomorphisms of* OM, the orthonormal frame bundle of M. This is defined by taking $V_0 = 0$ and $V_1, V_2, \ldots, V_d$ to be the canonical horizontal vector fields on OM. The one-point motions of this flow on OM perform the stochastic development (introduced by Itô in [It]); that is, they perform the stochastic version of parallel transport along Brownian paths in the base manifold M. However the flow on OM does not project to give a flow on M. For more details see [IW1], [CE], [El2], [El3] and [Li2].

2) If $M \subset \mathbf{R}^r$ is an isometric embedding there is an associated *gradient Brownian* stochastic flow on M. This arises by taking $V_0 = 0$ and $V_\alpha = \operatorname{grad} h_\alpha$, $1 \leq \alpha \leq r$, where $h_\alpha : M \to \mathbf{R}$ is the restriction to M of the αth coordinate function in $\mathbf{R}^r$. Some detailed calculations for the case where M is the unit sphere in $\mathbf{R}^{d+1}$ are given in Baxendale [Ba4]. For results on gradient Brownian stochastic flows in general see [El2], [El3] and references therein.

3) Suppose G is a finite-dimensional Lie group acting smoothly on M. Then G has a natural inclusion as a subgroup of Diff(M) and any Brownian motion process on G can be interpreted as a stochastic flow of diffeomorphisms of M. See for example Liao [Li1] for the case where G is the group of isometries of M. The support theorem for stochastic flow of diffeomorphisms (due to Ikeda and Watanabe [IW2], see also [Ku3, p283]) provides a way of recognising this situation; it implies that the stochastic flow of diffeomorphisms $\{\xi_t : t \geq 0\}$ corresponding to (1.1) takes values in a closed subgroup G of Diff(M) whenever the vector fields $V_0, V_1, \ldots, V_r$

all act as infinitesimal G-automorphisms of M.

4) Suppose M is a symmetric space with isometry group G. We say the stochastic flow $\{\xi_t : t \geq 0\}$ is *isotropic* if its law is G-invariant, i.e. if for all $g \in G$ the process $\{g^{-1}\xi_t g : t \geq 0\}$ has the same law as $\{\xi_t : t \geq 0\}$. A survey of isotropic stochastic flows on $\mathbf{R}^d$ is given by Darling [Da].

4 Stationary versions

Many of the results on the large time behavior of stochastic flows are based on ergodic theorems for stationary processes. In this section we construct the appropriate stationary process.

Let $\mathcal{P}(M)$ denote the space of probability measures on M (with the weak topology of weak convergence), and $\mathcal{B}(M)$ the Borel σ-algebra of M. We say $\rho \in \mathcal{P}(M)$ is *stationary* for the one-point motion $\{\xi_t(x) : t \geq 0\}$ if

$$\int_M \mathbf{P}\{\xi_t(x) \in A\}\, d\rho(x) = \rho(A) \tag{4.1}$$

for all $A \in \mathcal{B}(M)$ and $t \geq 0$. Notice that the definition depends only on the law of the diffusion process $\{\xi_t(x) : t \geq 0\}$ on M, and hence only on L. The question of the existence and uniqueness of such ρ is well studied. If M is compact there is at least one stationary $\rho \in \mathcal{P}(M)$; for the case $M = \mathbf{R}^d$ see for example Bhattacharya [Bh] or Khas'minskii [Kh2].

For $t \geq 0$ define $\rho_t = \rho\xi_t^{-1}$ (that is, $\rho_t(A) = \rho(\xi_t^{-1}(A))$ for all $A \in \mathcal{B}(M)$). Then $\{\rho_t : t \geq 0\}$ is a random process on $(\Omega, \mathcal{F}, \mathbf{P})$ with values in $\mathcal{P}(M)$. The equation (4.1) (asserting that ρ is stationary) can be rewritten as

$$\mathbf{E}(\rho_t) = \rho \tag{4.2}$$

for all $t \geq 0$. This is weaker than the statement that $\rho_t = \rho$ for all $t \geq 0$ w.p.1. If $\rho_t = \rho$ for all $t \geq 0$ w.p.1. we say that ρ is *invariant* under the stochastic flow $\{\xi_t : t \geq 0\}$. The distinction between the two concepts of stationarity and invariance for ρ is due directly to the randomness in the system; in the deterministic case the two concepts coincide.

Notice that an invariant probability measure ρ must be stationary. This immediately reduces the possible candidates for an invariant ρ. It is easy to construct examples of stochastic flows with a unique stationary ρ in which it can be directly checked that ρ is not invariant. See also Theorem 5.12 later.

In general there may be no $\rho \in \mathcal{P}(M)$ such that $\{\xi_t : t \geq 0\}$ forms a family of measure-preserving transformations of the measure space $(M, \mathcal{B}(M), \rho)$. Instead we consider an induced process on the product space $\Omega \times M$.

Thus far we have considered the stochastic flow $\{\xi_t : t \geq 0\}$ parametrized by non-negative time. We can interpret ξ_t as describing the random evolution of the state space M from time 0 to time t. It is convenient at this point to consider a larger family $\{\xi_{s,t} : -\infty < s \leq t < \infty\}$, where $\xi_{s,t}$ describes the random evolution of M from time s to time t. We do this as follows.

Firstly we may assume that the Brownian motion $\{(W_t^1, \ldots, W_t^r) : t \in \mathbf{R}\}$ is defined for all time. This can be done by joining two independent non-negative time copies 'back to back' (so that the important properties of independence of increments and time homogeneity are retained). Moreover we can assume that there exists a 1-parameter group $\{\theta_t : t \in \mathbf{R}\}$ of measure-preserving transformation of $(\Omega, \mathcal{F}, \mathbf{P})$ such that $W_t^\alpha(\theta_s\omega) = W_{t+s}^\alpha(\omega) - W_s^\alpha(\omega)$ for all $\omega \in \Omega$, $s, t \in \mathbf{R}$, $1 \leq \alpha \leq r$. For $-\infty < s \leq t < \infty$ let $\mathcal{F}_s^t$ denote the σ-algebra generated by $\{W_u^\alpha - W_v^\alpha : s \leq v \leq u \leq t, 1 \leq \alpha \leq r\}$. The process $\{\xi_{s,t} : -\infty < s \leq t < \infty\}$ in Diff(M) is then constructed by piecing together the one-point motions given by

$$\xi_{s,t}(x) = x + \int_s^t V_0(\xi_{s,u}(x))du + \sum_{\alpha=1}^{r} \int_s^t V_\alpha(\xi_{s,u}(x)) \circ dW_u^\alpha \tag{4.3}$$

for $s \leq t$ and $x \in M$. This can be done (see Kunita [Ku3]) in such a way that for $\mathbf{P}$-almost all $\omega \in \Omega$

$$\xi_{s,t}(\theta_u\omega, \cdot) = \xi_{s+u,t+u}(\omega, \cdot) \tag{4.4}$$

for all $s \leq t$ and all u, and

$$\xi_{t,u}(\omega, \cdot) \circ \xi_{s,t}(\omega, \cdot) = \xi_{s,u}(\omega, \cdot) \tag{4.5}$$

for all $s \leq t \leq u$. Here the random diffeomorphism $\xi_{s,t}$ is regarded as a mapping $\Omega \times M \to M$ and we write $\xi_{s,t}(\omega, \cdot)$ to emphasize the dependence of $\xi_{s,t}$ on $\omega \in \Omega$. Clearly $\xi_{s,t}$ is $\mathcal{F}_s^t$ measurable. Since each $\xi_{s,t}(\omega, \cdot)$ is invertible we may define $\xi_{s,t}(\omega, \cdot) = \xi_{t,s}(\omega, \cdot)^{-1}$ for $s > t$. Then (4.4) and (4.5) become valid for all $s, t, u \in \mathbf{R}$. Moreover if we define $\xi_t(\omega, \cdot) = \xi_{0,t}(\omega, \cdot)$ for $t \in \mathbf{R}$ then

$$\xi_s(\theta_t\omega, \cdot) \circ \xi_t(\omega, \cdot) = \xi_{s+t}(\omega, \cdot) \tag{4.6}$$

for all $s, t \in \mathbf{R}$. Notice that $\xi_t(\omega, \cdot) = \xi_{0,t}(\omega, \cdot)$ for $t \geq 0$ is consistent with our earlier notation. Equation (4.6) is the cocycle condition for the class of *random dynamical systems* studied by Crauel [Cr1]. Crauel's class of random dynamical systems contains solutions of (1.1) and much more besides. For example solutions of real noise stochastic differential equations and (in discrete time) products of random diffeomorphisms are included.

We define the *skew-product flow* on $\Omega \times M$ as follows. For $t \in \mathbf{R}$ define $\Theta_t : \Omega \times M \to \Omega \times M$ by

$$\Theta_t(\omega, x) = (\theta_t\omega, \xi_t(\omega, x)). \tag{4.7}$$

It follows from (4.6) that $\{\Theta_t : t \in \mathbf{R}\}$ is a one-parameter group of transformations of $\Omega \times M$.

If we were concerned solely with the future evolution of the system from some known present configuration then we would consider only $\{\xi_{s,t} : 0 < s \le t < \infty\}$ with corresponding 1-parameter semigroups $\{\theta_t : t \ge 0\}$ on $(\Omega, \mathcal{F}_0^\infty)$ and $\{\Theta_t : t \ge 0\}$ on $(\Omega \times M, \mathcal{F}_0^\infty \times \mathcal{B}(M))$. There would be corresponding restrictions on the time parameters s, t, and u in equations (4.4)–(4.7). This is the situation we will consider in Sections 5, 6 and 7. In Section 8 we return to the more general process $\{\xi_{s,t} : -\infty < s \le t < \infty\}$.

The skew product representation for products of independent identically distributed diffeomorphisms is studied by Ohno [Oh], Kifer [Ki] and Ledrappier and Young [LY1]. In particular [Ki] and [LW1] study the associated Lyapunov exponents and the entropy. The continuous time case is considered by Carverhill [Ca1] and Crauel [Cr1].

(4.8) Proposition. (i) *As a measure on $(\Omega \times M, \mathcal{F}_0^\infty \times \mathcal{B}(M))$, $\mathbf{P} \times \rho$ is preserved by the family $\{\Theta_t : t \ge 0\}$. Moreover ρ is ergodic (for the 1-point motion on M) if and only if $\mathbf{P} \times \rho$ is ergodic (for $\{\Theta_t : t \ge 0\}$ on $(\Omega \times M, \mathcal{F}_0^\infty \times \mathcal{B}(M))$).*
(ii) *As a measure on $(\Omega \times M, \mathcal{F} \times \mathcal{B}(M))$, $\mathbf{P} \times \rho$ is preserved by the family $\{\Theta_t : t \ge 0\}$ if and only if ρ is invariant under the flow $\{\xi_t : t \ge 0\}$.*

Proof. (i) See Ohno [Oh] or Carverhill [Ca1].
(ii) A direct calculation shows that, with respect to the projection $\Omega \times M \to \Omega$, the measure $(\mathbf{P} \times \rho)\Theta_t^{-1}$ on $(\Omega \times M, \mathcal{F} \times \mathcal{B}(M))$ has marginal $\mathbf{P}$ and regular conditional probability distribution $\rho\xi_{-t,0}(\omega, \cdot)^{-1} : \mathcal{F} \to \mathcal{P}(M)$. The result now follows since $\rho\xi_{-t,0}(\omega, \cdot)^{-1} = \rho\xi_{0,t}(\theta_{-t}\omega, \cdot)^{-1}$ which has the same distribution as ρ_t. □

Remark. The difference between these two results can be explained by observing that the regular conditional probability distribution of $(\mathbf{P} \times \rho)\Theta_t^{-1}$ as a measure on $(\Omega \times M, \mathcal{F}_0^\infty \times \mathcal{B}(M))$ is $\mathbf{E}(\rho\xi_{-t,0}(\omega, \cdot)^{-1} | \mathcal{F}_0^\infty) = \mathbf{E}(\rho\xi_{-t,0}(\omega, \cdot)^{-1}) = \mathbf{E}(\rho_t) = \rho$, where we use the fact that $\xi_{-t,0}$ is $\mathcal{F}_{-t}^0$ measurable and hence is independent of $\mathcal{F}_0^\infty$. We will pursue the matter of invariant measures for $\{\Theta_t : t \ge 0\}$ on $(\Omega \times M, \mathcal{F} \times \mathcal{B}(M))$ in Section 8. Crauel [Cr2] investigates the measurability properties of invariant measures for the skew-product representation of a random dynamical system.

5 Lyapunov exponents and the moment Lyapunov function

For any $x \in M$ we may linearize the stochastic flow $\{\xi_t : t \geq 0\}$ along the trajectory $\{\xi_t(x) : t \geq 0\}$ to obtain the process $\{D\xi_t(x) : t \geq 0\}$ where $D\xi_t(x)$ denotes the derivative of $\xi_t : M \to M$ at x. In this section we introduce the Lyapunov exponents which describe the almost sure growth rate of $D\xi_t(x)$ as $t \to \infty$. For an overview of the theory of Lyapunov exponents see the collections [AW] and [ACE].

Let M be given a Riemannian metric. Let $|v|$ denote the norm of $v \in T_xM$ and $\|A\|$ the operator norm of a linear mapping $A : T_xM \to T_yM$. Let $\log^+(r) = \max(\log(r), 0)$ for $r > 0$. We state the basic hypothesis we shall assume throughout the paper.

(5.1)

(i) $\{\xi_t : t \geq 0\}$ is the stochastic flow of diffeomorphisms corresponding to (1.1).

(ii) ρ is a stationary ergodic probability measure for the one-point motion $\{\xi_t(x) : t \geq 0\}$ on M.

(iii) $\mathbf{E} \int_M \sup_{0 \leq t \leq 1} (\log^+ \|D\xi_t(x)\| + \log^+ \|(D\xi_t(x))^{-1}\| + \log^+ \|D\xi_{t,1}(x)\| + \log^+ \|(D\xi_{t,1}(x))^{-1}\|) \, d\rho(x) < \infty.$

If M is compact (5.1)(iii) is automatically satisfied, see Baxendale [Ba6].

(5.2) Theorem. *Assume* (5.1). *For* $(\mathbf{P} \times \rho)$*-almost all* $(\omega, x) \in \Omega \times M$,

$$[D\xi_t(x)^* D\xi_t(x)]^{\frac{1}{2t}} \to \Lambda_{(\omega,x)} \text{ as } t \to \infty$$

where $\Lambda_{(\omega,x)}$ *is a random self-adjoint mapping of* T_xM *onto itself with non-random eigenvalues*

$$e^{\lambda_1} \geq e^{\lambda_2} \geq \cdots \geq e^{\lambda_d}.$$

Moreover if $\tilde\lambda_1 > \cdots > \tilde\lambda_{\tilde d}$ *(with* $\tilde d \leq d$*) denote the distinct* λ_j *and*

$$F^i_{(\omega,x)} = \bigoplus_{j=i}^{\tilde d} \ker(\Lambda_{(\omega,x)} - e^{\tilde\lambda_j} I) \subset T_xM$$

for $1 \leq i \leq \tilde d$ *and* $F^{\tilde d+1}_{(\omega,x)} = \{0\}$ *then*

$$\frac{1}{t} \log |D\xi_t(x)(v)| \to \tilde\lambda_i \text{ as } t \to \infty$$

whenever $v \in F^i_{(\omega,x)} \setminus F^{i+1}_{(\omega,x)}$.

This theorem is based on the multiplicative ergodic theorem of Oseledec [Os]. It was applied to (deterministic) dynamical systems by Ruelle [Ru1] and to stochastic flows of diffeomorphisms by Carverhill [Ca1]. The values $\lambda_1 \geq \lambda_2 \geq \cdots \geq \lambda_d$ are called the *Lyapunov exponents* for $\{\xi_t : t \geq 0\}$. The Riemannian structure on M is used both in (5.1)(iii) and also in the definition of $D\xi_t(x)^*$. However it is easily checked that the validity of (5.1)(iii) and the values of $\lambda_1, \ldots, \lambda_d$ are unchanged if the Riemannian metric is replaced by a uniformly equivalent one.

(5.3) Corollary. *For ρ-almost all $x \in M$ and Lebesgue-almost all $v \in T_xM$*

$$\lim_{t\to\infty} \frac{1}{t} \log |D\xi_t(x)(v)| = \lim_{t\to\infty} \frac{1}{t} \log ||D\xi_t(x)|| = \lambda_1 \text{ wp1}$$

and

$$\lim_{t\to\infty} \frac{1}{t} \log |\det D\xi_t(x)| = \lambda_\Sigma \text{ wp1},$$

where $\lambda_\Sigma = \lambda_1 + \lambda_2 + \cdots + \lambda_d$.

Remark. If the process $\{\xi_t : t \geq 0\}$ is sufficiently non-degenerate then the assertions of Corollary (5.3) will be valid for all $x \in M$. The question as to whether 'Lebesgue-almost all v' can be replaced by 'all non-zero v' is related to the randomness of the filtration

$$T_xM = F^1_{(\omega,x)} \supset F^2_{(\omega,x)} \supset \cdots \supset F^d_{(\omega,x)} \supset \{0\}.$$

If $\{D\xi_t(x)(v) : t \geq 0\}$ is sufficiently non-degenerate then the positive codimension subspaces $F^2_{(\omega,x)}, \ldots, F^d_{(\omega,x)}$ will contain no fixed non-zero vectors in T_xM. In this case the assertion of Corollary (5.3) would be valid for any fixed non-zero $v \in T_xM$. The question of the extent of randomness in the filtration is discussed in detail in Carverhill [Ca3] and Kifer [Ki].

Clearly the value of λ_1, the top Lyapunov exponent, is of great importance when discussing stability. We indicate briefly how it may be computed. The method is due to Carverhill [Ca2], extending a formula of Khas'minskii [Kh1] for linear stochastic differential equations. (The details of the following calculation are taken from [BS, Sect. 2]; a calculation based on an isometric embedding of M in some Euclidean space is given in [Ba2].) For non-zero $v \in T_xM$, write $x_t = \xi_t(x)$ and $v_t = D\xi_t(x)(v) \in T_{\xi_t(x)}M$. Then $\{x_t : t \geq 0\}$ satisfies (1.1) and $\{v_t : t \geq 0\}$ is the diffusion process in TM satisfying

$$dv_t = TV_0(v_t)dt + \sum_{\alpha=1}^{r} TV_\alpha(v_t) \circ dW_t^\alpha \qquad (5.4)$$

where TV_α denotes the natural lift of the vector field V_α on M to a vector field on TM. Using the Riemannian connection on M to decompose T_vTM into horizontal and vertical components, then $TV_\alpha(v)$ has horizontal and vertical components $V_\alpha(x)$ and $\nabla V_\alpha(x)(v)$ respectively. Hence $\{v_t : t \geq 0\}$ is the path in TM over $\{x_t : t \geq 0\}$ which is determined by

$$\nabla v_t = \nabla V_0(x_t)(v_t)dt + \sum_{\alpha=1}^{r} \nabla V_\alpha(x_t)(v_t) \circ dW_t^\alpha. \tag{5.5}$$

That is, (1.1) and (5.5) together are equivalent to (5.4).

Now write $\theta_t = v_t/|v_t| \in SM$, the unit sphere bundle of M. The stochastic calculus yields

$$d\theta_t = \tilde{V}_0(\theta_t)dt + \sum_{\alpha=1}^{r} \tilde{V}_\alpha(\theta_t) \circ dW_t^\alpha \tag{5.6}$$

and

$$d(\log|v_t|) = q_0(\theta_t)dt + \sum_{\alpha=1}^{r} q_\alpha(\theta_t) \circ dW_t^\alpha \tag{5.7}$$

where the vector fields $\tilde{V}_\alpha$ on SM have horizontal and vertical components $V_\alpha(x)$ and $\nabla V_\alpha(x)(\theta) - \langle \nabla V_\alpha(x)(\theta), \theta\rangle\theta$ respectively at $\theta \in S_x M$ and the functions q_α : $SM \to \mathbf{R}$ are given by $q_\alpha(\theta) = \langle \nabla V_\alpha(x)(\theta), \theta\rangle$. Thus $\{\theta_t : t \geq 0\}$ is a diffusion process on SM with generator

$$\tilde{L} = \tilde{V}_0 + \frac{1}{2}\sum_{\alpha=1}^{r} \tilde{V}_\alpha^2 \tag{5.8}$$

and

$$\log|v_t| = \log|v_0| + \int_0^t Q(\theta_s)ds + \sum_{\alpha=1}^{r}\int_0^t q_\alpha(\theta_s)dW_s^\alpha. \tag{5.9}$$

Here we have converted to Itô integrals and so $Q(\theta) = q_0(\theta) + \frac{1}{2}\sum_{\alpha=1}^{r}(\tilde{V}_\alpha \cdot q_\alpha)(\theta)$. Therefore

$$\begin{aligned} \lim_{t\to\infty} \frac{1}{t}\log|v_t| &= \lim_{t\to\infty}\frac{1}{t}\int_0^t Q(\theta_s)ds + \lim_{t\to\infty}\frac{1}{t}\sum_{\alpha=1}^{r}\int_0^t q_\alpha(\theta_s)dW_s^\alpha \\ &= \lim_{t\to\infty}\frac{1}{t}\int_0^t Q(\theta_s)ds \end{aligned}$$

since $M_t/t \to 0$ almost surely for any continuous martingale with $\frac{d}{dt}\langle M\rangle_t \leq C$ for a finite constant C. Since the unit sphere bundle $\pi : SM \to M$ has compact fiber there exists at least one stationary probability measure $\tilde{\rho}$ (with $\tilde{\rho} \circ \pi^{-1} = \rho$) for $\{\theta_t : t \geq 0\}$ on SM. Assuming $\tilde{\rho}$ is ergodic and $(\mathbf{P} \times \tilde{\rho})\{(\omega, \theta) : \theta \notin F^2_{(\omega,\pi\theta)}\} > 0$ then the ergodic theorem gives

$$\lambda_1 = \int_{SM} Q(\theta)\tilde{\rho}(d\theta). \tag{5.10}$$

This will be the situation if $\tilde{L}$ is hypoelliptic, in which case $\tilde{\rho}$ has a smooth density. (Notice that $F^2_{(\omega,x)}$ has positive codimension in T_xM.) Otherwise, see [Ca3] or [Ki].

A simple generalization of this argument yields a formula for $\lambda_1 + \cdots + \lambda_k$ for $1 \leq k \leq d$. The process $\{\theta_t : t \geq 0\}$ in SM is replaced by the process induced by $\{\xi_t : t \geq 0\}$ in $G_k(M)$, the Grassmannian bundle of k-dimensional subspaces of TM (see Baxendale [Ba2] for details). For the case $k = d$ the calculations simplify. We obtain

$$d(\log\det(D\xi_t(x))) = \mathrm{div}V_0(\xi_t(x))dt + \sum_{\alpha=1}^{r} \mathrm{div}V_\alpha(\xi_t(x)) \circ dW_t^\alpha$$

and hence

$$\lambda_\Sigma = \lim_{t\to\infty} \frac{1}{t}\int_0^t Q_0(\xi_s(x))ds = \int_M Q_0(x)\rho(dx) \tag{5.11}$$

where $Q_0(x) = \mathrm{div}V_0(x) + \frac{1}{2}\sum_{\alpha=1}^{r}(V_\alpha.\mathrm{div}V_\alpha)(x)$ for $x \in M$.

Let m denote the Riemannian measure on M. Then λ_Σ provides a measurement of the rate at which ξ_t distorts m. The following result extends this idea to the stationary probability measure ρ. For $\mu, \nu \in \mathcal{P}(M)$ let $h(\mu;\nu)$ denote the relative entropy of ν with respect to μ.

(5.12) Theorem. (i) *Assume* (5.1) *and that ρ is absolutely continuous with respect to m with* $\log(\frac{d\rho}{dm}) \in L^1(\rho)$. *Then for all* $t \geq 0$,

$$\mathbf{E}(h(\rho;\rho_t)) = -t\lambda_\Sigma.$$

(ii) *In particular* $\lambda_\Sigma \leq 0$ *with equality if and only if ρ is invariant under* $\{\xi_t : t \geq 0\}$.

Proof. See Baxendale [Ba6, Thm 4.2]. Part (ii) remains valid without the condition $\log(\frac{d\rho}{dm}) \in L^1(\rho)$, see Le Jan [Le5]. (In the special case when ρ has a strictly positive C^2 density with respect to m, part (ii) can be obtained directly from (5.11) by an integration by parts, see [Ba2].) □

Notice that $\mathbf{E}(h(\rho;\rho_t))$ is the relative entropy $h(\mathbf{P}\times\rho;(\mathbf{P}\times\rho)\Theta_t^{-1})$ of $(\mathbf{P}\times\rho)\Theta_t^{-1}$ relative to $\mathbf{P}\times\rho$ considered as probability measures on $(\Omega\times M, \mathcal{F}\times\mathcal{B}(M))$. Thus Theorem 5.12 may be regarded as a qualitative strengthening of Proposition 4.8.

For similar results characterizing stochastic flows $\{\xi_t : t \geq 0\}$ consisting of conformal tranformations (respectively isometries) by the condition $\lambda_1 = \lambda_2 = \cdots = \lambda_d$ (resp. $\lambda_1 = \lambda_2 = \cdots = \lambda_d = 0$) see Baxendale [Ba6].

The Lyapunov exponents, and especially the top Lyapunov exponent λ_1, describe the almost-sure exponential growth rate of $|D\xi_t(x)(v)|$ as $t \to \infty$. Corresponding information about the exponential growth rate of $\mathbf{E}|D\xi_t(x)(v)|^p$ is provided by the *moment Lyapunov function* g given by

$$g(p) = \lim_{t\to\infty} \frac{1}{t} \log \mathbf{E}|D\xi_t(x)(v)|^p, \qquad p \in \mathbf{R}. \tag{5.13}$$

The moment Lyapunov function was introduced by Arnold [Ar] for linear stochastic systems. See also papers by Arnold, Kliemann and Oeljeklaus [AKO] and Arnold, Oeljeklaus and Pardoux [AOP] in [AW]. The connection between the moment Lyapunov function and an associated large deviations problem was established for linear systems by Stroock [St]. See also Arnold and Kliemann [AK2]. For nonlinear systems a version of the moment Lyapunov function concerned with the growth rate of $\det(D\xi_t(x))$ was studied by Elworthy and Stroock [ES]. The moment Lyapunov function above is studied in Baxendale and Stroock [BS].

The remainder of the section will summarize some results about the moment Lyapunov function taken from [AOP] and [BS]. We need to impose stronger assumptions than those made so far. We will assume

(5.14)
(i) M is compact.
(ii) $\mathrm{Lie}(\tilde{V}_1, \tilde{V}_2, \ldots, \tilde{V}_r)(\theta) = T_\theta SM$ for all $\theta \in SM$.

In fact (5.14)(ii) is stronger than needed. For most of our results it can be replaced by the weaker (but more complicated) assumption (2.8) of [BS]. We note that (5.14) implies that there are unique stationary $\rho \in \mathcal{P}(M)$ and $\tilde{\rho} \in \mathcal{P}(SM)$ for the one-point motions $\{x_t : t \geq 0\}$ and $\{\theta_t : t \geq 0\}$ in M and SM respectively, and that ρ and $\tilde{\rho}$ have smooth positive densities.

We present the results on the moment Lyapunov function in two theorems; the first one concerns its intrinsic properties and the second one shows how it provides information about $\{D\xi_t(x)(v) : t \geq 0\}$.

(5.15) Theorem. *Assume* (5.14).
(i) $g : \mathbf{R} \to \mathbf{R}$ *is well-defined (i.e. the limit in* (5.13) *exists and does not depend on* v *so long as* $v \neq 0$*).*
(ii) g *is convex and analytic with* $g(0) = 0$.
(iii) $g(p)$ *is the largest eigenvalue of the operator*

$$\tilde{L}_p = \tilde{L} + p\sum_{\alpha=1}^{r} q_\alpha V_\alpha + pQ + \frac{p^2}{2}\sum_{\alpha=1}^{r} q_\alpha^2$$

on SM*, it is simple and the corresponding eigenfunction can be chosen to be strictly positive.*

(iv) $\lim_{p\to\infty} \frac{1}{p^2}g(p) = 0$ *if and only if* $\lim_{p\to-\infty} \frac{1}{p^2}g(p) = 0$ *if and only if there is a Riemannian structure on* M *with respect to which* $V_1, V_2, \ldots, V_r$ *are infinitesimal isometries.*

Remark. The eigenvalue representation of $g(p)$ in (iii) above is based on the formula

$$|v_t|^p = |v_0|^p \exp\left(p\int_0^t Q(\theta_s)ds + p\sum_{\alpha=1}^r \int_0^t q_\alpha(\theta_s)dW_s^\alpha\right)$$

from (5.9), together with a Girsanov transformation and the Feynman-Kac formula. If $\phi_p : SM \to (0,\infty)$ is a corresponding eigenfunction, then the eigenvalue equation on SM is equivalent to the equation

$$(TL - g(p))\left(|v|^p\phi_p(v/|v|)\right) = 0 \tag{5.16}$$

on TM, where TL denotes the generator for $\{v_t : t \geq 0\}$. It follows that $\{|v_t|^p\phi_p(v_t/|v_t|)e^{-g(p)t} : t \geq 0\}$ is a local martingale for each $p \in \mathbf{R}$. This family of local martingales is very useful when studying the process $\{v_t : t \geq 0\}$. For results in the case of a linear stochastic differential equation (where TM reduces to $\mathbf{R}^d$ and SM to S^{d-1}) see Baxendale [Ba5].

(5.17) Theorem. *Assume* (5.14).
(i) $g'(0) = \lambda_1 = \lim_{t\to\infty} \frac{1}{t}\log|D\xi_t(x)(v)|$ *wp1 for all* $v \in T_xM \setminus \{0\}$.
(ii) $(\log|D\xi_t(x)(v)| - \lambda_1 t)/\sqrt{t}$ *converges in distribution to* $N(0, g''(0))$ *as* $t \to \infty$.
(iii) *Define* $I(y) = \sup\{py - g(p) : p \in \mathbf{R}\}$*; then* I *is the rate function for the large deviation principle associated with the distribution of* $\frac{1}{t}\log|D\xi_t(x)(v)|$ *as* $t \to \infty$.
(iv) $g(-d) \geq 0$ *with equality if and only if* ρ *is invariant for* $\{\xi_t : t \geq 0\}$.

6 Local stability

In this section we consider stability of the original non-linear stochastic flow of diffeomorphisms in the sense of stability along trajectories. That is, for nearby initial points $x \neq y \in M$ we ask whether $\mathrm{dist}(\xi_t(x), \xi_t(y)) \to 0$ as $t \to \infty$. The previous section gave information about the behavior of tangent vectors under the stochastic flow. The principles of linearization indicate that we should obtain similar information about the behavior of the two-point motion $\{(\xi_t(x), \xi_t(y)) : t \geq 0\}$ on M^2 during periods when the two points are close together.

We establish some notation. Let $B(x,\delta)$ and $\overline{B}(x,\delta)$ denote the open and closed balls in M with center x and radius δ. Let $\widehat{M} = \{(x,y) \in M^2 : x \neq y\}$ and $U^\delta = \{(x,y) \in M^2 : 0 < \mathrm{dist}(x,y) < \delta\}$. We write $r_t(x,y) = \mathrm{dist}(\xi_t(x), \xi_t(y))$.

The first result is based on the local stable manifold theorem for stochastic flows, see Ruelle [Ru1] and Carverhill [Ca1]. Roughly stated, the local stable manifold theorem says that if $\mu < 0$ and $\tilde{\lambda}_{i-1} > \mu > \tilde{\lambda}_i$ then for $(\mathbf{P} \times \rho)$-almost all $(\omega, x) \in (\Omega \times M)$ there exist $r(\omega, x) > 0$ and $C(\omega, x) < \infty$ such that the set

$$\{y \in \overline{B}(x, r(\omega, x)) : \mathrm{dist}(\xi_t(\omega, x), \xi_t(\omega, y)) \leq C(\omega, x)e^{\mu t} \text{ for all } t \geq 0\} \tag{6.1}$$

is a smooth submanifold of $\overline{B}(x, r(\omega, x))$ which is tangent to $F^i_{(\omega,x)}$ at x. Observe that the location of the submanifold given by (6.1) depends on the future evolution $\{\xi_t : t \geq 0\}$ of the stochastic flow (except in certain cases of degenerate noise). Accordingly the major usefulness of this result in predicting the future behavior of the stochastic flow occurs in the case where the submanifold is of full dimension in M. In this case we obtain the following.

(6.2) Theorem. *Assume* (5.1) *and* $\lambda_1 < 0$. *If* $0 > \mu > \lambda_1$, *then for* $(\mathbf{P} \times \rho)$*-almost all* $(\omega, x) \in (\Omega \times M)$ *there exist (measureable)* $r(\omega, x) > 0$ *and* $C(\omega, x) < \infty$ *such that*

$$B(x, r(\omega, x)) \subset \{y \in M : \mathrm{dist}(\xi_t(\omega, x), \xi_t(\omega, y)) \leq C(\omega, x)e^{\mu t} \textit{ for all } t \geq 0\}$$

In particular for ρ*-almost all* $x \in M$ *and all* $\varepsilon > 0$ *there exists* $\delta > 0$ *such that*

$$\mathbf{P}\{\mathrm{diam}(\xi_t(B(x, \delta)) \to 0 \textit{ as } t \to \infty\} > 1 - \varepsilon. \tag{6.3}$$

(6.4) Remark. If the one-point motion is sufficiently non-degenerate that its transition probabilities $P(t, x, \cdot)$ are absolutely continuous with respect to ρ for all $x \in M$ and all $t > 0$, then 'ρ-almost all $x \in M$' may be replaced by 'all $x \in M$' in the conditions for (6.3) above. If furthermore M is compact then $\delta > 0$ may be chosen uniformly for $x \in M$, and depending only on ε. For an example of a stochastic flow on the circle where the precise dependence of δ on ε is given (the gradient Brownian flow on the unit circle in $\mathbf{R}^2$) see Baxendale [Ba3].

In deterministic dynamical systems, unstable manifolds can be regarded as stable manifolds for the time-reversed flow. This viewpoint is much less useful for stochastic flows. Whereas for a deterministic flow the families $\{\phi_{-t,0} : t \geq 0\}$ and $\{\phi_{0,t} : t \geq 0\}$ are identical, for stochastic flows the families $\{\xi_{-t,0} : t \geq 0\}$ and $\{\xi_{0,t} : t \geq 0\}$ are independent. In particular knowledge of the local unstable manifold (when defined in terms of $\{\xi_{-t,0} : t \geq 0\}$) is of no direct use in predicting the future (i.e. $t \geq 0$) behavior of the flow. See however Boxler [Bo] and Arnold and Boxler [AB] where a stochastic center manifold is constructed (using knowledge of ξ_t for all $t \in \mathbf{R}$) in the case when $\lambda_i = 0$ for some $1 \leq i \leq d$ and is then used to develop a stochastic bifurcation theory.

The situation when $\lambda_1 \geq 0$ is more complicated than when $\lambda_1 < 0$ because now the linearized process $\{v_t : t \geq 0\}$ does not decay, and so it cannot be such a good approximation to the two-point motion $\{(\xi_t(x), \xi_t(y)) : t \geq 0\}$ on M^2. For example on a compact manifold it is impossible for $r_t(x,y)$ to grow at an exponential rate even if $\lambda_1 > 0$. We adopt a different approach and describe the method of Baxendale and Stroock [BS] which uses Lyapunov functions to describe the behavior of $\{r_t(x,y) : t \geq 0\}$ during the time that the two-point motion lies in some U^δ.

If δ is sufficiently small then for $(x,y) \in U^\delta$ there exists a unique $\overline{v}(x,y) \in T_xM$ such that the geodesic curve with initial velocity $\overline{v}(x,y)$ goes from x to y in time 1. In particular $|\overline{v}(x,y)| = \text{dist}(x,y)$. This provides an isomorphism between U^δ and $\{v \in TM : 0 < |v| < \delta\}$ which allows us to compare the processes $\{D\xi_t(x)(v) : t \geq 0\}$ and $\{(\xi_t(x), \xi_t(y)) : t \geq 0\}$. More precisely we can compare their generators TL and $L^{(2)}$. The following theorem [BS, Thm 3.18] provides a non-linear version of equation (5.16).

(6.5) Theorem. *Assume* (5.14) *and that g is not identically zero.*
(i) *For each $p \in \mathbf{R}$ there exist $\delta > 0$, $K < \infty$, and smooth real-valued functions ϕ_p^+ and ϕ_p^- on U^δ such that*

$$\left(L^{(2)} - g(p)\right)\phi_p^+ \geq 0 \geq \left(L^{(2)} - g(p)\right)\phi_p^-$$

and

$$\frac{1}{K}\text{dist}(x,y)^p \leq \phi_p^\pm \leq K\text{dist}(x,y)^p.$$

(ii) *There exist $\delta > 0$, $K < \infty$, and smooth real-valued functions ψ^+ and ψ^- on U^δ such that*

$$L^{(2)}\psi^+ \geq \lambda_1 \geq L^{(2)}\psi^-$$

and

$$|\psi^\pm(x,y) - \log \text{dist}(x,y)| \leq K.$$

Under assumptions (5.14), g is identically zero if and only if $\{\xi_t : t \geq 0\}$ consists of isometries of some smooth Riemannian structure on M (see [BS, Cor 2.14]). As a consequence of Theorem 6.5, for each $p \in \mathbf{R}$ and while $\{(\xi_t(x), \xi_t(y)) : t \geq 0\}$ remains inside U^δ, the process $\{(r_t(x,y))^p e^{-g(p)t} : t \geq 0\}$ behaves, modulo multiplicative constants, like both the local submartingale $\{\phi_p^+(\xi_t(x), \xi_t(y))e^{-g(p)t} : t \geq 0\}$

and the local supermartingale $\{\phi_p^-(\xi_t(x),\xi_t(y))e^{-g(p)t} : t \geq 0\}$. The processes $\{\phi_p^\pm(\xi_t(x),\xi_t(y))e^{-g(p)t} : t \geq 0\}$ may be regarded as non-linear approximations to the local martingale $\{|v_t|^p\phi_p(v_t/|v_t|)e^{-g(p)t} : t \geq 0\}$ which occurred in the previous section. Similar observations may be made for the process $\{\log r_t(x,y)-\lambda_1 t : t \geq 0\}$. The local submartingales and supermartingales can now be used to deduce information about the stepping times

$$\tau_r(x,y) = \inf\{t > 0 : r_t(x,y) = r\}$$

for $r > 0$.

This method is valid for all $\lambda_1 \in \mathbf{R}$. The case $\lambda_1 < 0$ has already been treated using the local stable manifold theorem. Under the extra assumption (5.14) the explicit dependence of δ on ε may be calculated, see Baxendale [Ba7, Thm 4.6]. We concentrate here on the cases $\lambda_1 > 0$ and $\lambda_1 = 0$. In the case of $\lambda_1 > 0$ the convexity of g together with $g(0) = 0$, $g'(0) = \lambda_1$, and $g(-d) \geq 0$ implies the existence of a unique $\gamma \in (0,d]$ such that $g(-\gamma) = 0$.

(6.6) Theorem. *Assume* (5.14) *and* $\lambda_1 > 0$. *There exist* $\delta > 0$, $K < \infty$ *and* $k \in (0,1)$ *such that the following is true.*

(i) $\mathbf{P}\{\tau_R(x,y) < \infty\} = 1$ *and*

$$\frac{1}{\lambda_1}\left(\log(\frac{R}{\text{dist}(x,y)}) - K\right) \leq \mathbf{E}(\tau_R(x,y)) \leq \frac{1}{\lambda_1}\left(\log(\frac{R}{\text{dist}(x,y)}) + K\right)$$

whenever $0 < \text{dist}(x,y) < R < \delta$.

(ii)

$$\frac{1}{K}\left(\frac{\varepsilon}{\text{dist}(x,y)}\right)^\gamma \leq \mathbf{E}\left(\int_0^{\tau_R(x,y)} \chi_{(0,\varepsilon)}(r_s(x,y))\,ds\right) \leq K\left(\frac{\varepsilon}{\text{dist}(x,y)}\right)^\gamma$$

whenever $0 < \varepsilon < \text{dist}(x,y) < kR < k\delta$.

Proof. See Theorem 3.19 and Corollary 3.24 of [BS]. □

In the case $\lambda_1 = 0$, we define $V = g''(0)$. The convexity of g implies $V \geq 0$. It is shown in Baxendale [Ba7, Prop 5.1] that $V = 0$ if and only if g is identically zero.

(6.7) Theorem. *Assume* (5.14) *and* $\lambda_1 = 0$ *and* $V > 0$. *There exist* $\delta > 0$, $K < \infty$ *and* $k \in (0,1)$ *such that the following is true.*

(i) $\mathbf{P}\{\tau_R(x,y) < \infty\} = 1$ *whenever* $0 < \text{dist}(x,y) < R < \delta$.

(ii) $\mathbf{E}(\tau_R(x,y)) = \infty$ *whenever* $0 < \text{dist}(x,y) < kR < \delta$.

(iii)

$$\begin{aligned}\frac{2}{V}\left[\log\left(\frac{R}{\operatorname{dist}(x,y)}\right)-K\right] &\le \liminf_{\varepsilon\to 0}\frac{1}{|\log\varepsilon|}\mathbf{E}\left(\int_0^{\tau_R(x,y)}\chi_{[\varepsilon,\infty)}(r_s(x,y))\,ds\right)\\ &\le \limsup_{\varepsilon\to 0}\frac{1}{|\log\varepsilon|}\mathbf{E}\left(\int_0^{\tau_R(x,y)}\chi_{[\varepsilon,\infty)}(r_s(x,y))\,ds\right)\\ &\le \frac{2}{V}\left[\log\left(\frac{R}{\operatorname{dist}(x,y)}\right)+K\right]\end{aligned}$$

whenever $0 < \operatorname{dist}(x,y) < R < \delta$.

Proof. See Corollary 5.4 and Proposition 5.6 of [Ba7]. □

Observe that in both cases $\lambda_1 > 0$ and $\lambda_1 = 0$ there is a δ such that for any $x \neq y$ with $\operatorname{dist}(x,y) < \delta$ then almost surely the points $\xi_t(x)$ and $\xi_t(y)$ are at least distance δ apart at some future time. This is a form of local instability.

7 The two-point motion

The previous section described the behavior of the two-point motion within $U^\delta = \{(x,y) \in M^2 : 0 < \operatorname{dist}(x,y) < \delta\}$ for some $\delta > 0$. Here we consider the two-point motion $\{(\xi_t(x),\xi_t(y)) : t \geq 0\}$ on the invariant set $\widehat{M} = \{(x,y) \in M^2 : x \neq y\}$. Notice that although the two-point motion is defined on all of M^2, its restriction to the diagonal $M^2 \setminus \widehat{M}$ looks like a copy of the one-point motion; the joint motion of two distinct points takes place on $\widehat{M}$.

Even if M is compact, the space $\widehat{M}$ will be non-compact. The statement that $\operatorname{dist}(\xi_t(x),\xi_t(y)) \to 0$ can be interpreted as the statement that $(\xi_t(x),\xi_t(y))$ leaves compact subsets of $\widehat{M}$, i.e. that $(\xi_t(x),\xi_t(y))$ goes to infinity in $\widehat{M}$. Thus a statement about stability/instability can be reinterpreted as a statement about transience/recurrence of the two-point motion on $\widehat{M}$.

We make the following assumptions.

(7.1)
(i) M is compact and there is no non-empty compact $C \in \widehat{M}$ which is invariant for the two-point motion.
(ii) $\operatorname{Lie}(TV_1, TV_2, \ldots, TV_r)(v) = T_vTM$ for all $v \in TM$ with $|v| \neq 0$.

Notice that (7.1) implies (5.14). (7.1)(i) implies that for all $\delta > 0$ the two-point motion from any starting position in $\widehat{M}$ will almost surely hit U^δ (for details see [BS, Prop. 4.1]), at which time the estimates from the previous section may be used.

(7.2) Theorem. *Assume* (7.1).
(i) *If* $\lambda_1 < 0$ *then* $\mathbf{P}\{r_t(x,y) \to 0 \text{ as } t \to \infty\} = 1$ *for all* $(x,y) \in \widehat{M}$.
(ii) *If* $\lambda_1 > 0$ *there exists a unique* $\rho^{(2)} \in \mathcal{P}(\widehat{M})$ *such that*

$$\mathbf{P}\left\{\frac{1}{t}\int_0^t f(\xi_s(x),\xi_s(y))\,ds \to \int f\,d\rho^{(2)} \text{ as } t \to \infty\right\} = 1$$

for all bounded measurable $f : \widehat{M} \to \mathbf{R}$ *and all* $(x,y) \in \widehat{M}$. *In particular* $\rho^{(2)}$ *is the unique stationary measure for the two-point motion on* $\widehat{M}$. *Moreover there exist* $\gamma \in (0,d]$ *and* $K < \infty$ *such that*

$$\frac{1}{K}r^\gamma \le \rho^{(2)}\{U^r\} \le Kr^\gamma$$

for all sufficiently small r.
(iii) *If* $\lambda_1 = 0$ *there exists a* σ*-finite measure* $\rho^{(2)}$ *on* $\widehat{M}$, *unique up to a multiplicative constant, such that*

$$\mathbf{P}\left\{\frac{\int_0^t f(\xi_s(x),\xi_s(y))\,ds}{\int_0^t g(\xi_s(x),\xi_s(y))\,ds} \to \frac{\int f\,d\rho^{(2)}}{\int g\,d\rho^{(2)}} \text{ as } t \to \infty\right\} = 1$$

for all bounded measurable $\rho^{(2)}$*-integrable* $f,g : \widehat{M} \to \mathbf{R}$ *with* $\int g\,d\rho^{(2)} \neq 0$ *and all* $(x,y) \in \widehat{M}$. *In particular* $\rho^{(2)}$ *is the unique, up to multiplicative constant, stationary measure for the two-point motion on* $\widehat{M}$. *Moreover there exists* $a \in (0,\infty)$ *such that*

$$\rho^{(2)}\{\widehat{M} \setminus U^r\}/|\log r| \to a \quad \text{as} \quad r \to 0.$$

Proof. See [BS, Thm 4.6] and [Ba7, Thm 4.10 and 4.14]. We note that for (i) the assumption (7.1)(ii) can be replaced by some weaker assumption on the one-point motion, see Remark 6.4. □

(7.3) Remark. Suppose in addition to assumption (7.1) that the two-point motion is non-degenerate on $\widehat{M}$. Then using recurrence in the sense of hitting open sets, Theorem 7.2 gives a complete classification of the two-point motion on $\widehat{M}$ as transient, or null-recurrent, or positive recurrent according as $\lambda_1 < 0$, or $\lambda_1 = 0$, or $\lambda_1 > 0$.

(7.4) Remark. The exponent γ in Theorem 7.2(ii) is characterized by $g(-\gamma) = 0$. It follows from Theorem 5.17(iv) that the following are equivalent.

(i) $\gamma = d$.

(ii) $\rho \times \rho$ is stationary for the two-point motion (whether regarded as a process on M^2 or on $\widehat{M}$).

(iii) ρ is invariant under $\{\xi_t : t \geq 0\}$.

The equivalence of (ii) and (iii) is noted by Harris [Ha] and Kunita [Ku3].

8 Statistical equilibrium

Recall from Section 4 that the probability measure ρ, although stationary for the one-point motion, is often not preserved by the stochastic flow of diffeomorphisms $\{\xi_t : t \geq 0\}$. In this case the process $\{\rho_t : t \geq 0\}$ given by $\rho_t = \rho\xi_t{}^{-1}$ is a random process with values in $\mathcal{P}(M)$.

In this section we will discuss the limiting behavior of $\{\rho_t : t \geq 0\}$ and of the closely related process $\{\rho\xi_{-t,0}{}^{-1} : t \geq 0\}$. We can interpret ρ_t as the random distribution induced by the stochastic flow at time t when starting at time 0 with distribution ρ; similarly $\rho\xi_{-t,0}{}^{-1}$ can be interpreted as the random distribution obtained at time 0 by starting at time $-t$ with distribution ρ. For each fixed $t \geq 0$, ρ_t and $\rho\xi_{-t,0}{}^{-1}$ have the same distribution (as random elements in $\mathcal{P}(M)$). However as processes $\{\rho_t : t \geq 0\}$ and $\{\rho\xi_{-t,0}{}^{-1} : t \geq 0\}$ have their increments on different sides, and this will cause a significant difference in the behavior of the two processes.

Various authors have studied the effect of stochastic flows on measures on M. See for example Harris [Ha], Baxendale and Harris [BH] and Kunita [Ku3]. Here we concentrate on the concept of the statistical equilibrium associated with $\{\xi_t : t \geq 0\}$, as developed by Le Jan [Le1–5] and Young [Yo].

The crucial observation to be made is that $\{\rho\xi_{-t,0}{}^{-1} : t \geq 0\}$ is a $\mathcal{P}(M)$-valued martingale relative to the filtration $\{\mathcal{F}_{-t}^0 : t \geq 0\}$. In fact, if $f : M \to \mathbf{R}$ is any bounded continuous function and $M_t^f = \int_M f\, d(\rho\xi_{-t,0}{}^{-1})$, then

$$\begin{aligned}
\mathbf{E}\left(M_{t+s}^f \,|\, \mathcal{F}_{-t}^0\right) &= \mathbf{E}\left(\int_M (f \circ \xi_{-t-s,0})\, d\rho \,|\, \mathcal{F}_{-t}^0\right) \\
&= \mathbf{E}\left(\int_M (f \circ \xi_{-t,0} \circ \xi_{-t-s,-t})\, d\rho \,|\, \mathcal{F}_{-t}^0\right) \\
&= \mathbf{E}\left(\int_M (f \circ \xi_{-t,0})\, d(\rho \circ \xi_{-t-s,-t}{}^{-1}) \,|\, \mathcal{F}_{-t}^0\right) \\
&= \int_M (f \circ \xi_{-t,0})\, d\rho = M_t^f.
\end{aligned}$$

We have used the fact that $\xi_{-t-s,-t}$ is independent of $\mathcal{F}_{-t}^0$ and has the same distribution as ξ_s, together with the fact that $\mathbf{E}(\rho_s) = \rho$. Notice that the process $\{\xi_{-t,0} : t \geq 0\}$ has independent increments on the right. This fact is crucial to the calculation above, and is the reason why we consider $\{\xi_{-t,0} : t \geq 0\}$ rather than $\{\xi_{0,t} : t \geq 0\}$. (It is instructive to attempt the calculation above with M_t^f and $\mathcal{F}_{-t}^0$ replaced by $\int f\, d\rho_t$ and $\mathcal{F}_0^t$, and to understand why the attempt fails.) Thus

$\{(M_t^f, \mathcal{F}_{-t}^0) : t \geq 0\}$ is a bounded continuous martingale which converges almost surely and in $L^1(\Omega, \mathcal{F}, \mathbf{P})$ to M_∞^f, say, where $\mathbf{E}(M_\infty^f) = \int f \, d\rho$. By restricting f to run through a suitable countable set of bounded continuous functions, we deduce that with probability 1

$$\rho\xi_{-t,0}(\omega, \cdot)^{-1} \to \mu_\omega \text{ as } t \to \infty \tag{8.1}$$

for some $\mu_\omega \in \mathcal{P}(M)$, where the convergence is in the weak topology on $\mathcal{P}(M)$. This result is due to Le Jan [Le4, Le5] and Young [Yo] in the case of independent identically distributed random diffeomorphisms and Le Jan [Le1, Le3] for certain isotropic stochastic flows; the extension to the case of a general stochastic flow of diffeomorphisms is automatic.

In view of our interpretation of $\rho\xi_{-t,0}{}^{-1}$ above, we can think of μ_ω as the measure obtained at time 0 by starting in the infinite past with measure ρ and letting the system evolve randomly up to time 0.

Let Q denote the distribution of the random measure μ_ω. That is, Q is the image of $\mathbf{P}$ under the mapping $\Omega \to \mathcal{P}(M), \omega \mapsto \mu_\omega$. Thus $Q \in \mathcal{P}(\mathcal{P}(M))$. Q is called the ***statistical equilibrium*** associated with the stochastic flow of diffeomorphisms $\{\xi_t : t \geq 0\}$. Heuristically, a typical $\nu \in \mathcal{P}(M)$ chosen according to the distribution Q is the typical probability measure occurring at time 0 in a system which has evolved randomly starting from measure ρ an infinite length of time ago.

For each fixed t, the random measures $\rho\xi_{-t,0}{}^{-1}$ and ρ_t have the same distribution. Since $\rho\xi_{-t,0}(\omega, \cdot)^{-1} \to \mu_\omega$ as $t \to \infty$ for $\mathbf{P}$-almost all $\omega \in \Omega$, it follows that $\rho_t \to \mu_\omega$ as $t \to \infty$ in distribution. (Some authors write $\rho_t \to Q$ as $t \to \infty$ in distribution.) However, except in some special cases, the processes $\{\rho\xi_{-t,0}{}^{-1} : t \geq 0\}$ and $\{\rho_t : t \geq 0\}$ evolve with different laws (since the increments are composed on different sides) and $\{\rho_t : t \geq 0\}$ does not converge in an almost-sure sense.

The existence of the family $\{\mu_\omega : \omega \in \Omega\}$ given by (8.1) is very closely connected with the existence of an invariant probability for $\{\Theta_t : t \in \mathbf{R}\}$ on $(\Omega \times M, \mathcal{F} \times \mathcal{B}(M))$. Recalling that $\{\rho\xi_{-t,0}(\omega, \cdot)^{-1} : \omega \in \Omega\}$ is the regular conditional probability distribution for $(\mathbf{P} \times \rho)\Theta_t^{-1}$ (see Prop. 4.8), we see that (8.1) can be reinterpreted as

$$(\mathbf{P} \times \rho)\Theta_t^{-1} \to \overline{\mathbf{P}} \text{ as } t \to \infty \tag{8.2}$$

where $\overline{\mathbf{P}}$ denotes the probability measure on $(\Omega \times M, \mathcal{F} \times \mathcal{B}(M))$ with marginal $\mathbf{P}$ and regular conditional probability distribution $\{\mu_\omega : \omega \in \Omega\}$. It is clear from (8.1) and (4.4) that

$$\mu_\omega \xi_{0,t}(\omega,\cdot)^{-1} = \mu_{\theta_t\omega} \text{ for all } t \in \mathbf{R} \tag{8.3}$$

for **P**-almost all $\omega \in \Omega$. Moreover, (8.3) is equivalent to the statement that $\overline{\mathbf{P}}$ is preserved by the family $\{\Theta_t : t \in \mathbf{R}\}$. (This also follows directly from (8.2).)

The definition (8.1) of μ_ω, together with the L^1 martingale convergence imply that μ_ω is $\mathcal{F}^0_{-\infty}$ measurable and $\mathbf{E}(\mu_\omega|\mathcal{F}_0^\infty) = \mathbf{E}(\mu_\omega) = \rho$. The corresponding fact about $\overline{\mathbf{P}}$ is that the restriction of $\overline{\mathbf{P}}$ to $\mathcal{F}_0^\infty \times \mathcal{B}(M)$ is $(\mathbf{P}|_{\mathcal{F}_0^\infty}) \times \rho$ (see Prop. 4.8).

We give results on the nature of the statistical equilibrium. The first result is due to Le Jan [Le5], arguing directly from the multiplicative ergodic theorem and the local stable manifold theorem.

(8.4) Theorem. *Assume* (5.1) *and that M is compact.*
(i) *If $\lambda_\Sigma < 0$ then μ_ω is* **P***-almost surely singular with respect to the Riemannian measure m.*
(ii) *If $\lambda_1 < 0$ then there exists $n \geq 1$ such that μ_ω consists* **P***-almost surely of n atoms of mass $1/n$.*

Recall our earlier result (Theorem 5.12) asserting, under mild non-degeneracy conditions on ρ, that $\lambda_\Sigma \leq 0$ with equality if and only if ρ is invariant under $\{\xi_t : t \geq 0\}$, in which case clearly $\mu_\omega \equiv \rho$.

In Theorem 8.4(ii) all positive integer values of n can occur (for example for $\mathbf{Z}_n$-invariant flows on the circle). However under a non-degeneracy condition such as (7.1)(i), it can be shown that $n = 1$. Once it is known that μ_ω consists of a single atom, we may use the fact that $\mathbf{E}(\mu_\omega) = \rho$ to assert that the position of the atom is randomly distributed with distribution ρ.

An alternative approach to studying μ_ω is via its 'second moment' $\mathbf{E}(\mu_\omega \times \mu_\omega)$. The idea of using the two-point motion to study μ_ω is due to Le Jan [Le1, Le2]. It is easy to verify that $\mathbf{E}(\mu_\omega \times \mu_\omega)$ is a stationary probability measure for the two-point motion $\{(\xi_t(x), \xi_t(y)) : t \geq 0\}$ on M^2. Indeed it is the limiting probability measure for the two-point motion on M^2 started with initial distribution $\rho \times \rho$. The question of whether μ_ω contains atoms is clearly related to the value of $\mathbf{E}(\mu_\omega \times \mu_\omega)(D)$, where D denotes the diagonal $\{(x,y) \in M^2 : x = y\} = M^2 \setminus \widehat{M}$. Furthermore, if $\mathbf{E}(\mu_\omega \times \mu_\omega)(\widehat{M}) > 0$ then (after rescaling) $\mathbf{E}(\mu_\omega \times \mu_\omega)|_{\widehat{M}}$ provides a stationary probability measure for the two-point motion on $\widehat{M}$. Putting these ideas together with the results of section 7 we obtain the following.

(8.5) Theorem. *Assume* (7.1).
(i) *If* $\lambda_1 \leq 0$ *then* μ_ω *consists* **P**-*almost surely of a single atom.*
(ii) *If* $\lambda_1 > 0$ *then* μ_ω *is* **P**-*almost surely atomless. Moreover there exist* $\gamma \in (0, d]$ *and* $K < \infty$ *such that*

$$\frac{1}{K} r^\gamma \leq \mathbf{E}^{\overline{\mathbf{P}}}(\mu_\omega(B(x,r))) \leq K r^\gamma$$

for all sufficiently small $r > 0$.

Proof. See [Ba7, Thm 4.10, 4.18, and 5.12] The result for $\lambda_1 < 0$ follows easily from Theorem 8.4. □

Here, as earlier, the exponent γ is characterized by $g(-\gamma) = 0$. Let $\{X_r : r > 0\}$ be the family of random variables on $(\Omega \times M, \mathcal{F} \times \mathcal{B}(M), \overline{\mathbf{P}})$ defined by $X_r(\omega, x) = \mu_\omega(B(x,r))$. Thus X_r measures how much mass a typical μ_ω assigns to the ball $B(x,r)$ whose center x is chosen with distribution μ_ω. In the case when $\lambda_1 > 0$ then μ_ω is almost surely atomless and so $X_r \to 0$ as $r \to 0$ almost surely $(\overline{\mathbf{P}})$. One way of describing a typical μ_ω would be to describe the rate at which $X_r \to 0$ as $r \to 0$. The result above shows that $X_r \to 0$ in $L^1(\overline{\mathbf{P}})$ like r^γ. The following result, due to Ledrappier and Young [LY2], gives the almost sure convergence rate. The Lyapunov dimension σ, say, of $\{\xi_t : t \geq 0\}$ is defined by

$$\sigma = k - \frac{1}{\lambda_{k+1}} \sum_{j=1}^{k} \lambda_j$$

where k is the largest integer such that $\sum_{j=1}^{k} \lambda_j > 0$.

(8.6) Theorem. *Assume that* M *is compact,* $\lambda_1 > 0$, $\lambda_j \neq 0$ *for all* j *and*

(8.7) $$\mathrm{Lie}(GLV_0, GLV_1, \ldots, GLV_r)(f) = T_f GLM \text{ for all } f \in GLM.$$

Then

$$\lim_{r \to 0} \frac{\log X_r}{\log r} = \sigma \textit{ almost surely } (\overline{\mathbf{P}}).$$

In particular, μ_ω *almost surely* (**P**) *has Hausdorff dimension* σ.

Here GLM denotes the bundle of linear frames over M and GLV_α is the natural lift of V_α to a vector field on GLM. The result of Ledrappier and Young remains valid under various alternative non-degeneracy conditions, see [LY2] for more details.

It follows from Jensen's inequality that $\sigma \geq \gamma$ whenever the hypotheses of both Theorems 8.5 and 8.6 are valid. A related inequality is given by Le Jan [Le4]. Examples are known in which $\sigma > \gamma$ (see [Ba7]).

9 Random sources and sinks

We will assume throughout this section that M is compact and that

$$\mathrm{Lie}(V_1, V_2, \ldots, V_r)(x) = T_xM \text{ for all } x \in M.$$

This implies that the one-point motion $\{\xi_t(x) : t \geq 0\}$ has a unique stationary $\rho \in \mathcal{P}(M)$, and that ρ has a smooth, strictly positive density with respect to the Riemannian measure m.

Suppose that the stochastic flow of diffemorphisms $\{\xi_t : t \geq 0\}$ has property the that

$$\mathbf{P}\{\mathrm{dist}(\xi_t(x), \xi_t(y)) \to 0 \text{ as } t \to \infty\} = 1 \tag{9.1}$$

for all $x, y \in M$. We have seen earlier that this will be the case if $\lambda_1 < 0$ and certain non-degeneracy conditions are satisfied. Then for any fixed $x \in M$ the trajectory $\{\xi_t(x) : t \geq 0\}$ can be regarded as a (wandering) *random sink* in the sense that

$$m\{y \in M : \xi_t(y) \in B(\xi_t(x), \delta)\} \to \mathrm{vol}(M) \text{ as } t \to \infty \tag{9.2}$$

for each fixed $\delta > 0$. Moreover, for any $\nu \in \mathcal{P}(M)$,

$$\nu\xi_t^{-1} - \delta(\xi_t(x)) \to 0 \text{ as } t \to \infty \tag{9.3}$$

where $\delta(x)$ denotes the unit measure at x, and the convergence is in the weak topology in the space of bounded signed measures on M.

Both (9.2) and (9.3) express the fact that as t increases, ξ_t clusters more and more of M in a small neighborhood of $\xi_t(x)$. Of course as t increases the position of this cluster moves around M like the one-point motion. In particular this shows that in this case ρ_t does not achieve an almost sure limit.

Topological considerations in the case that M is compact lead us to believe that if most of M is being squashed together by ξ_t then there must be some small piece of M which is being greatly expanded by ξ_t. Thus we are interested in the behavior of $m(\xi_t(A))$ as $t \to \infty$, for $A \in \mathcal{B}(M)$. This question has been considered for stochastic flows in Euclidean spaces by Baxendale and Harris [BH], Le Jan [Le1, Le3] and Kunita [Ku3]. We proceed by considering the stochastic flow of diffeomorphisms $\{\xi_{-t,0}{}^{-1} : t \geq 0\}$. Notice that $\{\xi_{0,t} : t \geq 0\}$ has independent left increments, so $\{\xi_{-t,0} : t \geq 0\}$ has independent right increments, and $\{\xi_{-t,0}{}^{-1} : t \geq 0\}$ again has independent left increments. It is shown by Kunita [Ku1] that the process $\{\xi_{-t,0}{}^{-1} : t \geq 0\}$ arises (in distribution) from a version of equation (1.1) in which V_0 is replaced by $-V_0$. In particular the generator of the one-point motion $\{\xi_{-t,0}{}^{-1}(x) : t \geq 0\}$ in M is given by

(9.4) $$L^- = \frac{1}{2}\sum_{\alpha=1}^{r} V_\alpha^2 - V_0.$$

Suppose that ρ^- is stationary for the one-point motion $\{\xi_{-t,0}{}^{-1}(x) : t \geq 0\}$. Thus $\mathbf{E}(\rho^-\xi_{-t,0}) = \rho^-$, or equivalently $\mathbf{E}(\rho^-\xi_{0,t}) = \rho^-$. The martingale convergence argument of Section 8 can now be adapted to show that $\{\rho^-\xi_{0,t} : t \geq 0\}$ is a $\mathcal{P}(M)$ valued martingale relative to $\{\mathcal{F}_0^t : t \geq 0\}$, and so

(9.5) $$\rho^-\xi_{0,t}(\omega,\cdot) \rightarrow \mu_\omega^- \text{ as } t \rightarrow \infty$$

for some $\mu_\omega^- \in \mathcal{P}(M)$, where the convergence is in the weak topology on $\mathcal{P}(M)$.

Clearly $\{\mu_\omega^- : \omega \in \Omega\}$ is the statistical equilibrium for the reversed stochastic flow $\{\xi_{-t,0}{}^{-1} : t \geq 0\}$, so we can make many of the same sort of statements about μ_ω^- as we did earlier about μ_ω. Notice that the change of V_0 to $-V_0$ will entail a new computation of the Lyapunov exponents. The statements about μ_ω^- have to be made in terms of the Lyapunov exponents $\lambda_1^- \geq \lambda_2^- \geq \cdots \geq \lambda_d^-$ for $\{\xi_{-t,0}{}^{-1} : t \geq 0\}$. There appears to be no a priori reason why the flows $\{\xi_{0,t} : t \geq 0\}$ and $\{\xi_{-t,0}{}^{-1} : t \geq 0\}$ should have the same Lyapunov exponents. (Notice however that $\lambda_1 + \lambda_2 + \cdots + \lambda_d = 0$ if and only if ρ is invariant under $\{\xi_t : t \geq 0\}$ if and only if ρ^- is invariant under $\{\xi_{-t,0}{}^{-1} : t \geq 0\}$ if and only if $\lambda_1^- + \lambda_2^- + \cdots + \lambda_d^- = 0$.) One important difference is that whereas μ_ω is $\mathcal{F}_{-\infty}^0$ measurable, the measure μ_ω^- depends on the forward evolution of the stochastic flow from time 0 to time ∞.

Suppose now in addition that λ_1^-, the top Lyapunov exponent for the reversed stochastic flow $\{\xi_{-t,0}{}^{-1} : t \geq 0\}$, is negative. Then (under a version of assumption (7.1)(i) for the reversed flow) we may deduce that μ_ω^- consists almost surely of a single atom. That is, $\mu_\omega^- = \delta(x_\omega)$ for some $\mathcal{F}_0^\infty$ measurable random point $x_\omega \in M$. If A is an open neighborhood of x_ω in M then $\rho^-(\xi_t(A)) \rightarrow 1$ as $t \rightarrow \infty$; and if A is closed with $x_\omega \notin A$ then $\rho^-(\xi_t(A)) \rightarrow 0$ as $t \rightarrow \infty$. In other words for any $A \in \mathcal{B}(M)$,

$$m(\xi_t(A)) \rightarrow \text{vol}(M) \text{ as } t \rightarrow \infty \qquad \text{if} \quad x_\omega \in \text{int}(A)$$

and

$$m(\xi_t(A)) \rightarrow 0 \text{ as } t \rightarrow \infty \qquad \text{if} \quad x_\omega \notin \text{cl}(A).$$

Thus we may interpret x_ω as the position of the *random source* of the particular evolution $\{\xi_t(\omega,\cdot) : t \geq 0\}$.

Notice that we may find the position of x_ω corresponding to a particular evolution $\{\xi_t(\omega,\cdot) : t \geq 0\}$ by observing which sets expand in volume and which sets contract.

More precisely, if $\{A_i : i \geq 1\}$ denotes a countable basis of open sets for the topology of M, then

$$x_\omega = \bigcap_{i \in I(\omega)} A_i$$

where

$$I(\omega) = \{i \geq 1 : m(\xi_t(A_i)) \to \text{vol}(M) \text{ as } t \to \infty\}.$$

(9.6) Remark. Since $E(\mu_\omega^-) = \rho^-$, it follows that the position of the random source x_ω has distribution ρ^-. At time 0 this is all we can say about x_ω; it is only after any particular evolution of the stochastic flow that we can pinpoint x_ω.

(9.7) Remark. If the increments $\{\xi_{n-1,n} : n \geq 1\}$ of the stochastic flow were commutative then for each continuous function $f : M \to \mathbf{R}$ the value $\int_M f \, d\mu_\omega^-$ would be an exchangeable function of the family $\{\xi_{n-1,n} : n \geq 1\}$, and the Hewitt-Savage 0–1 law would apply. In this case we could deduce that μ_ω^- would be a fixed probability measure, and so $\mu_\omega^- = \rho^-$. In particular ρ^- would be invariant for $\{\xi_{-t,0}{}^{-1} : t \geq 0\}$, implying in turn that $\rho^- = \rho$ and that ρ is invariant for $\{\xi_t : t \geq 0\}$. We conclude that the existence of random point sources is a consequence of the lack of commutativity in the increments of $\{\xi_t : t \geq 0\}$. Similar remarks may be made about the randomness of the filtration $\{F^i_{(\omega,x)} : 1 \leq i \leq \tilde{d}\}$ and of the local stable manifolds.

(9.8) Remark. Even if we drop the assumption that $\lambda_1^- < 0$, then (9.1) implies that $\lambda_1 + \lambda_2 + \cdots + \lambda_d < 0$ which implies in turn that $\lambda_1^- + \lambda_2^- + \cdots + \lambda_d^- < 0$ so that μ_ω^- is almost surely singular with respect to m. Therefore there is a random $B_\omega \in \mathcal{B}(M)$ such that $m(B_\omega) = 0$ and $\mu_\omega^-(B_\omega) = 1$. The set B_ω now plays the role of a generalized random source.

REFERENCES

[Ar] L. Arnold (1984). A formula connecting sample and moment stability of linear stochastic systems. *SIAM J. Appl. Math.* **44** 793–802.

[AB] L. Arnold and P. Boxler (1989). Eigenvalues, bifurcation and center manifolds in the presence of noise. In *Equadiff '87* (C. Dafermos et al, eds) Lecture Notes Pure & Applied Math. Ser. **118**. Dekker, New York.

[ACE] L. Arnold, H. Crauel and J.-P. Eckmann, eds *Lapunov exponents, Proc. Oberwolfach 1990* Lect. Notes Math. Springer, Berlin Heidelberg New York. (To appear)

[AK1] L. Arnold and W. Kliemann (1983). Qualitative theory of stochastic systems. In *Probabilistic Analysis and Related Topics* (A. T. Bharucha-Reid, ed.) **3** 1–79. Academic Press, New York.

[AK2] L. Arnold and W. Kliemann (1987). Large deviations of linear stochastic differential equations. In *Stochastic differential systems* (H.J. Engelbert, W. Schmidt, eds) Lect. Notes Control Inf. Sci. **96** 117–151. Springer, Berlin Heidelberg New York.

[AKO] L. Arnold, W. Kliemann and E. Oeljeklaus (1986). Lyapunov exponents of linear stochastic systems. In [AW] 85–125.

[AOP] L. Arnold, E. Oeljeklaus and E. Pardoux (1986). Almost sure and moment stability for linear Itô equations. In [AW] 129–159.

[AW] L. Arnold and V. Wihstutz, eds (1986). *Lyapunov exponents, Proc. Bremen 1984* Lect. Notes Math. **1186** Springer, Berlin Heidelberg New York.

[Ba1] P.H. Baxendale (1984). Brownian motions in the diffeomorphism group I. *Compositio Math.* **53** 19–50.

[Ba2] P.H. Baxendale (1986). The Lyapunov spectrum of a stochastic flow of diffeomorphisms. In [AW] 322–337.

[Ba3] P.H. Baxendale (1986). Asymptotic behaviour of stochastic flows of diffeomorphisms. In *Stochastic processes and their applications. Proc. Nagoya 1985.* (K. Itô and T. Hida, eds) Lect. Notes Math. **1203** 1–19. Springer, Berlin Heidelberg New York.

[Ba4] P.H. Baxendale (1986). Asymptotic behaviour of stochastic flows of diffeomorphisms: two case studies. *Probab. Th. Rel. Fields* **73** 51–85.

[Ba5] P.H. Baxendale (1987). Moment stability and large deviations for linear stochastic differential equations. In *Proc. Taniguchi Symposium on Probabilistic Methods in Mathematical Physics. Katata and Kyoto 1985.* (N. Ikeda, ed.) 31–54. Kinokuniya, Tokyo.

[Ba6] P.H. Baxendale (1989). Lyapunov exponents and relative entropy for a stochastic flow of diffeomorphisms. *Probab. Th. Rel. Fields* **81** 521–554.

[Ba7] P.H. Baxendale (1991). Statistical equilibrium and two-point motion for a stochastic flow of diffeomorphisms. In *Spatial Stochastic Processes: Festschrift* *for T. E. Harris* (K. Alexander, J. Watkins, eds) Birkhauser, Boston Basel Stuttgart (in press).

[BH] P.H. Baxendale and T.E. Harris (1986). Isotropic stochastic flows. *Ann. Probab.* **14** 1155–1179.

[BS] P.H. Baxendale and D.W. Stroock (1988). Large deviations and stochastic flows of diffeomorphisms. *Probab. Th. Rel. Fields* **80** 169–215.

[Bh] R. N. Bhattacharya (1978). Criteria for recurrence and existence of invariant measures for multidimensional diffusions. *Ann. Probab.* **6** 541–553.

[Bo] P. Boxler (1989). A stochastic version of center manifold theory. *Probab. Th. Rel. Fields* **83** 509–545.

[Ca1] A.P. Carverhill (1985). Flows of stochastic dynamical systems: ergodic theory. *Stochastics* **14** 273–317.

[Ca2] A.P. Carverhill (1985). A formula for the Lyapunov numbers of a stochastic flow. Application to a perturbation theorem. *Stochastics* **14** 209–226.

[Ca3] A.P. Carverhill (1986). A non-random Lyapunov spectrum for nonlinear stochastic dynamical systems. *Stochastics* **17** 253–287.

[CE] A.P. Carverhill and K. D. Elworthy (1986). Lyapunov exponents for a stochastic analogue of the geodesic flow. *Trans. Amer. Math. Soc.* **295** 85–105.

[Cr1] H. Crauel (1987). *Random dynamical systems: positivity of Lyapunov exponents, and Markov systems.* Dissertation, Universität Bremen.

[Cr2] H. Crauel (1989). Markov measures for random dynamical systems. Preprint, Universität Bremen.

[Da] R.W.R. Darling (1991). Isotropic stochastic flows: a survey. (In this volume)

[El1] K.D. Elworthy (1982). *Stochastic differential equations on manifolds.* Cambridge University Press.

[El2] K.D. Elworthy (1989). Geometric aspects of diffusions on manifolds. In *École d'Été de Probabilités de Saint-Flour XV–XVII.* (P. Hennequin, ed.) Lect. Notes Math. **1362** 276–425. Springer, Berlin Heidelberg New York.

[El3] K.D. Elworthy (1991). Stochastic flows on Riemannian manifolds. (In this volume)

[ES] K.D. Elworthy and D.W. Stroock (1986). Large deviation theory for mean exponents of stochastic flows. In *Stochastic processes, mathematics and physics.* (S. Albeverio, P. Blanchard, L. Streit, eds) Lect. Notes Math. **1158** 72–80. Springer, Berlin Heidelberg New York.

[FK] T. Fujiwara and H. Kunita (1985). Stochastic differential equations of jump type and Lévy processes in diffeomorphisms group. *J. Math. Kyoto Univ.* **25** 71–106.

[Ha] T.E. Harris (1981). Brownian motions on the homeomorphisms of the plane. *Ann. Probab.* **9** 232–254.

[IW1] N. Ikeda and S. Watanabe (1981). *Stochastic differential equations and diffusion processes.* North-Holland/Kodansha, Amsterdam/Tokyo.

[IW2] N. Ikeda and S. Watanabe (1984). Stochastic flows of diffeomorphisms. *Advances in Probability* **7** (Stochastic analysis and applications, M.Pinsky ed.) 179–198.

[It] K. Itô (1963). The Brownian motion and tensor fields on a Riemannian manifold. In *Proc. Internat. Congr. Math. Stockholm 1962.* 536–539. Inst. Mittag-Leffler, Djursholm.

[Kh1] R. Z. Khas'minskii (1967). Necessary and sufficient conditions for the asymptotic stability of linear stochastic systems. *Theory Probab. Appl.* **12** 144–147.

[Kh2] R. Z. Khas'minskii (1980). *Stochastic stability of differential equations.* Sijthoff and Noordhoff, Alphen aan den Rijn.

[Ki] Y. Kifer (1986). *Ergodic theory of random transformations.* Birkhäuser, Boston.

[Ku1] H. Kunita (1982). On bachward stochastic differential equations. *Stochastics* **6** 293–313.

[Ku2] H. Kunita (1984). Stochastic differential equations and stochastic flow of diffeomorphisms. In *École d'Été de Probabilités de Saint-Flour XII.* (P. Hennequin, ed.) Lect. Notes Math. **1097** 143–303. Springer, Berlin Heidelberg New York.

[Ku3] H. Kunita (1990). *Stochastic flows and stochastic differential equations.* Cambridge University Press.

[Kuo] H.H. Kuo (1975). *Gaussian measures in Banach spaces.* Lect. Notes Math. **463**. Springer, Berlin Heidelberg New York.

[Kus] H. Kushner (1967). *Stochastic stability and control.* Academic Press, New York.

[LY1] F. Ledrappier and L.-S. Young (1988). Entropy formula for random transformations. *Probab. Th. Rel. Fields* **80** 217–240.

[LY2] F. Ledrappier and L.-S. Young (1988). Dimension formula for random transformations. *Commun. Math. Phys.* **117** 529–548.

[Le1] Y. Le Jan (1984). Équilibre et exposants de Lyapounov de certains flots browniens. *C. R. Acad. Sci. Paris Sér. A* **298** 361–364.

[Le2] Y. Le Jan (1985). Equilibrium state for turbulent flows of diffusion. In *Infinite dimensional analysis and stochastic processes.* Pitman Research Notes **124** 83–93. Longman, London.

[Le3] Y. Le Jan (1985). On isotropic stochastic flows. *Z. Wahrscheinlichkeitstheorie verw. Gebiete* **70** 609–620.

[Le4] Y. Le Jan (1986). Hausdorff dimension for the statistical equilibrium of stochastic flows. In *Stochastic processes — mathematics and physics.* (S. Albeverio, Ph. Blanchard, L. Streit, eds) Lect. Notes Math. **1158** 201–207.

[Le5] Y. Le Jan (1987). Équilibre statistique pour les produits de difféomorphismes aléatoires indépendants. *Ann. Inst. Henri Poincaré (Probabilités et Statistiques)* **23** 111–120.

[LW] Y. Le Jan and S. Watanabe (1984). Stochastic flows of diffeomorphisms. In *Proc. Taniguchi Symposium on Stochastic Analysis. Katata and Kyoto 1982.* (K. Itô, ed.) 307–332. Kinokuniya, Tokyo.

[Li1] M. Liao (1990). The existence of isometric stochastic flows for Riemannian Brownian motions. (In this volume)

[Li2] M. Liao (1991). The Brownian motion and the canonical stochastic flow on a symmetric space. (To appear)

[Oh] T. Ohno (1983). Asymptotic behaviours of dynamical systems with random parameters. *Publ. R.I.M.S. Kyoto Univ.* **19** 83–98.

[Os] V. I. Oseledec (1968). A multiplicative ergodic theorem. Lyapunov characteristic numbers for dynamical systems. *Trans. Moscow Math. Soc.* **19** 197–231.

[RW] L.C.G. Rogers and D. Williams (1987). *Diffusions, Markov processes, and Martingales. Volume 2: Itô calculus.* Wiley, Chichester.

[Ru1] D. Ruelle (1979). Ergodic theory of differential dynamical systems. *Publ. Math. IHES* **50** 275–306.

[Ru2] D. Ruelle (1989). *Elements of Differentiable Dynamics and Bifurcation Theory.* Academic Press, San Diego.

[St] D.W. Stroock (1986). On the rate at which a homogeneous diffusion approaches a limit, an application of the large deviation theory of certain stochastic integrals. *Ann. Probab.* **14** 840–859.

[WZ] E. Wong and M. Zakai (1969). Riemann-Stieltjes approximations of stochastic integrals. *Z. Wahrscheinlichkeitstheorie verw. Gebiete* **12** 87–97.

[Yo] L.-S. Young (1986). Stochastic stability of hyperbolic attractors. *Ergod. Th. & Dynam. Sys.* **6** 311–319.

Peter H. Baxendale
Department of Mathematics
University of Southern California
Los Angeles, CA 90089-1113, USA

Stochastic Flows on Riemannian Manifolds

K.D. Elworthy

Introduction

The global analytic approach to flows of stochastic differential equations can simplify and clarify many proofs, especially by reducing the repitition of similar estimates. It is not generally appreciated that there are several different probabilistic solutions to the heat equations for differential forms. Given a gradient vector field ∇h on a torus with respect to some smooth Riemannian metric, there is no way of adding noise (large or small) to give a moment stable stochastic flow with corresponding diffusion generator $\frac{1}{2}\alpha\Delta + \nabla h$ where δ is the Laplace–Beltrami operator and $\alpha > 0$.

The scheme of this article is to briefly describe the global analytic approach, and especially Baxendale's integrability theorem in Sections 1 and 2. This is used in Section 3 to look at the various semigroups on differential forms obtained using derivatives of stochastic flows. Gradient flows have especially good properties, described in Section 4 where the details of Kusuoka's result in [32] about how they can be used to solve the heat equation for p-forms are given. Somewhat better known solutions to these heat equations are described in Section 5 although the treatment of the heat equation for Witten type deformed Laplacians does not seem well known. In Section 6, moment exponents are described and it is shown that there are topological obstructions to moment stability for flows with generator of the form $\frac{1}{2}\Delta + \nabla h$. For gradient flows, more detailed results are given. At the end of Section 6 there is some discussion of Bougerol's characterization of gradient Brownian flows on spheres [12]. In the last section the earlier results are used to give a preliminary look at some reformulations of Elliott–Kohlmann 'homogeneous chaos expansions' and to get a new solution of the heat equation for closed 1-forms (with deformed Laplacian) which on exact forms essentially reduces to Bismut's celebrated formula for $\nabla \log p_t(x, y)$.

We deal mainly with the compact case. However the non-compact case is especially interesting and a discussion of it on the lines of [25] and [27] is expected to give more detailed results on toplogocial obstructions to stability than we have been able to give here in both the compact and the non-compact case.

1. Construction

A. Consider a stochastic differential equation

$$dx_t = X(x_t) \circ dB_t + A(x_t)dt \tag{1}$$

on a connected n-dimensional manifold M. We will assume that M, X and A are C^∞. For simplicity, $\{B_t : t \geq 0\}$ is a Brownian motion for some $\mathbb{R}^m$ with m not necessarily equal to n. Then A is a vector field and for $y \in M$

$$X(y) : \mathbb{R}^m \to T_yM$$

is a linear map into the tangent space at y to M.

It is sometimes convenient to introduce the vector fields $X^1, \ldots, X^m$ on M given by

$$X^i(y) = X(y)(e_i)$$

for $e_1, \ldots, e_m$ the standard basis for $\mathbb{R}^m$. Equation (1) can then be written

$$dx_t = \sum_{i=1}^{m} X^i(x_t) \circ dB_t^i + A(x_t)dt \tag{1$'$}$$

where $B_t^1, \ldots, B_t^m$ are the components of B_t.

It will be convenient later to take the canonical representation of our basic Brownian motion B, so $\Omega = C_0([0,\infty); \mathbb{R}^m)$, the space of continuous paths in $\mathbb{R}^m$ starting at the origin, and $\mathbb{P}$ is a Wiener measure, with $\mathcal{F}, \{\mathcal{F}_t : t \geq 0\}$, the Borel σ-algebra and standard filtration (augmented so that it satisfies the usual conditions).

B. A *solution flow* for (1) will be taken to be some

$$F_t : M \times \Omega \to M \qquad \{t \geq 0\}$$

such that each ω in Ω

$$F_{\cdot}(\cdot, \omega) : [0, \infty) \times M \to M$$

is continuous, and such that $\{F_t(y) : t \geq 0\}$ is a solution to (1) with $F_0(y) = y$, for each $y \in M$. For a thorough treatment of stochastic flows there is the recent book by Kunita [31]; the sketch which follows is designed to emphasize the aspects needed here.

When M is compact the most direct method of proving the existence of a solution flow is to take any smooth embedding of M in some $\mathbb{R}^p$, extend X and A to give a stochastic differential equation on $\mathbb{R}^p$ whose coefficients have compact support. The Totoki–Kolomogorov criterion for the existence of sample continuous versions can then be used to obtain a flow for this equation on $\mathbb{R}^p$, following Blagovescenskii and Freidlin e.g. see [23] and [31]. This restricts to give the required flow on M (there are details in [23]).

However to get more information about the flow, for example its composition properties or its inverse, a lot of unnecessary estimation can be avoided, following [23], by working on a suitable Sobolev space of diffeomorphisms of M. This is described in detail in [23] with a version in [13] which avoids most of the infinite dimensional manifold theory and gives more applications. The idea is that for sufficiently large s the space $H^s(M;M)$ of maps locally in the Sobolev class H^s forms a separable Hilbert manifold and the group $\mathcal{D}^s$ of diffeomorphisms in $H^s(M;M)$ is an open subset, thereby inheriting a Hilbert manifold structure. The tangent space $T_{id}\mathcal{D}^s$ to $\mathcal{D}^s$ at the identity can be identified with the H^s vector fields on M. Define

$$\widetilde{X}(id) : \mathbb{R}^m \to T_{id}\mathcal{D}^s$$

by

$$\widetilde{X}(id)(e) = X(\cdot)(e)$$

and let

$$\widetilde{X}(h)(e) = T_(id)R_h[\widetilde{X}(id)(e)]$$

for $h \in \mathcal{D}^s$ where $R_h : \mathcal{D}^s \to \mathcal{D}^s$ is right translation i.e.

$$g \mapsto g \circ h \quad \text{and} \quad T_{id}R_h : T_{id}\mathcal{D}^s \to T_h\mathcal{D}^s$$

is the induced (derivative) map on tangent spaces.

Similarly $\widetilde{A}$ is defined to be the the vector field on $\mathcal{D}^s$ given by $\widetilde{A}(h) = T_{id}R_h[A]$. The stochastic differential equation

$$dF_t = \widetilde{X}(F_t) \circ dB_t + \widetilde{A}(F_t) \tag{2}$$

is $\mathcal{D}^s$ is right invariant and has a solution $\{F_t : t \geq 0\}$ with $F_0 = id$. it is easily seen that this solution is a solution flow for (1). Moreover it consists of diffeomorphisms and since $\mathcal{D}^\infty = \bigcap_s \mathcal{D}^s$ is the space of C^∞ diffeomorphisms it can be chosen so that $F_t(y,\omega)$ is C^∞ in y and continuous in t into the C^∞ topology of $\mathcal{D}^\infty$.

C. This can be generalized to the case of "infinite dimensional noise." The simplest way to do this is to observe that our S. D. E. (1) determines an element A of $T_{id}\mathcal{D}^\infty$ and a Gaussian measure γ^x on $T_{id}\mathcal{D}^\infty$, considered as the space of C^∞ vector fields on M, by letting γ^x be the image under $e \mapsto X(\cdot)e$ of the standard Gaussian measure on $\mathbb{R}^n$. Naturally associated to (γ^x, A) is the process $\{W_t : t \geq 0\}$

$$W_t = X(\cdot)B_t + A(\cdot)t \tag{3}$$

which is just the Wiener process of γ^x with constant drift A. Equation (2) is then equivalent to

$$dF_t = (T_{id}R_{F_t})(\circ\, dW_t). \tag{2'}$$

From this, two points emerge:

(a) to obtain more general flows simply take a mean zero Gaussian measure γ on a suitable space E of vecotr fields on M and a vector field A in E (or equivalently a Gaussian measure with mean A for some A) and let $\{W_t : t \geq 0\}$ be the E-valued process which is the Wiener process $\{W_t^A : t \geq 0\}$ of γ with constant drift A. The flow is obtained by solving (2′). There is then a corresponding stochastic differential equation on M of which $\{F_t : t \geq 0\}$ will be the solution flow, namely

$$dx_t = X(x_t) \circ dW_t^A.$$

or equivalently $dx_t = X(x_t) \circ dW_t^0 + A(x_t)dt$ where

$$X(y) : E \mapsto T_y M$$

is the evaluation map.

(b) The flow is determined by γ^x and A. Since γ^x is determined by its covariance Γ^x,

$$\Gamma^x : \bigcup_{x,y} T_x^* M \otimes T_y^* M \to \mathbb{R}$$

with

$$\Gamma^x(\alpha \otimes \beta) = \int_E \alpha(V(x))\beta(V(y))d\gamma^x(V)$$

for $\alpha \otimes \beta \in T_x^* M \otimes T_y^* M$, the flow is determined by Γ^x and A.

The point (a) above brings the 'global analytic' approach here into line with Baxendale's approach in [5] where he started with Gaussian measures on vector fields. It also relates it to the extensions of Le Jan and Watanabe [33] and Kunita, see [31], using new forms of stochastic integrals. The global analytic approach demains a certain amount of regularity e.g. Γ^x being supported on the space of H^s vector fields for s sufficiently big ($s > \frac{1}{2}n + 1$ would suffice). The more direct methods do not need so much, and this is important in the analysis of general stochastic flows, a topic we next consider briefly.

D. In [6] Baxendale considered random processes $\{\varphi_t : t \geq 0\}$ taking the values in a Polish topological group G (i.e. one which is separable and admits a complete metric inducing its topology). These were called *Brownian motions* on G if they satisfy

(i) for $0 \leq s < t \leq u < v$ there is independence of $\varphi_t\varphi_s^{-1}$ from $\varphi_v\varphi_u^{-1}$ i.e. "independent increments on the left"

(ii) time homogeneity: the distribution of $\varphi_t\varphi_s^{-1}$ depends only on $t - s$

(iii) almost sure sample continuity
(iv) $\varphi_0 = e$ almost surely for e the identity element of G.

For example our process $\{W_t : t \geq 0\}$ given by (3) is a Brownian motion on $T_{id}\mathcal{D}^s$ considered as an additive group. In general [6] if $\{\varphi_t : t \geq 0\}$ is a Brownian motion on G and μ_t the distribution at time t then $\{\mu_t : t \geq 0\}$ forms a convolution semigroup of probability measures on G satisfying

$$(*) \quad \lim_{t \downarrow 0} \frac{1}{t}\mu_t(G - U) = 0 \quad \text{for every neighborhood } U \text{ of } e \text{ in } G.$$

Conversely every such convolution semigroup corresponds to a Brownian motion on G.

The solution flow $\{F_t : t \geq 0\}$ forms Brownian motions on the groups $\mathcal{D}^s, \mathcal{D}^\infty$ and the groups $\mathrm{Diff}^r(M)$ of C^r diffeomorphisms of M. One of the principle results of [6], extended later by Kunita (see [31]), was that to any Brownian motion on C^r Diff$M$ for M compact and $r \geq 3$ there corresponds a C^{r-3} vector field A and a Gaussian measure γ on the space of such vector fields, related to the Brownian motion in essentially the same way as the γ^x and A above were to the flow $\{F_t : t \geq 0\}$. This is an infinite dimensional extension of Hunt's classification of Brownian motions on finite-dimensional Lie groups.

E. The distributions of the solutions to (1) are determined by the generator $\mathcal{A}$ where

$$\mathcal{A}f = Af + \frac{1}{2}\sum_{i=1}^{m} X^i X^i(f)$$

for C^2 functions f. Different coefficients can give the same $\mathcal{A}$ and the solution flows for these different stochastic differential equations can have radically different properties, e.g. see [16] or the end of Chapter I of [26]. However in [6] it was shown that the "two-point motion" $\{(F_t(y_1), F_t(y_2)) : t \geq 0\}$ on $M \times M$ does determine the flow: in the sense that the law of the 2-point motion determines the law of the flow as an infinte dimensional Brownian motion on $\mathrm{Diff}^\infty M$. Indeed let $\mathcal{A}^{(2)}$ be the generator of the 2-point motion, so $\mathcal{A}^{(2)}$ operates on smooth $f : M \times M \to \mathbb{R}$. For such f

$$\mathcal{A}^{(2)}f(y) = \mathcal{A}_1 f(y) + \mathcal{A}_2(f)(y) + \frac{1}{2}\sum_{i,j=1 i\neq j} \Gamma^x(d_i d_j f)_y$$

where $\mathcal{A}_1, \mathcal{A}_2$ denote $\mathcal{A}$ acting on the first and second coordinates of $y = (y_1, y_2) \in M \times M$ and d_1, d_2 denote the differentiation with respect to the corresponding coordinate so that, for example

$$(d_1 f)_{y_1} : M \to T^*_{y_1} M$$

giving

$$(d_2 d_1 f)_{(y_1, y_2)} \in \mathbb{L}(T_{y_2}M; T^*_{y_1}M) \approx T^*_{y_2}M \otimes T^*_{y_1}M$$

(fortunately Γ^x is symmetric, so we do not need to be careful about the order). Thus $\mathcal{A}^{(2)}$ determines Γ^x and A and so determines the flow.

2. Baxendale's Integrability Theorem

A. Another important result from [6] is the following:

Suppose $\{\mu_t : t \geq 0\}$ is a convolution semigroup of probability measures on a Polish group G satisfying condition (*) above. For some *Banach* space F suppose

$$\alpha : G \times F \to F$$

is a continuous group action such that

$$\alpha(g, -) : F \to F$$

is linear for each g in G. Then defining

$$P_t f = \int_G g f \, d\mu_t(g)$$

the integral exists for each f in F and $\{P_t\}_{t\geq 0}$ is a semigroup of continuous linear operators on E with $P_t f \to f$ as $t \downarrow 0$ and $\|P_t\| \leq ce^{dt}$ for some constants c and d.

Even when M is not compact the space C^r Diff$M$ of C^r diffeomorphisms of M with the C^r compact open topology is a Polish group ($0 \leq r \leq \infty$), e.g. see [6]. The theorem may therefore be applied when μ_t is the distribution at time t of the solution flow of a stochastic differential equation on M provided the flow exists and consists of diffeomorphisms. In the non-compact case the flow may not exist, even when the solutions exist for all time [23], and there are very few results giving conditions for existence outside the globally Lipschitz case on $\mathbb{R}^n$.

3. Stochastic flows and the heat equation for forms

A. Let $\{F_t : t \geq 0\}$ be a flow for (1). The derivative flow $TF_t : TM \to TM$ is the solution flow to an equation

$$dv_t = \delta X(v_t) \circ dB_t + \delta A(x_t)dt \tag{5}$$

on the tangent bundle TM. When M is an open subset of $\mathbb{R}^n$ we can write $TU = U \times \mathbb{R}^n$ and $v_t = (x_t, v_t^0)$ so that (5) becomes

$$\begin{aligned} dv_t^0 &= DX(x_t)(v_t^0) \circ dB_t + DA(x_t)(v_t^0)dt \\ dx_t &= X(x_t) \circ dB_t + A(x_t)dt. \end{aligned} \tag{5'}$$

When M is given a connection (5) can be more neatly written as the covaraint equation

$$Dv_t = \nabla X(v_t) \circ dB_t + \nabla A(v_t)dt. \tag{5''}$$

See [23] for details.

It must be emphasized that just as the flow itself is not determined by the generator $\mathcal{A}$ of the 'one point motion' so the diffusion $\{v_t : t \geq 0\}$ depends on more than $\mathcal{A}$. In particular a geometrically constructed S. D. E. (1) can be expected to give rise to derivative flows with geometrically interesting properties.

Let $G = C^\infty \operatorname{Diff} M$ with the C^∞ compact-open topology. Let G^1 be the subgroup of $C^\infty \operatorname{Diff} TM$ consisting of vector bundle maps $\xi : TM \to TM$. Thus there are maps $p_1 : G^1 \to G$ and $T : G \to G^1$ with ξ a vector bundle map over $p_1(\xi)$:

$$\begin{array}{ccc} TM & \xrightarrow{\xi} & TM \\ \downarrow & & \downarrow \\ M & \xrightarrow{p_1\xi} & M \end{array}$$

and $p_1(TH) = h$.

Set $T^{(1)}M = TM$ and $T^{(r)}M = T(T^{(r-1)}M)$ for $r = 2, 3, \dots$. Inductively define G^r to be the subgroup of $C^\infty \operatorname{Diff}(T^{(r)}M)$ consisting of vector bundle maps over elements of $G^{(r-1)}$, so there are maps $p_r : G^r \to G^{r-1}$ and $T : G^{r-1} \to G^r$ as for $r = 1$, taking $G^0 = G$. Set $T^{(p)} = T \circ T \circ \cdots \circ T : G^r \to G^{r+p}$ for any $r = 0, 1, 2, \dots$. A smooth solution flow $\{F_t : t \geq 0\}$ of (1) determines Brownian motions $\{T^{(p)}F_t : t \geq 0\}$ on each G^p, $p = 1, 2, \dots$ and corresponding convolution semigroups $\{\mu_t^p : t \geq 0\}$ for $p = 0, 1, 2, \dots$.

B. Let $C^r\Omega^1$ be the space of C^r 1-forms on M. Consider such a 1-form as a map $\varphi : TM \to \mathbb{R}$ such that each restriction $\varphi_x : T_xM \to \mathbb{R}$ is linear. For example if $f : M \to \mathbb{R}$ is C^{r+1} then its differential df is in $C^r\Omega^1$. Then G^1 acts on $C^r\Omega^1$ on the right by composition

$$\alpha_1 : C^r\Omega^1 \times G^1 \to C^r\Omega^1 \qquad \alpha_1(\varphi, \xi) = \varphi \circ \xi.$$

Set

$$\delta P_t(\varphi) = \mathbb{E}\varphi \circ (TF_t) = \int_G \alpha_1(\varphi, \xi) d\mu_t^1(\xi)$$

and $P_t f = \mathbb{E} f \circ F_t$ whenever they exist. The following is well known e.g. see [23].

Proposition 3B. *Suppose M is compact and $f : M \to \mathbb{R}$ is C^1 then*

$$\delta P_t(df) = d(P_t f). \tag{6}$$

Proof. There is the commutative diagram

$$\begin{array}{ccc} C^1(M;\mathbb{R}) \times G & \xrightarrow{\alpha} & C^1(M;\mathbb{R}) \\ \downarrow d \times T & & \downarrow d \\ C^0\Omega^1 \times G^1 & \xrightarrow{\alpha} & C^0\Omega^1 \end{array}$$

where α is composition. For M compact, $C^1(M;\mathbb{R})$, with the C^1 topology, and $C^0\Omega^1$ are Banach spaces. Baxendale's integrability theorem applies equally well to right actions so we have immediately that P_t maps $C^1(M;\mathbb{R})$ continuously to itself and so does δP_t to $C^0\Omega^1$. Since d is continuous and linear

$$\begin{aligned}\delta P_t(df) &= \int_{G^1} df \circ \xi d\mu_t^1(\xi)\\ &= \int_G df \circ Th d\mu_t^0(h)\\ &= \int_G d(f\circ h)d\mu_t^0(h)\\ &= d\int_G f\circ h d\mu_t^0(h)\\ &= d(P_t(f)).\end{aligned}$$

C. The spaces $T^{(r)}M$ are unnecessarily large for our purposes when $r>1$. For example if M is an open set U of $\mathbb{R}^n$ we can identify $T^{(2)}M$ with $(U\times\mathbb{R}^n)\times(\mathbb{R}^n\times\mathbb{R}^n)$ and $T^{(2)}F_t$ with

$$(x,v)(u,V))\mapsto((F_t(x),DF_t(x)v),\ (DF_t(x)u,\ D^2F_t(x)(u,v)+DF_t(x)V)). \tag{7}$$

A connection on M determines a splitting of $T^{(2)}M$ into the sum $HT^{(2)}M\oplus VT^{(2)}M$ of horizontal and vertical subundles both natrually isomorphic to the pull-back of TM by the projection of TM onto M. For $v\in T_xM$ we can therefore write T_vTM as $T_xM\times T_xM$. Using this an element of $T^{(2)}M$ is determined by a triple (v,u,V) of vectors T_xM for some $x\in M$. In this representation $T^{(2)}F_t$ can be written

$$(v,u,V)\mapsto(TF_t(v),TF_t(u),TF_t(V)+\nabla_1TF_t(u,v)) \tag{7'}$$

where

$$\nabla_1TF_t(u,v)\in T_{F_t(x)}M$$

is bilinear in (u,v). A good reference is [20] (but the notation there reverses the order of u and v). For a smooth $\psi: T^{(r)}M\to\mathbb{R}$ there is $d\psi: T^{(r+1)}M\to\mathbb{R}$. To avoid confusion with exterior derivation write $d^{(q)}\psi$ for $d(d^{(q-1)}\psi): T^{(q+r)}M\to\mathbb{R}, q\geq 1$, with $d^{(0)}\psi=\psi$. In terms of the splitting of $T^{(2)}M$ above, for $f: M\to\mathbb{R}$

$$d^{(2)}f(v,u,V)=df(V)+\nabla df(u,v). \tag{8}$$

We can now extend Proposition 3B as in [23] Proposition 8C.

Proposition 3C. *For M compact and $f: M\to\mathbb{R}$ of class $C^{(r)}$*

$$d^{(r)}(P_tf)=\mathbb{E}\{d^{(r)}f\circ T^{(r)}F_t\}\quad r=1,2,\dots. \tag{9}$$

In particular, for $r = 2$ for $u, v \in T_{x_0}M$

$$\nabla d(P_t f)(u,v) = \mathbb{E}\{df \circ \nabla_1 T_{x_0}F_t(u,v) + \nabla df(T_{x_0}F_t(u), T_{x_0}F_t(v)). \quad (10)$$

Proof. As for Proposition 3B, using (7′) and (8) to deduce (10).

D. There is a natural right action of G on $C^r\Omega^p$ by

$$(\varphi, h) \mapsto h^*(\varphi)$$

where

$$h^*(\varphi)(v_1, \ldots, v_p) = \varphi(Th(v_1), \ldots, Th(v_p))$$

for $v_1, \ldots, v_p$ in T_xM. It commutes with exterior differentiation (this is just the invariance of exterior differentiation under diffeomorphisms):

$$\begin{array}{ccc} C^r\Omega^p \times G & \longrightarrow & C^r\Omega^p \\ d \times id \downarrow & & \downarrow d \\ C^{r-1}\Omega^{p+1} \times G & \longrightarrow & C^{r-1}\Omega^{p+1} \end{array}$$

For $\varphi \in C^r\Omega^p$ define $\widehat{P}_t(\varphi) = \mathbb{E}(F_t)^*(\varphi)$. Since exterior differentiation $d : C^r\Omega^p \to C^{r-1}\Omega^{p+1}$ is continuous and linear we can apply Baxendale's integrability result as before to obtain:

Proposition 3D. *For M compact and $\varphi \in C^1\Omega^p$*

$$d(\widehat{P}_t(\varphi)) = \widehat{P}_t(d\varphi).$$

E. We will need Itô's formula for the action of TF_t on a differential form. Before doing that we will set down some of the notation and interrelationship between various operations on differential forms. Note, and beware, that different constants appear in the various sources, e.g. in [30]. We are following Abraham and Marsden [1] (second edition) who follow Bourbaki. The definitions are given for q-forms but the $q = 1$ case will suffice for much of this article:

Let $\{h_t : t \in \mathbb{R}\}$ be a solution flow for a smooth vector field Y on M. Then the *Lie derivative* of φ by Y is the q-form

$$L_Y\varphi = \frac{d}{dt} h_t^*(\varphi)|_{t=0}. \quad (11)$$

This definition easily extends to the case where only a partial flow exists. There is also the *interior product* of φ by Y which is the $(q-1)$-form $i_Y\varphi$ given by

$$i_Y\varphi(v_1, \ldots, v_{q-1}) = \varphi(Y(x), v_1, \ldots, v_{q-1})$$

for $v_1, \ldots, v_{q-1}$ in T_xM. When $q = 0$ define $i_Y\varphi \equiv 0$. There are the following formulae.

For Z a vector field:

$$L_Y(\varphi(Z(\cdot)) = L_Z\varphi(Y) - \varphi([Y, Z]) \tag{12}$$

i.e.

$$L_Y i_Z\varphi = i_Y L_Z\varphi + i_{[Y,Z]}\varphi;$$

and

$$L_Y\varphi = i_Y d\varphi + d i_Y\varphi. \tag{13}$$

Lemma 3E. *For $\varphi \in C^2\Omega^q$ and a C^1 vecotr field Y with partial flow h_t*

$$\frac{d^2}{dt^2} h_t^*\varphi|_{t=0} = L_Y L_Y\varphi.$$

Proof. We can assume the flow is global (e.g. by modifying Y outside a compact set). Then

$$\begin{aligned}\frac{d^2}{dt^2} h_t^*\varphi|_{t=0} &= \frac{d}{dt}\left(\frac{d}{ds} h_{t+s}^*\varphi|_{s=0}\right)|_{t=0}\\ &= \frac{d}{dt}(h_t^* L_Y\varphi)|_{t=0} = L_Y L_Y\varphi.\end{aligned}$$

Proposition 3E. *Suppose $\varphi \in C^2\Omega^q$ and $\{F_t : t \geq 0\}$ is a smooth flow for (1). Then for $v_0^i \in T_{x_0}M$, $i = 1$ to q writing $v_s^i = T_{x_0}F_s(v_0^i)$ and $x_s = F_s(x_0)$*

$$\begin{aligned}\varphi(v_t^1, \ldots, v_t^q) &= \varphi(v_0^1, \ldots, v_0^q) + \int_0^t \sum_{j=1}^m L_{X^j}\varphi(v_s^1, \ldots, v_s^q) dB_s^j\\ &+ \int_0^t L_A\varphi(v_s^1, \ldots, v_s^q) ds + \frac{1}{2}\int_0^1 L_{X^j} L_{X^j}\varphi(v_s^1, \ldots, v_s^q) ds.\end{aligned} \tag{14}$$

Proof. Immediate from the general Itô formula in the form given in [23] or [26] Proposition 3A, using the Lemma.

Corollary 3E1 [23]. *For $q = 1$, given a torsion free linear connection on M and $v_0 \in T_{x_0}M$*

$$\varphi(v_t) = \varphi(v_0) + \int_0^t \nabla\varphi(X(x_s)dB_s)v_s + \int_o^t \varphi(\nabla X(v_s)dB_s)$$

$$+\int_0^t L_{A^x}\varphi(v_s)ds + \int_0^t \text{trace } \nabla\varphi(X(x_s)-)\nabla X(v_s)(-)ds$$
$$+\int_0^t \frac{1}{2}\{\text{trace } \nabla^2\varphi(X(x_s)-, X(x_s)-)(v_s)$$
$$+\varphi(\text{trace } R(X(x_s)-, v_s)X(x_s)-)\}ds. \tag{15}$$

Here A^X is the vector field

$$A^X(x) = A(x) = \frac{1}{2}\text{trace } \nabla X(X(x)-)(-) \tag{16}$$

and traces are taken by treating their subjects as bilinear in the arguments denoted by "-." For the smooth vector fields Y and q-forms φ we consider the covariant derivatives as maps

$$\nabla Y : TM \to TM$$
$$\nabla\varphi : TM \to \Lambda^q T^* M$$
$$\nabla^2 Y : TM \oplus TM \to TM$$
$$\nabla^2\varphi : TM \oplus TM \to \Lambda^q T^* M$$

and

$$R : (TM \oplus TM) \oplus TM \to TM$$

is the curvature tensor with sign convention such that

$$\nabla^2 Y(v_1, v_2) = \nabla^2 Y(v_2, v_1) + R(v_1, v_2)Y(x) \tag{17}$$

for $v_1, v_2 \in T_xM$; see II, Section 1 of [26].

For example, when $q = 1$, and $v \in T_xM$

$$d(i_Y\varphi)(v) = \nabla\varphi(v)(Y(x)) + \varphi(\nabla Y(v))$$

Remark. In [23] a term $\frac{1}{2}\nabla\varphi(\text{trace } \nabla x(X(x_s)-)-)$ has been missed.

Proof. By the Proposition it is simply a matter of evaluating the Lie derivatives. With the notation above

$$L_Y\varphi(v_0) = \frac{d}{dt}\varphi(Th_t(v_0)|_{t=0}$$
$$= \nabla\varphi(Y(x_0))v_0 + \varphi(\nabla Y(v_0))$$

so that

$$L_Y L_Y(v_0) = \nabla(L_Y\varphi)(Y(x_0))v_0 + L_Y\varphi(\nabla Y(v_0))$$
$$= \nabla^2\varphi(Y, Y)(v_0) + \nabla\varphi(\nabla Y(Y))(v_0) + 2\nabla\varphi(Y)(\nabla Y(v_0))$$
$$+ \varphi(\nabla^2 Y(Y, v_0) + (\nabla Y(v_0)).$$

However

$$\nabla^2 Y(Y(x_0), v_0) + \nabla Y(\nabla Y(v_0)) = \nabla(\nabla Y(Y(\cdot))(v_0) + R(Y(x_0), v_0)(x_0)$$

by (17) and also

$$\varphi(\nabla A^X(v_0)) + \nabla\varphi(A^X)(v_0) = d\varphi(A^X, v_0) \\ + d(\varphi(A^X(\cdot))(v_0) = L_{A^X}(\varphi)(v_0)$$

by (13).

An important special case is when M is Riemannian and the solutions to (1) are Brownian motions with drift, i.e. the differential generator of the corresponding Markov process $\mathcal{A}$ is given by

$$\mathcal{A}f \equiv \sum_i X^i(X^i(f) + A(f) = \frac{1}{2}\Delta f + Z(f)$$

where Δ denotes the Laplace–Beltrami operator on M and Z is a vector field (the drift). Remember that

$$Z(f)(x) = < Z(x), \nabla f(x) >_x = df(Z(x)).$$

This holds if and only if
(i) $X(x) : \mathbb{R}^m \to T_xM$
is an orthogonal projection for each $x \in M$, i.e. iff

$$X(x)X(x)^* v = v \qquad \text{all} \quad v \in T_xM$$

and (ii) $A^X = Z$ for A^X defined by (16) above. Every non-degenerate system (1) with smooth coefficients has this form for some Riemannian metric on M.

Corollary 3E2. *For this case, using the Levi–Civita connection, (15) reduces to*

$$\varphi(v_t) = \varphi(v_0) + \int_0^t \nabla\varphi(X(x_s)dB_s)v_s + \int_0^t \varphi(\nabla X(v_s)dB_s) + \int_0^t L_z\varphi(v_s) \\ + \frac{1}{2}\int_0^t \mathit{trace}\, d\varphi(X(x_s)-, \nabla X(v_s)-)ds + \frac{1}{2}\int_0^t \Delta^1\varphi(v_s)ds \tag{18}$$

where Δ^1 is the Laplace–Beltrami operator on 1-forms.

Proof of 3E2. Using the Weitzenböck formula

$$\Delta^1\varphi(v) = \text{trace}\, \nabla^2\varphi(-,-)(v) - \text{Ric}_x(v, \varphi_x^\#) \tag{19}$$

for $v \in T_xM$ where $\varphi_x^\# \in T_xM$ defined by

$$< \varphi_x^\#, v >_x = \varphi_x(v) \qquad v \in T_xM$$

and

$$\mathrm{Ric}_x : T_xM \oplus T_xM \to \mathbb{R}$$

is

$$\mathrm{Ric}_x(v_1, v_2) = \mathrm{trace} < R(-, v_1)v_2, - >_x$$

(e.g. see [1] or [16], the only problem is the term

$$\int_0^t \mathrm{trace}\, \nabla\varphi(X(x_s)-)\, \nabla X(v_s)(-)ds$$

in (15).

To deal with this, fix x in M and choose an orthonormal basis $e_1, \ldots, e_m$ for $\mathbb{R}^m$ such that $X(x)e_j = 0$ for $j = n+1, \ldots, m$.

Set $f_j = X(x)e_j$, so $e_j = X(x)^* f_j$ for $j = 1, \ldots, n$. Then, is $v \in T_xM$

$$\mathrm{trace}\, \nabla\varphi(X(x)-)\, \nabla X(v)(-) = \sum_{j=1}^n \nabla\varphi(f_j)\nabla X(v)X(x)^* f_j.$$

Now $\nabla X(v)X(x)^* : T_xM \to T_xM$ is seen to be skew adjoint since $\nabla(X(\cdot)X(\cdot)^*) = 0$. Therefore by elementary linear algebra and the fact that

$$\nabla\varphi(v_1)(v_2) = -d\varphi(v_2, v_1) + \nabla\varphi(v_2)(v_1)$$

for $v_1, v_2 \in T_xM$, we have

$$\mathrm{trace}\, \nabla\varphi(X(x)-)\nabla X(v)(-) = \frac{1}{2}\sum_j d\varphi(f_j, \nabla X(v)X(x)^* f_i)$$

e.g. see Lemma VII 12D of [23] where Kobayashi and Nomizu's conventions for d were used, giving twice the result here.

Remark. Of course (14), (15) and (19) remain valid (if necessary taking an explosion time into account) for non-compact M even if there is no smooth flow provided TF_t is interpreted as a derivative in probability, as in [23].

F. By definition (with our sign conventions)

$$\Delta^1 = -(dd^* + d^* d)$$

while

$$\Delta = -d^*d$$

where d^* is the formal L^2 adjoint of exterior differentiation, extended to give maps

$$d^* : C^r\Omega^p \to C^{r-1}\Omega^{p-1}.$$

Given that the Ricci curvature is bounded below and M is complete there is a semigroup $\{P_t^1 : t \geq 0\}$ with a generator agreeing with $\frac{1}{2}\Delta^1$ on compactly supported C^2 forms acting on each of the spaces $L^p\Omega^1$ of L^p 1-forms for $1 \leq p \leq \infty$. For $p = 2$ this can be defined as the L^2 semigroup $\{e^{t\Delta^1} : t \geq 0\}$, using a self-adjoint extension of Δ^1, [19], and for $p = \infty$ it can be defined using the Ricci flow as in Section 5 below. These two semigroups agree on their common domain of definition, e.g. see [25] and determine semigroups on the other L^p spaces.

Since $d^2 = 0$ we have $d\Delta f = \Delta^1 df$ for all C^2 functions f and also the L^2 closure od Δ^1 restricts to a self-adjoint operator on the closed subspace of closed forms in $L^2\Omega^1$. Thus P_t^1 maps closed forms in L^2 to closed forms. Also for any L^2 function f with $df \in L^2\Omega^1$ we have

$$dP_tf = P_t^1 df \tag{21}$$

Proposition 3F1. *Suppose the solutions of (1) are Brownian motions and M is compact. Then, for $\varphi \in C^2\Omega^1$ with $d\varphi = 0$ we have*

$$\delta P_t(\varphi) = P_t^1(\varphi). \tag{22}$$

Proof. Apply (18) to the time dependent 1-form $\{P_{t-s}^1\varphi : 0 \leq s \leq t\}$.

Remark. By (21) and Proposition 3B we know already that (22) holds when $\varphi = df$.

Proposition 3F1 can be generalized to the case where Z is a gradient by using Bismut's modification, [1], of Witten's deformed Laplacian as follows.

Suppose $h : M \to \mathbb{R}$ is C^∞ and $Z = \nabla h$. Let μ_h be the measure on M given by $e^{2h(x)}dx$ where dx refers to the Riemannian measure of M. Let $L^2\Omega^q(M, \mu_h)$ be the Hilbert space of q-forms which are L^2 for μ_h with inner product

$$< \varphi, \psi >^h = \int_M < \varphi_x, \psi_x > e^{2h(x)}dx. \tag{22}$$

Let δ^h denote the adjoint of d in these spaces, M being assumed complete. Define δ^h on q-forms by

$$\Delta^{h,q} = -(d + \delta^h)^2 = -(d\delta^h + \delta^h d)$$

As observed in [11]

$$\Delta^h = e^{-h}\Box^h e^h \tag{23}$$

where

$$\Box^h = -(d^h + (d^h)^*)^2$$

for $d^h = e^h de^{-h}$, is Witten's deformed Laplacian. Moreover, using Lie differentiation

$$\Delta^h = \Delta + 2L_{\nabla h}. \tag{24}$$

Let $\{P_t^h : t \geq 0\}$ be the corresponding semigroup on the space $\oplus_q L^2\Omega^q(M, \mu_h)$ of L^2 forms for μ_h with $P_t^{h,q}$ the restriction to q-forms. (By (23) the analytical details can be taken over from those for $\Box^h$, treated carefully in [19] for example.) As for the case $h = 0$ these semigroups map closed forms to closed forms and commute with exterior differentiation:

$$dP_t^h = P_t^h d \tag{25}$$

In Section 5 below the 'Hessian flow' is used to obtain L^∞ semigroups.

Proposition 3F2. *If M is compact and $\mathcal{A} = \frac{1}{2}\Delta + \nabla h$ then*

$$\delta P_t(\varphi) = P_t^{h,1}(\varphi)$$

for $\varphi \in C^2\Omega^1$ with $d\varphi = 0$.

Proof. As for 3F1.

G. To look at $P_t^q : C^\infty\Omega^q \to C^\infty\Omega^q$ for $q > 1$ it will be convenient to introduce some more notation and to consider the value φ_x of a q-form at x as a linear map

$$\varphi_x : \Lambda^q T_x M \to \mathbb{R}.$$

In general if E is a vector space and $A : E \to E$ is linear there are the two induced linear maps

$$\Lambda^q A : \Lambda^q E \to \Lambda^q E$$

and

$$(d\Lambda)^q A : \Lambda^q E \to \Lambda^q E$$

determines by

$$\Lambda^q A(v^1 \wedge \ldots \wedge v^q) = Av^1 \wedge AV^2 \wedge \ldots \wedge Av^q$$

and

$$(d\Lambda)^q A(v^1 \wedge \ldots \wedge v^q) = \sum_{j=1}^{q} v^1 \wedge \ldots \wedge v^{j-1} \wedge Av^j \wedge v^{j+1} \wedge \ldots \wedge v^q$$

(so they are the same when $q = 1$) e.g. see [19] p. 250. It will also be useful to use the notation

$$(\delta^2\Lambda)^q A = ((D\Lambda)^q A) \circ (d\Lambda)^q A - (d\Lambda)^q A^2$$

e.g.

$$(\delta^2\Lambda)^2 A(v^1 \wedge v^2) = \Lambda^2 A(v^1 \wedge v^2) = Av^1 \wedge Av^2.$$

Proposition 3G. *Suppose $\psi \in C^2\Omega^q$ and $(F_t : t \geq 0)$ is a smooth flow for (1). Then for a torsion free linear connection on M and any $V_0 \in \Lambda^q T_{x_0} M$, writing $x_s = F_s(x_0)$ and $V_s = \Lambda^q(TF_s)V_0$:*

$$\begin{aligned}(F_t)^*\psi(V_0) = \psi(V_t) &= \psi(V_0) + \int_0^t \nabla\psi(X(x_s)dB_s)V_s \\ &+ \int_0^t \psi((d\Lambda)^q(\nabla X(\cdot)dB_s)V_s) + \int_0^t L_{A^X}\psi(V_s)ds \\ &+ \int_0^t trace\{\nabla\psi(X(x_s)-)((d\Lambda)^q)(\nabla X(\cdot)(-))V_s)\}ds \\ &+ \int_0^t \frac{1}{2} trace\{\nabla^2\psi(X(x_s)-, X(x_s)-)V_s \\ &+ \psi(\delta^2\Lambda)^q \nabla X(\cdot)(-)V_s) \\ &+ \psi(d\Lambda)^q(R(X(x_s)-, \cdot)X(x_s)-)V_s)\}ds. \end{aligned} \tag{26}$$

Here the traces are taken with respect to the variables indicated by '–' and those indicated by '·' are the relevant ones for $d\Lambda$ and $\delta^2\Lambda$.

Proof. By Proposition 3E it is simply a computation of Lie derivatives as for 3E1. Note that for a vector field Y and $V \in \Lambda^q T_x M$

$$L_Y\psi(V) = \nabla\psi(Y(x))V + \psi((d\Lambda)^q(\nabla Y)V)$$

and in particular this holds for $Y = A, Y = X^i$, and $Y = A^X$. The curvature term comes from the use of equation (17).

Equation (30) will be seen to have especially simple form for gradient Brownian flows in the next section.

4. Example One: Gradient Brownian Flows with Drift

A. One of the first classes of non-linear stochastic flows to be studied from the point of view of ergodic theory was that of gradient Brownian flows, [16],[24]. Here there is an isometric immersion of M into a Euclidean space $\mathbb{R}^m$

$$f : M \to \mathbb{R}^m$$

(for example the standard inclusion of the sphere S^n into $\mathbb{R}^{n+1}$) and $X(x) : \mathbb{R}^m \to T_xM$ is the orthogonal projection of $\mathbb{R}^m$ onto the tangent space at x to M considered as a subset of $\mathbb{R}^m$ by using T_xf as an identification). If $f(x) = (f^1(x), \ldots, f^m(x))$ then

$$X^i = \nabla f^i$$

when $e_1, \ldots, e_m$ is the standard base for $\mathbb{R}^m$. Hence the *gradient flows.*

Let $\nu_xM = (T_xM)^\perp$, the space of vectors in $\mathbb{R}^m$ normal to M at x. The second fundamental form is a bilinear symmetric map, (see [30]),

$$\alpha_x : T_xM \times T_xM \to \nu_xM \qquad x \in M.$$

Correspondingly there is the map

$$A_x : T_xM \times \nu_xM \to T_xM$$

with

$$< A_x(u, \xi), v)_{\mathbb{R}^m} =< \alpha_x(u, v), \xi >_{\mathbb{R}^m}$$

for all $u, v \in T_xM$, $\xi \in \nu_xM$.

Let $Y(x)$ be the orthogonal projection onto ν_xM

$$Y(x)(e) = e - X(x)(e) \quad x \in M, e \in \mathbb{R}^m. \tag{27}$$

Then

$$\nabla X(v)(e) = A_x(v, Y(x)e) \quad v \in T_xM, \quad e \in \mathbb{R}^m \tag{28}$$

For any x in M, if we take the orthonormal base $e_1, \ldots, e_m$ for $\mathbb{R}^m$ such that $e_1, \ldots, e_m \in T_xM$ then for $v \in T_xM$

$$\textit{either} \qquad \nabla X(v)e_i = 0 \quad \textit{or} \quad X(x)e_i = 0 \tag{29}$$

In particular

$$\nabla X(X(x)e_i)e_i = 0$$

for each i. Thus the solutions to (1) are Brownian motions with drift A, e.g. see [23],[26].

B. For our flow $\{F_t : t \geq 0\}$ for this system there are important simplifications to the Itô formulae (18), (26). The first follows immediately from (18) by using (29):

Lemma 4B. *Let $v_t = TF_t(v_0)$ for $v_0 \in T_{x_0}M$. Then if φ is a C^2 1-form:*

$$\varphi(v_t) = \varphi(v_0) + \int_0^t \nabla\varphi(X(x_s)dB_s)v_s$$
$$+ \int_0^t \varphi(A_{x_s}(v_s, Y(x_s)dB_s)) + \int_0^t (\frac{1}{2}\Delta^1 + L_A)(\varphi)(v_s)ds \tag{30}$$

In particular when $A = \nabla h$ and M is compact

$$\delta P_t(\varphi) = P_t^{h,1}(\varphi) \tag{31}$$

(without the assumption that $d\varphi = 0$).

Proposition 4B. *Let $V_0 \in \Lambda^q T_{x_0} M$. Set $V_t = \Lambda^q(TF_t)(V_0)$. Then for $\psi \in C^2\Omega^q$, $1 \le q \le n$,*

$$\begin{aligned}\psi(V_t) = \psi(V_0) &+ \int_0^t \nabla\psi(X(x_s)dB_s)(V_s)\\ &+ \int_0^t \psi((d\Lambda)^q A_{x_s}(\cdot, Y(x_s)dB_s)(V_s))\\ &+ \int_0^t (\frac{1}{2}\Delta^q + L_A)(\psi)(V_s)ds\end{aligned} \tag{32}$$

Proof. Assume first that $A \equiv 0$ and M is compact. From the lemma and Proposition 3D, for $\varphi \in C^2\Omega^1$

$$\begin{aligned}\mathbb{E}(d\varphi)(V_t) \equiv \widehat{P}_t(d\varphi)(V_0) &= d(\delta P_t)(\varphi)(V_0)\\ &= d(P_t^1\varphi)(V_0) = P_t^2(d\varphi)(V_0)\end{aligned}$$

where $\{P_t^q : t \ge 0\}$ is the heat flow on q-forms, generator

$$\frac{1}{2}\Delta^q = -\frac{1}{2}(dd^* + d^*d). \tag{33}$$

It follows that for $\varphi \in C^3\Omega^2$

$$\frac{d}{dt^+}\widehat{P}_t(d\varphi)(V_0)|_{t=0} = \frac{1}{2}\Delta^2(d\varphi)(V_0). \tag{34}$$

Now in our situation all the terms in (26) which are of bounded variation vanish identically except for those under the last intergal sign. Thus

$$\frac{d}{dt^+}\widehat{P}_t(\psi)|_{t=0} = \frac{1}{2}\text{trace}\,\nabla^2\psi - \frac{1}{2}\widetilde{\mathcal{R}}^q(\psi) \tag{35}$$

where

$$\widetilde{\mathcal{R}}^q(\psi)(V_0) = -\psi(\text{trace}\,(\delta^2\Lambda)^q(A_{x_0})(V_0) - (d\Lambda)^q(\text{Ric}^\#)V_0) \tag{36}$$

for $\text{Ric}^\#(v) = \text{Ric}(v, -)^\# \in T_xM$, when $v \in T_xM$, see (45) below. Thus $\widetilde{\mathcal{R}}^q$ is a zero order operator, acting pointwise. However there is the Weitzenböck formula for q-forms:

$$\Delta^q = \text{trace}\,\nabla^2 - \mathcal{R}^q \tag{37}$$

where $\mathcal{R}^q$, the *Weitzenböck term*, is a zero order operator acting pointwise. From (34) and (35) we have $\widetilde{\mathcal{R}}^2(d\varphi) = \mathcal{R}^2(d\varphi)$. For any 2-form ψ we can write ψ_{x_0} as $(d\varphi)_{x_0}$ for a fixed x_0 and some $\varphi \in C^3\Omega^1$. Thus $\widetilde{\mathcal{R}}^2 = \mathcal{R}^2$ and (32) holds for $q = 2$, $A \equiv 0$, and M compact.

Keeping $A \equiv 0$ and M compact, we see from (25) as in the proof of 3F1 that $\widehat{P}_t\psi = P_t^2\psi$ for all $\psi \in C^2\Omega^2$. Using Proposition 3D again yields $\widehat{P}_t(d\psi) = P_t^3(d\psi)$. But then the same argument as above shows that $\widetilde{\mathcal{R}}^3 = \mathcal{R}^3$ and $\widehat{P}_t(\psi) = P_t^3\psi$ for 3-forms. Iterating this gives (32) for all q when $A \equiv 0$ and M is compact, and also

$$\widetilde{\mathcal{R}}^q = \mathcal{R}^q \qquad q = 1, 2, \ldots, n. \tag{38}$$

On the other hand (38) is a pointwise identity independent of global properties like compactness and independent of the vanishing of A. Consequently, (32) is valid without these assumptions.

The following can be found in Kusuoka [32], (for $h \equiv 0$), the proof is as for Proposition 3F2:

Corollary 4B1. *When M is compact and $A = \nabla h$,*

$$P_t^{h,q}\psi = \widehat{P}_t^q\psi$$

for all $\psi \in C^2\Omega^q$.

C. In the proof of (38) we could have used expression for $\mathcal{R}^q$ in terms of the curvature tensor (e.g. in [9]) together with Gauss' theorem:

$$\begin{aligned} < \alpha_x(Y,Z), \alpha_x(X,W) >_{\mathbb{R}^n} - < \alpha_x(X,Z), \alpha_x(Y,W) >_{\mathbb{R}^n} \\ = < R(X,Y)Z, W >_x \end{aligned}$$

for X, Y, Z, W in T_xM, e.g. see [30], p. 23. This would give a direct proof of (38) for $q = 2$ and presumably in general. However the usual expression for $\mathcal{R}^p$ are rather opaque and the 'probabilistic' proof seems simpler. In fact we can argue backwards to obtain the following rephrasing:

Addendum 4C. *For $\psi \in C^2\Omega^q$ and any Riemannian manifold isometrically immersed in $\mathbb{R}^m$*

$$\Delta^q\psi = \sum_{i=1}^m L_{X^i}L_{X^i}\psi \tag{39}$$

Moreover for each $x \in M$

$$\mathcal{R}_x^q = -(Q_x^q)^*$$

where

$$Q_x^q : \Lambda^q T_xM \to \Lambda^q T_xM$$

is given by

$$Q^q_x(V) = \sum_{i=1}^{m} \frac{D^2}{\partial t^2} \Lambda^q(TS^i_t)(V)|_{t=0} \tag{40}$$

for $S^i_t : M \to M$ the flow of X^i, (using a partial flow if a full one does not exist).

5. Canonical, Ricci and Hessian Flows

A. Let $\pi : OM \to M$ be the orthonormal frame bundle to M, so if $x \in M$ an element $u \in \pi^{-1}(x)$ is an isometry

$$u : \mathbb{R}^n, <,>_{\mathbb{R}^n} \to T_x M, \quad <,>_x .$$

Our s.d.e (1) has a *horizontal lift*

$$du_t = \widetilde{X}(u_t) \circ dB_t + \widetilde{A}(u_t)dt \tag{41}$$

where $\widetilde{X}$ and $\widetilde{A}$ are the horizontal lifts of X and A e.g. see [23],[29]. If a flow $\{F_t : t \geq 0\}$ exists for (1) then there is a flow $\{\widetilde{F}_t : t \geq 0\}$ on OM for (41) satisfying

$$\pi \widetilde{F}_t(u) = F_t(\pi u) \tag{42}$$

and

$$\widetilde{F}_t(u.g) = \widetilde{F}_t(u) \cdot g \tag{43}$$

for $t \geq 0$, $u \in OM$ and $g \in O(n)$, the n-dimensional orthogonal group, almost surely, where the right action of $O(n)$ on $O(M)$ is just that of composition. Indeed a partial flow for (41) could be chosen to satisfy the analogue of (42) and (43) e.g. as shown in [14], and from this strong completeness follows.

From this there is a flow of *parallel translations* along $\{F_t : t \geq 0\}$:

$$//_t(\omega) : TM \to TM \qquad \omega \in \Omega$$

given by

$$//_t(\omega)(v_0) = \widetilde{F}_t(u_0)u_0^{-1}(v_0) \in T_{F_t(x_0)}M$$

where $v_0 \in T_{x_0}M$ and $u_0 \in \pi^{-1}(x_0)$. By (43) the choice of such u_0 is irrelevant.

B. Suppose $A = \frac{1}{2}\Delta$. By the *Ricci flow* we will mean $\{W_t : t \geq 0\}$ with

$$W_t(\omega) : TM \to TM \qquad \omega \in \Omega$$

over $\{F_t : t \geq 0\}$ given by the covariant equation for random vectors

$$\frac{DW_t}{\partial t}(v_0) = -\frac{1}{2}\mathrm{Ric}(W_t(v_0), -)^{\#} \quad v_0 \in TM \tag{44}$$

where Ric: $TM \oplus TM \to \mathbb{R}$ is the Ricci curvature and $\mathrm{Ric}(v,-)^{\#} \in T_xM$ if $v \in T_xM$ is defined by

$$< \mathrm{Ric}(v,-)^{\#}, w >_x = \mathrm{Ric}(v,w) \qquad w \in T_xM. \tag{45}$$

N.B. This has nothing to do with R. Hamilton's Ricci flow which deterministically deforms the metric of M.

For a C^2 1-form φ on M there is well known Itô formula

$$\varphi(v_t) = \varphi(v_0) + \int_0^t \nabla\varphi(X(x_s)dB_s)(v_s) + \frac{1}{2}\int_0^t \Delta^1\varphi(v_s)ds \tag{46}$$

where $v_s = W_s(v_0)$ and $x_s = F_s(x_0)$ for $v_0 \in T_{x_0}M$. This gives the standard solution to the heat equation for 1-forms:

$$P_t^1\varphi(v_0) = \mathbb{E}\,\varphi(W_t(v_0)) \tag{47}$$

when the Ricci curvature is bounded below.

The Ricci flow gives the "geodesic deviation" of Dohrn and Guerra in stochastic mechanics [35].

C. There are corresponding flows $\{W_t^q : t \geq 0\}$ on $\Lambda^q TM$ along $\{F_t : t \geq 0\}$ given by

$$\frac{DW_t^q}{\partial t} = -\frac{1}{2}\mathcal{R}^q(W_t^q(-))$$

where $\mathcal{R}^q$ is the Wetzenböck term of the previous section. These given the solutions to the heat equation for q-forms e.g. see [29],[26] and Section 5E below.

D. The *canonical* flow is the flow on OM given by the canonical S.D.E. on OM. We refer to [23],[26],[29] for details. If it is denoted by $\{H_t : t \geq 0\}$ then for φ a *closed* form on M

$$P_t^1\varphi(v_0) = \mathbb{E}\,\varphi(T\pi \circ TH_t(\tilde{v}_0)) \tag{48}$$

where $v_0 \in T_{x_0}M$, $u_0 \in \pi^{-1}(x_0)$ and $\tilde{v}_0 \in T_{u_0}OM$ is the horizontal lift of v_0, see [23], [34], [35].

E. When $\mathcal{A} = \frac{1}{2}\Delta + Z$ we can define a flow $W_t^{Z,q}$ on $\Lambda^q TM$ over $\{F_t : t \geq 0\}$ by

$$\frac{DW_t^{Z,q}}{\partial t}(V_0) = -\frac{1}{2}\mathcal{R}^q(W_t^{Z,q}(V_0)) + (d\Lambda)^q(\nabla Z(\cdot))(W_t^{Z,q}(V_0)) \tag{49}$$

for $V_0 \in \Lambda^q T_{x_0} M$. We shall call this the *Hessian flow* when Z is a gradient vector field. Writing $V_t = W_t^{Z,q}(V_0)$ and $x_t = F_t(x_0)$ the Weitzenböck formula (37) and general Itô formula give

$$\psi(V_t) = \psi(V_0) + \int_0^t \nabla\psi(X(x_s)dB_s)V_s + \int_0^t L_Z(\psi)(V_s)ds + \frac{1}{2}\int_0^t \Delta^q(\psi)(V_s)ds \tag{50}$$

for $\psi \in C^2\Omega^q$. The existence of a continuous flow is of course irrelevant for this. In fact if ξ_x is the composition time of (1) for solutions starting from a point x and ξ_x is finite there are the obvious modifications to (50).

We will say that $\frac{1}{2}\mathcal{R}^q - (d\Lambda)^q(\nabla Z)(\cdot))$ is *bounded below* if

$$C^q(Z) \equiv \inf \frac{1}{2} < \mathcal{R}^q(V), V > - < (d\Lambda)^q(\nabla Z)(\cdot))V, V >: V\in \Lambda^q TM \text{ and } |V| = 1 > -\infty, \tag{51}$$

where we are using the induced Riemannian metric on $\Lambda^q TM$. A standard proof now gives the following proposition:

Proposition 5E. *Suppose $\frac{1}{2}\mathcal{R}^q - d\Lambda^q(\nabla Z(\cdot))$ is bounded below. For $t \geq 0$ define $Q_t^q : L^\infty\Omega^q \to L^\infty\Omega^q$ by*

$$Q_t^q(\psi)(V) = \mathbb{E}\chi_{\{t<\xi_x\}}\psi(W_t^{Z,q}(V)) \tag{52}$$

for $V \in \Lambda^q T_x M$, $\psi \in L^\infty\Omega^q$. Then $\{Q_t^q : t \geq 0\}$ is a semigroup which preserves the subspace of bounded continuous forms and is strongly continuous there with generator agreeing with

$$\frac{1}{2}\Delta^q + L_Z$$

on C^2 forms with compact support. When $Z = \nabla h$, $\psi \in L^2\Omega^q(M, \mu_h) \cap L^\infty$ and M is compact

$$Q_t^q\psi = P_t^{h,q}\psi \tag{53}$$

for $P_t^{h,q}$ as defined in Section 3F.

Remarks. (1) In [4] Bakry shows that the lower bound (51) for $C^1(\nabla h)$ implies non-explosion. Note

$$C^1(\nabla h) = \inf\{\frac{1}{2}\mathrm{Ric}^\# - \nabla\nabla h\}$$

(2) There is the 'semigroup domination' by (51) and the definition of $W_t^{Z,q}$:

$$|Q_t^q(\psi)(V)| \leq e^{-C^q(Z)t} Q_t^0(|\psi|.)(x)|V|_x$$

for $\psi \in L^\infty \Omega^q$, $V \in \Lambda^q T_x M$, and $x \in M$. For $q = 1$ and $Z = \nabla h$ this is essentially that proved and used by Bakry in [4] in order to obtain his non-explosion result.

(3) More generally if we have a diffusion on M with differential generator $\mathcal{A} = \frac{1}{2}\Delta + Z$, not necessarily given by an S.D.E. on M, we can take its horizontal lift to give a process $\{u_t : t \geq 0\}$ in OM, starting at $u_0 \in \pi^{-1}(x_0)$, over the paths $\{x_t : t \geq 0\}$ of the diffusion starting from a given point x_0 e.g. see [28]. Then $W_t^{Z,q}$ can be defined by (49) as before but (50) becomes

$$\psi(V_t) = \psi(V_0) + \int_0^t \nabla\psi(u_s dB_s)V_s + \int_0^t \left(\frac{1}{2}\Delta^q + L_Z\right)(\psi)(V_s)ds \quad (54)$$

cf. [26] Chapter IV Section 1, [23] 13.

6. Moment exponents and topological obstructions to stability

A. Let $\{F_t : t \geq 0\}$ be a solution flow for (1). Assume M is compact. Also assume (1) is non-degenerate so the associated differential generator is elliptic with a unique invariant measure given by a C^∞ density $\rho : M \to \mathbb{R}(> 0)$.

For $v_0 \in T_{x_0}M$ with $\|v_0\| = 1$ and for $p \in \mathbb{R}$ define

$$\mu(v_0, p) = \overline{\lim}_{t\to\infty}\frac{1}{t}\log \mathbb{E}\,|TF_t(v_0)|^p \quad (55)$$

and

$$\mu_{x_0}(p) = \overline{\lim}_{t\to\infty}\frac{1}{t}\log\mathbb{E}\,\|T_{x_0}F_t\|^p. \quad (56)$$

The first of these form the *moment exponents*, discussed for linear systems in [2], and given a detailed analysis for the manifold case in [9]. In fact in our compact case the upper limit is a limit and given a hypoellipticity condition (on an induced S.D.E. on the sphere bundle of M) $\mu(v_0, p)$ is independent of v_0 and x_0, [9], and can be shown to be equal to $\mu_{x_0}(p)$ for $p > 0$ as in [3].

There is also the *sample exponent*

$$\mu_1 = \lim_{t\to\infty} \frac{1}{t}\log\|T_{x_0}F_t\|$$

for almost all $(x_0, \omega) \in M \times \Omega$. This is easily seen to be the same as the highest Lyapunov exponent as defined in [14].

Following [2] or [3], as [26], but not assuming the additional hypoellipticity conditions of [3], [9], we have

(i) $p \mapsto \mu_{x_0}(p)$ is convex
(ii) $p \mapsto \frac{1}{p}\mu_{x_0}(p)$ is increasing
(iii) $\frac{1}{-p}\mu_{x_0}(-p) \le \mu_1 \le \frac{1}{p}\mu_{x_0}(p)$ if $p > 0$
(iv) $\frac{d}{dp^-}\mu_{x_0}(p)|_{p=0} \le \mu_1 \le \frac{d}{dp^+}\mu_{x_0}(p)|_{p=0}$.

B. We will say that a flow is *moment stable* if $\mu_{x_0}(1) < 0$ for almost all x_0 in M (with respect to each invariant measure for the case of a general flow). We show now that there exist toplological obstructions to moment stability. An important point is that

$$\Delta^{q,h}\varphi = 0 \leftrightarrow d\varphi = 0 \quad \text{and} \quad \delta^h\varphi = 0$$

because

$$< \Delta^{q,h}\varphi, \varphi >_{L^2} = < d\varphi, d\varphi >_{L^2} + < \delta^h\varphi, \delta^h\varphi >_{L^2} .$$

Theorem 6B. *Suppose M is compact with $H^1(M;\mathbb{R})$ non-zero. Then if (1) has generator $\frac{1}{2}\Delta + \nabla h$, for some smooth Riemannian metric on M and some C^2 map h: $M \to \mathbb{R}$, its flow will not be moment stable.*

Proof. Let $\rho_1 = \dim(H^1;\mathbb{R})$, so ρ_1 is the dimension of the space harmonic 1-forms by Hodge's theorem. According to Witten $\rho_1 = \dim \ker \square^{h,1}$, in the notation of Section 3F, e.g. see [19] Theorem 11.12. Using (23)

$$\rho_1 = \dim \ker \Delta^{h,1}$$

as observed by Bismut [11]. Thus our hypothesis implies there exists $\varphi \in C^\infty\Omega^1$ with $\Delta^{h,1}\varphi = 0$ and with $x_0 \in M$ and $v_0 \in T_{x_0}M$ such that $|v_0| = 1$ and $\varphi(v_0) > 0$. Then, using Proposition 3F2 for a flow for (1) when $A = \frac{1}{2}\Delta + \nabla h$

$$\begin{aligned} 0 &= \lim_{t\to\infty} \frac{1}{t} \log \varphi(v_0) = \lim \frac{1}{t} \log P_t^1 \varphi(v_0) \\ &= \lim_{t\to\infty} \frac{1}{t} \log \mathbb{E}\,\varphi(TF_t(v_0)) \le \mu(v_0, 1) \le \mu_{x_0}(1). \end{aligned}$$

The classification of surfaces gives the following corollary.

Corollary 6B. *If a compact 2-dimensional Riemannian manifold admits an S.D.E. (1) with $A = \frac{1}{2}\Delta + \nabla h$ for some C^2 map $h: M \to \mathbb{R}$ which has a moment stable flow then it is diffeomorphic either to the sphere S^2 or the real projective space $\mathbb{P}^2$.*

Remarks. (i) The question as to whether diffeomorphism can be replaced by an isometric isomorphism perhaps with the assumption, or conclusion, that also $h \equiv 0$.

(ii) The intention is to treat higher dimensional analogues of Corollary 6B in a sequel by the methods of [25], [27], at least for gradient flows.

(iii) Note that if $\mathcal{A} = \frac{1}{2}\Delta + A$ for A a vector field then moment stability is impossible, by essentially the same argument, if there is a harmonic 1-form φ invariant under the flow A (so that $L_A\varphi = 0$).

(iv) Given unformity in x_0 of the limit in (56) it is possible to obtain *homotopy* vanishing theorems given p-th moment stability, for solution flows of otherwise arbitrary stochastic differential equations (1) with smooth coefficients. This uniformity is observed to be true in [9], after equation (2.9), assuming their non-degeneracy conditions on the tangent bundle. Homotopy vanishing results are stronger than homology vanishing theorems. The proofs particularly for $\pi_p M$ when $p > 1$, depend on deep results taken from global analysis and will be treated in a sequel.

C. For gradient flows Theorem 6B has a quantitative refinement. Let $\lambda^{h,q}$ be the highest eigenvalue of $\Delta^{h,q}$. Then $\lambda^{h,q} \leq 0$ with equality if and only if $H^q(M;\mathbb{R}) \neq 0$, as can be seen by the proof for the case $q = 1$ given while proving Theorem 6B.

Theorem 6C. *For a gradient system with generator $\mathcal{A} = \frac{1}{2}\Delta + \nabla h$ on a compact Riemannian manifold M*

$$\lambda^{h,q} \leq \mu_{x_0}(q) \qquad q = 1, \ldots, n$$

for almost all x_0 in M. In particular $\mu_{x_0}(q) < 0$ implies $H^q(M;\mathbb{R}) = 0$.

Proof. Take a non-zero q-form φ with $\Delta^{h,q}\varphi = \lambda^{h,q}\varphi$. If $\varphi_{x_0} \neq 0$ take $V_0 \in \Lambda^q T_{x_0}M$ with $\varphi(V_0) > 0$. Using Corollary 4B1

$$\begin{aligned}
\lambda^{h,q} &= \lim_{t\to\infty} \frac{1}{t}\log|P_t^{h,q}(\varphi)(V_0)| \\
&= \lim_{t\to\infty} \frac{1}{t}\log|\mathbb{E}\varphi(\Lambda^q(TF_t)(V_0))| \\
&\leq \lim_{t\to\infty} \frac{1}{t}\log\mathbb{E}|\Lambda^q(TF_t)(V_0)| \leq \mu_{x_0}(q).
\end{aligned}$$

Remark. We would get more refined results by considering the exponents

$$\mu^q_{x_0}(p) = \overline{\lim}_{t\to\infty}\frac{1}{t}\log\mathbb{E}\|\Lambda^q T_{x_0}F_t\|p \tag{57}$$

for $1 \leq q \leq n$, $p \in \mathbb{R}$. There is the formula, almost surely on $M \times \Omega$,

$$\mu_1 + \mu_2 + \ldots + \mu_q = \lim_{t\to\infty} \frac{1}{t}\log\|\Lambda^q T_{x_0}F_t\|$$

for the sum of the q highest sample exponents, taking multiplicity into account, e.g. see [7]. When $q = n$ this reduces to the well studied case of the *mean exponents* [16],[17],[8],[9], [18],24].

$$\mu^n_{x_0}(p) = \overline{\lim}_{t\to\infty}\frac{1}{t}\log\mathbb{E}|\det T_{x_0}F_t|^p. \qquad (58)$$

The corresponding results to (i), (ii), (iii), (iv) holds as before. From the definition of $\|\Lambda^q T_{x_0}F_t\|$, (and luckily any sensible choice of norm will do) it is clear that

$$\mu^q_{x_0}(p) \le \mu_{x_0}(qp) \qquad p > 0 \qquad (59)$$

c.f. [9] equation (2.13). Moreover the proof of Theorem 6C works equally well to give the strongest result

$$\lambda^{h,q} \le \mu^q_{x_0}(1). \qquad (60)$$

Since any compact orientable manifold has $H^n(M;\mathbb{R}) \neq 0$ (for example because the volume form

$$(v_1, \ldots, v_n) \mapsto \det[v^j_i]$$

is harmonic), (60) has the corollary that $\mu^n_{x_0}(1) \ge 0$. For non-orientable manifolds this also holds by going to the orientable double cover and observing that this does not change the exponents. In fact for any compact M and gradient flow

$$\mu_{x_0}(\dim M) \ge \mu^n_{x_0}(1) = 0. \qquad (61)$$

This was observed independently by Chappell [17] and L. Arnold, at least for $h \equiv 0$. For non-zero h the proof of [17], which is a direct computation, yields (61) by an additional use of the ergodic theorem and the divergence theorem.

By the properties of μ_{x_0} in Section A, Theorem 6C gives strong topological restrictions to equality in (61) unless in the exceptional situation of $\mu_{x_0} \equiv 0$ on $[0, \dim M]$, c.f. [9] Corollary (2.14) but remembering the hypoellipticity assumptions there.

D. Now we look at the calculations involved to estimate the exponents for gradient Brownian flows. First for a general submanifold of $\mathbb{R}^m$ and then for spheres.

Take $v_0 \in T_{x_0}M$ with $|v_0| = 1$. Set $v_t = TF_t(v_0)$ and write $\eta_s = v_s/_{|v_s|}$. Then, for $x_s = F_s(x_0)$,

$$\log|v_t| = \int_0^t < \eta_s, A(\eta_s, dB_s - X(x_s)dB_s) >$$
$$+ \int_0^t \left\{\frac{1}{2}|\alpha_{x_s}(\eta_s, -)|^2 - |\alpha_{x_s}(\eta_s,\eta_s)|^2 - \frac{1}{2}\mathrm{Ric}(\eta_s,\eta_s)\right\} ds \quad (62)$$

e.g. see [26] formula (151). From this

$$|v_t|^p = 1 + p\int_0^t |v_s|^p < \eta_s, A(\eta_s, dB_s - X(x_s)dB_s) > \tag{63}$$
$$+p\int_0^t |v_s|^p \left\{\frac{1}{2}|\alpha_{x_s}(\eta_s, -)|^2 + \left(\frac{1}{2}p - 1\right)|\alpha_{x_s}(\eta_s, \eta_s)|^2 - \frac{1}{2}\text{Ric}(\eta_s, \eta_s)\right\} ds$$

For any tangent vector v to M, Gauss' theorem gives

$$\text{Ric}(v, v) = -|\alpha_x(v, -)|^2 + < \alpha_x(v, v), \text{trace } \alpha_x > .$$

Therefore

$$\mathbb{E}|v_t|^p = 1 + p\mathbb{E}\int_0^t |v_s|^p \{|\alpha_{x_s}(\eta_s, -)|^2$$
$$+ \left(\frac{1}{2}p - 1\right)|\alpha_{x_s}(\eta_s, \eta_s)|^2 - \frac{1}{2} < \alpha_{x_s}(\eta_s, \eta_s), \text{trace}\, \alpha_{x_s} > \} ds \tag{64}$$
$$= 1 + p\mathbb{E}\int_0^t |v_s|^p \text{Ric}(\eta_s, \eta_s) + \frac{1}{2} < \alpha_{x_s}(\eta_s, \eta_s), \text{trace}\, \alpha_{x_s} >$$
$$+ \left(\frac{1}{2}p - 1\right)|\alpha_{x_s}(\eta_s, \eta_s)|^2\} ds. \tag{65}$$

In general it does not seem easy to find conditions on the embedding of the submanifold e.g. conditions on the second fundamental form, which ensure that particular exponents are negative (see [26]). For hypersurfaces certain convexity conditions were shown by Chappell [18] to imply that $\lambda_1 < 0$ which implies sample stability [14]. These are conditions on the eigenvalues of the second fundamental form. The guess that $\lambda_1 < 0$ when M is the boundary of a convex domain seems to still remain a guess.

For the sphere $S^n(r)$ of radius r in $\mathbb{R}^{n+1}$

$$\alpha_x(u, v) = -\frac{1}{r} < u, v > \frac{x}{r} \qquad u, v \in T_x S^n(r)$$

so that

$$\mathbb{E}|v_t|^p = 1 + \frac{1}{2r^2}p(p-n)\int_0^t \mathbb{E}|v_s|^p ds$$

giving

$$\mu_{x_0}(p) = \mu(v_0, p) = \frac{1}{2r^2}p(p-n)$$

as is well known.

E. Another simple but important example is that of the Clifford torus [16], [18]. This refers to the embedding of the flat torus

$$f : S^1\left(\frac{1}{\sqrt{2}}\right) \times S^1\left(\frac{1}{\sqrt{2}}\right) \to \mathbb{R}^4$$

$$f(x,y) = \left(\frac{1}{\sqrt{2}}\cos\sqrt{2}x, \frac{1}{\sqrt{2}}\sin\sqrt{2}x, \frac{1}{\sqrt{2}}\cos\sqrt{2}y, \frac{1}{\sqrt{2}}\sin\sqrt{2}y\right)$$

where x, y parametrize $S^1\left(\frac{1}{\sqrt{2}}\right)$ by arc length. It gives an isometric embedding as a minimal submanifold in $S^3(1)$ and the general theory described in [16] can be applied [8] to give the asymptotic behavior of $\det TF_t$, including its rate function for large deviations [24]. It is also easy to proceed directly: for $(b^1, b^2, b^3, b^4) \in \mathbb{R}^4$

$$X(x,y)(b^1,b^2,b^3,b^4) = (-b^1\sin\sqrt{2}x + b^2\cos\sqrt{2}x, -b^3\sin\sqrt{2}y + b^4\cos\sqrt{2}y)$$

so that if $(x_t, y_t) = F_t(x_0, y_0)$ and $(u_t, v_t) = TF_t(u_0, v_0)$ for (u_0, v_0) tangent at (x_0, y_0)

$$du_t = -\sqrt{2}((\sin\sqrt{2}x_t)dB_t^1 + (\cos\sqrt{2}x_t)dB_t^2)u_t$$
$$dv_t = -\sqrt{2}((\sin\sqrt{2}y_t)dB_t^3 + (\cos\sqrt{2}y_t)dB_t^4)v_t.$$

Thus u_t and v_t are exponential martingles

$$u_t = u_0\exp(\sqrt{2}\beta_t^1 - t),$$
$$v_t = v_0\exp(\sqrt{2}\beta_t^2 - t)$$

for one-dimensional Brownian motions $\{\beta_t^1, t \geq 0\}$, $\{\beta_t^2 : t \geq 0\}$.

We have, for non-zero (u_0, v_0)

$$\mu(x_0,y_0)(p) = \mu((u_0,v_0),p) = \overline{\lim}_{t\to\infty}\frac{1}{t}\log\mathbb{E}(|u_t|^p + |v_t|^p)$$
$$= \overline{\lim}_{t\to\infty}\frac{1}{t}\log((|u_0|^p + |v_0|^p)\exp(p^2t - pt)) = p(p-1).$$

While

$$\mu_{x_0}^2 = \overline{\lim}_{t\to\infty}\frac{1}{t}\log\mathbb{E}\exp(p\sqrt{2}(\beta_t^1 + \beta_t^2) - 2_pt)$$
$$= \overline{\lim}_{t\to\infty}\frac{1}{t}2(p^2t - pt) = 2p(p-1).$$

It is one of Chappell's counterexamples [18] to the extension to higher codimension of Bougerol's result [12] that for a hypersurface the Lyapunov exponents cannot all be equal unless the hypersurface is a sphere. Indeed it is easy to see that all the exponents are -1. Under hypoellipticity hypotheses equality of all exponents is equivalent to equality of $\mu_x^n(p)$ with $\mu_x(np)$ for all $p \in \mathbb{R}$, $x \in M$, [9]. The latter equality is not true for the Clifford torus, and the considerations above suggest that it could imply that M is a sphere, at least toplogically, for arbitrary codimension, c.f. Corollary 6B above and also Remark (iv), in Section 6B.

7. Elliott–Kohlmann type expansions and another expression for the heat kernel on 1-forms

A. Let M be compact and $\{F_t : t \geq 0\}$ a smooth solution flow for (1). Mimicking Elliott and Kohlmann [21],[22], if $f : M \to \mathbb{R}$ is C^∞, application of Itô's to $(t,x) \mapsto P_{T-t}f(x)$ for $0 \leq t \leq T$ with fixed $T > 0$ yields

$$\begin{aligned} P_{T-s}f(x_s) &= P_T f(x_0) + \int_0^s d(P_{T-t}f)X(x_t)dB_t \\ &= P_T f(x_0) + \int_0^s \delta P_{T-t}(df)X(x_t)dB_t \end{aligned} \tag{66}$$

by Proposition 3B.

However, similarly, for $\varphi \in C^2\Omega^1$, using (15), as a $T^*_{x_0}M$ valued process

$$\begin{aligned} \delta P_{T-t}(\varphi)T_{x_0}F_t = \delta P_T(\varphi) &+ \int_0^t \nabla\delta P_{T-s}(\varphi)(X(x_s)dB_s)T_{x_0}F_s \\ &+ \int_0^t \delta P_{T-s}(\varphi)(\nabla X(T_{x_0}F_s(-))dB_s) \end{aligned} \tag{67}$$

Here $x_s = F_s(x_0)$ and $v_s = TF_s(v_0)$. Writing

$$\begin{aligned} \delta P_{T-t}(df)X(x_t) &= \delta P_{T-t}(df)(TF_t(TF_t^{-1}X(x_t)(-))) \\ &= \delta P_{T-t}(df)(TF_t(F_t^*(X)(x_0)(-))) \end{aligned}$$

in (66) with $s = T$ we can substitute $\varphi = df$ in (67) and apply (67) to $F_t^*(X)(x_0)(-)$ then substitute in (66) to obtain

$$\begin{aligned} f(x_T) = P_T f(x_0) &+ \delta P_T(df)\int_0^T (F_s)^*(X)(x_0)dB_s \\ &+ \int_{t=0}^T \int_{s=0}^t \{\nabla\delta P_{T-s}(df)(X(x_s)dB_s)TF_sF^*(X)(x_0)dB_t) \\ &+ \delta P_{T-s}(df)\nabla X((TF_sF_t^*(X)(x_0)dB_t)dB_s\}. \end{aligned} \tag{68}$$

The process can then be continues, for example using (10) and associated Itô formulae. In general this looks as if it will get complicated very quickly. The simplest expansion can be obtained by just using $\delta^{(r)}P_t$ together with

(9) although this is geometrically unappealing. For the Brownian case considerable simplifications occur by using $\delta P_{T-s}(df) = P^1_{T-s}(df)$ and then using the Ricci flow.

B. A minor variation of Elliott and Kohlmann's technique for getting an integration by parts formula will give us a new, and rather surprising, expression for the heat kernel for closed 1-forms. Suppose M is Riemannian and (1) has $\mathcal{A} = \frac{1}{2}\Delta + \nabla h$, with solution flow $\{F_t : t \geq 0\}$. For $x_t = F_t(x_0)$ as usual, there is the Stratonovich integral of $\varphi \in C^2\Omega^1$ along the paths $\{x_s : 0 \leq s \leq t\}$

$$\begin{aligned}\int_0^t \varphi \circ dx_s &= \int_0^t \varphi(X(x_s) \circ dB_s) + \int_0^t \varphi(A(x_s))ds \\ &= \int_0^t \varphi(X(x_s)dB_s) - \frac{1}{2}\int_0^t \delta^h\varphi(x_s)ds \end{aligned} \tag{69}$$

using the notation of Section 3F and noting that for $\varphi \in C^1\Omega^1$

$$\delta^h\varphi = -\text{trace}\,\nabla\varphi(-)(-) - 2i_{\nabla h}(\varphi).$$

When $\varphi = df$ for a C^2 function f this reduces to $f(x_t) - f(x_0)$.

For $t \geq 0$ define $Q_t(\varphi) : M \to \mathbb{R}$ by

$$Q_t(\varphi)(x_0) = \mathbb{E}\int_0^t \varphi \circ dx_s.$$

Restricting to the case of M compact for simplicity (69) gives

$$\frac{\partial}{\partial t}Q_t(\varphi)(x_0) = -\frac{1}{2}P_t^h(\delta^h\varphi)(x_0) = -\frac{1}{2}\delta^h(P_t^{h,1}\varphi)(x_0)$$

since $\delta^h\Delta^h = \Delta^{h,1}\delta^h$. For $v_0 \in T_{x_0}M$

$$\begin{aligned} d(Q_t(\varphi))(v_0) &= -\frac{1}{2}\int_0^t d\delta^h(P_s^{h,1}\varphi)ds(v_0) \\ &= \frac{1}{2}\int_0^t \Delta^{h,1}P_s^{h,1}(\varphi)(v_0)ds \end{aligned}$$

if $d\varphi = 0$. Thus for a *closed form*

$$d(Q_t(\varphi)) = P_t^{h,1}\varphi - \varphi \tag{70}$$

and consequently

$$\Delta^h Q_t(\varphi) = -P_t^h(\delta^h\varphi) + \delta^h\varphi. \tag{71}$$

Thus for $d\varphi = 0$, by Itô's formula, if $0 \le t \le T$

$$\begin{aligned} Q_{T-t}(\varphi)(F_t(x_0)) &= Q_T(\varphi)(x_0) + \int_0^t (P^{h,1}_{T-s}\varphi - \varphi)(X(x_s)dB_s) \\ &+ \frac{1}{2}\int_0^t \{\delta^h\varphi(x_s) - P^h_{T-s}(\delta^h\varphi)(x_s)\}ds \\ &+ \frac{1}{2}\int_0^t \delta^h P^h_{T-s}(\varphi)(x_s)ds \\ &= Q_T(\varphi)(x_0) - \int_0^t \varphi\circ dx_s + \int_0^t P^{h,1}_{T-s}(\varphi)(X(x_s)dB_s) \end{aligned}$$

by (69).

Taking $t = T$:

$$\int_0^T \varphi\circ dx_s = Q_T(\varphi)(x_0) + \int_0^T P^{h,1}_{T-s}(\varphi)(X(x_s)dB_s). \tag{72}$$

Now let $\{W^h_t : 0 \le t < \infty\}$ be the Hessian flow on TM over $\{F_t : t \ge 0\}$ of Section 5E, i.e. $W^h_t = W^{\nabla h,1}_t$. For $v_0 \in T_{x_0}M$ by (72)

$$\begin{aligned} &\mathbb{E}\left\{\int_0^T \varphi\circ dx_t \int_0^T < W^h_t(v_0), X(x_t)dB_t >\right\} \\ &= \mathbb{E}\left\{\int_0^T P^{h,1}_{T-t}(\varphi)(X(x_t)dB_t) \int_0^T < W^h_t(v_0), X(x_t)dB_t >\right\} \\ &= \mathbb{E}\int_0^T P^{h,1}_{T-t}(\varphi)(W^h_t(v_0))dt \end{aligned}$$

using Itô's formula for the product of the two martingales obtained by letting the range of integration vary (between 0 and T) in each of the integrals. By the semigroup property of $\{P^{h,1}_t : t \ge 0\}$ and Proposition 5E this gives:

Proposition 7B. *Suppose M is compact and $h : M \to \mathbb{R}$ is C^2. Then, for any $\varphi \in C^2\Omega^1$ with $d\varphi = 0$,*

$$P^{h,1}_t(\varphi)(v_0) = \frac{1}{t}\mathbb{E}\left\{\int_0^t \varphi\circ dx_s \int_0^t < W^h_s(v_0), X(x_s)dB_s >\right\} \tag{73a}$$

all $v_0 \in T_{x_0}M$.

Remarks. (1) Strictly speaking, flows are not relevant here. The most natural formula would be obtained by taking the canonical construction of a diffusion with $\mathcal{A} = \frac{1}{2}\Delta + \nabla h$. Then (73) holds with $X(x_s)dB_s$ replaced by $u_s dB_s$ for $\{u_s : 0 \le s < \infty\}$ the horizontal lift of $\{x_s : 0 \le s < \infty\}$ and $\{B_s : 0 \le s < \infty\}$ an n-dimensional Brownian motion. Alternatively, Proposition 3F2 could be used to give

$$P_t^{h,1}(\varphi)(v_0) = \frac{1}{t}\mathbb{E}\left\{\int_0^t \varphi \circ dx_s \int_0^t < TF_s(v_0), X(x_s)dB_s >\right\} \qquad (73b)$$

(2) Extensions of these formulae to the case of q-forms, $q > 1$, have been worked out by X.-M.Li.

(3) The formula is certainly not true for non-closed forms. Indeed if $\delta^h\varphi = 0$ then $\{\int_0^t \varphi \circ dx_s : t \ge 0\}$ is a martingale and the right hand side of (73) is just $\frac{1}{t}\int_0^t P_s^{h,1}(\varphi)(v_0)$ by (69) and Proposition 3F2. This is the same as the left hand side for $t \ge 0$ and all v_0 is and only if $\Delta^h\varphi = 0$ or equivalently $\delta^h\varphi = 0$ and $d\varphi = 0$

(4) When $\varphi = df$ for some $f : M \to \mathbb{R}$ the formulae reduces to

$$d(P_t^h f)(v_0) = P_t^{h,1}(df)(v_0) = \frac{1}{t}\mathbb{E}\left\{f(x_t)\int_0^t < w_s^h(v_0), X(x_s)dB_s >\right\} \qquad (74)$$

which has a slightly more straightfoward proof on the same lines. It is special, and explicit, case of Malliavin's integration by parts formula. Indeed if $p_t^h(x,y)$ represents the fundamental solution for $\frac{1}{2}\Delta + \nabla h$ then

$$d(P_t^h f)(v_0) = \int_M < \nabla P_t^h(-,y) >, v_0 >_{x_0} f(y)dy$$

while the right hand side if (74) is

$$\int_M p_t^h(x_0,y)f(y)\mathbb{E}\left\{\frac{1}{t}\int_0^t < W_s^h(v_0), X(x_s)dB_s > |x_t = y\right\}dy$$

Thus (74) is essentially Bismut's formula, [11], Theorem 2.71,

$$\begin{aligned}\nabla \log p_t^h(-,y) &= \mathbb{E}\left\{\frac{1}{t}\int_0^t (W_s^h)^*(X(x_s)dB_s)|x_t = y\right\}\\ &= \mathbb{E}\left\{\frac{1}{t}\int_0^t (W_s^h)^*(u_s dB_s)|x_t = y\right\} \qquad (75)\end{aligned}$$

using the notation of Remark (1). (In [11, $h \equiv 0$). Of course the work still remains to be done, that Bismut did to verify that the right-hand side of (74) is continuous in y so that (75) is true for all y, rather than only almost all y. using (73′) there is the variant in terms of flows:

$$\nabla(\log p_t^h(-,y))(x_0) = \mathbb{E}\{\frac{1}{t}\int_0^t (TF_s)^*(X(x_s)dB_s)|x_t = y\} \qquad (75')$$

See [36] for an alternative proof and substantial generalizations of Bismut's formula.

(4) Note that (70) gives an explicit cohomology between φ and $P_t^{h,1}\varphi$.

(5) Equation (72) can be written as

$$\int_0^t \varphi \circ dx_s = -\frac{1}{2}\int_0^t \delta^h P_s^{h,1}(\varphi)(x_s)ds + \int_0^t P_{t-s}^{h,1}(\varphi)(X(x_s)dB_s) \tag{76}$$

valid for closed $\varphi \in C^1\Omega^1$ when M is compact. However for each $t > 0$ the right-hand side is continuous as a linear map from $L^2\Omega^1(M, \mu_h)$ into the space of L^2 random variables $L^2(\Omega, \mathcal{F}_t, \mathbb{P}; \mathbb{R})$. Thus (76) can be used to extend the definition of $\int_0^t \varphi \circ dx_s$ to arbitrary closed φ in $L^2\Omega^1(M, \mu_h)$ and hence to all φ in $L^2\Omega^1(M, \mu_h)$ using (69) and the Hodge decomposition, c.f. the foward and backward martingale decomposition of Lyons and Zheng [37].

Acknowledgment. The initial formulation of this article was made during a visit to the Statistical Laboratory, Paris VI in the spring of 1990, organized by Professor M. Yor. My introduction to the Elliott–Kohlmann approach to integration by parts formulae was made especially pleasant by the restaurants of Rome and the company of Professor Elliott and the Bernoulli Conference of 1988, and extended by a seminar by E. Kopp on financial options. I am grateful to Professor H. Kunita and other Japanese colleagues, especially Professor N. Ikeda, for the opportunity to describe some of these ideas in the Nagoya Conference on Topics in Probability Theory, August 1990, and for useful discussions. This was supported by the Japanese Society for the Promotion of Science. Comments and corrections by X.-M. Li have been very helpful. This work can be considered as part of a joint project with S. Rosenberg, partially supported by NATO Collaborative Research Grants Programme 0232/87.

REFERENCES

[1] Abraham R. and Marsden J. E., *Foundations of Mechanics,* (Second Edition) (1978), The Benjamin/Cummings Publishing Co. Inc.

[2] Arnold L., *A formula connecting sample and moment stability of linear stochastic systems,* SIAM J. Appl. Math., **44**(1984), 793–802.

[3] Arnold, L., Oeljeklaus E. and Pardoux E., *Almost sure and moment stability for linear Itô equations,* In: Lyapunov Exponents, Proceedings Bremen 1984, eds. L. Arnold and V. Wihstutz. Lecture Notes in Math., **1186**(1986), Springer–Verlag.

[4] Bakry D., *Un critère de non-explosion pour certaines diffusions sur une variété riemannienne complète,* C. R. Acad. Sci. Paris, **303**(1), (1986), 23–26.

[5] Baxendale P., *Weiner processes on manifolds of maps,* Proc. Royal Soc. Edinburgh, **87A**(1986), 127–152.

[6] Baxendale P., *Brownian motions in the diffeomorphism group I.*, Compositio Math, **53**(1984), 19–50.

[7] Baxendale P., *The Lyapunov spectrum of a stochastic flow of diffeomorphisms,* In: Lyapunov Exponents, Proceedings, Bremen 1984, eds. L. Arnold and V. Wihstutz. Lecture Notes in Math., **1186**(1986), 322–337.

[8] Baxendale P., *Lyapunov exponents and relative entropy for a stochastic flow of diffeomorphisms,* Probab. Th. Rel. Fields, **81**(1989), 521–554.

[9] Baxendale P. and Stroock D. W., *Large deviations and stochastic flows of diffeomorphisms,* Probab. Th. Rel. Fields, **80**(1988), 169–215.

[10] Bismut J. -M., *Large deviations and the Malliavin calculus,* Progress in Mathematics, **45**(1984), Birkhäuser: Boston, Basel, Berlin.

[11] Bismut J. -M., *The Witten complex and the degenerate Morse inequalities,* J. Diff. Geom., **23**(1986), 207–240.

[12] Bougerol P., *Comparaison des exposants de Lyupanov des processus Markoviens multiplicatifs,* Ann. Inst. H. Poincaré Probab. Statist., **24**(4), (1986), 439–489.

[13] Caverhill A. P. and Elworthy K. D., *Flows of stochastic dynamical systems: the functional analytic approach,* Z. für wahrscheinlichkeitstheorie verw. geb., (1983), 245–267.

[14] Carverhill A. P., *Flows of stochastic dynamical systems: Ergodic Theory,* Stochastics, **14**(1985), 273–317.

[15] Carverhill A. P. and Elworthy K. D., *Lyupanov exponents for a stochastic analogue of the geodesic flow,* Trans. A. M. S., **295**(1), (1986), 85–105.

[16] Carverhill A. P., Chappell M. and Elworthly K. D., *Characteristic exponents for stochastic flows,* In: Stochastic Processes - Mathematics and Physics. Proceedings, Bielefeld 1984. Ed. S. Albeverio et al. pp. 52-762. Lecture Notes in Mathematics, **1158**(1986), Springer–Verlag.

[17] Chappell M. J., *Bounds for average Lyapunov exponents of gradient stochastic systems,* In: Lyapunov Exponents, Proceedings, Bremen 1984, eds. L. Arnold and V. Wihstutz. Lecture Notes in Math., **1186** (1986), 308–321.

[18] Chappell M. J., *Lyapunov exponents for certain stochastic flows,* Ph. D. Thesis. Mathematics Institute, University of Warwick, Coventry CV4 7Al, England, (1987).

[19] Cycon H., Froese R., Kirsch W. and Simon B., *Schrödinger operators with applications to quantum mechanics and global geometry,* Texts and Monographs in Physics, Springer–Verlag, (1987).

[20] Eliasson H., *Geometry of manifolds of maps,* J. Diff. Geom., **1**(1967), 169–194.

[21] Elliott R. J. and Kohlmann M., *Martingale Representation and the Malliavin Calculus,* Appl. Math. Optim., **20**(1967), 105–112.

[22] Elliott R. J. and Kohlmann M., *Integration by parts, homogeneous chaos expansions and smooth densities,* Ann. Probab., **17**(1), (1989), 194–207.

[23] Elworthy K. D., *Stochastic differential equations on manifolds,* London Mathematical Society Lecture Notes, Cambridge: Cambridge University Press, **70**(1989).

[24] Elworthy K. D. and Stroock D., *Large deviation theory for mean exponents of stochastic flows,* (1984), Appendix to [17] above.

[25] Elworthy K. D. and Rosenberg S., *Generalized Bochner theorems and the spectrum of complete manifolds,* Acta. Appl. Math., **12**(1988), 1–33.

[26] Elworthy K. D., *Geometric aspects of diffusions on manifolds,* In: Ecole d'Eté Probabilités de Saint-Flour XV - XVII 1985, 1987, ed. P. L. Hennequin, pp. 276–425. Lecture Notes in Maths., **1362**(1989), Springer–Verlag.

[27] Elworthy K. D. and Rosenberg S., *Manifolds with wells of negative curvature,* (1990), to appear in Invent. Math.

[28] Emery M., *Stochastic Calculus on Manifolds,* (1989), Springer–Verlag Universitext.

[29] Ikeda N., and Watanabe S., *Stochastic Differential Equations and Diffusion Processes,* (1981), Tokyo: Kodansha. Amsterdam, New York, Oxford: North-Holland.

[30] Kobayashi S. and Nomizu K., *Foundations of differential gometry,* **II** (1969), New York, Chichester, Brisbane, Toronto: Interscience Publishers, John Wiley & Sons.

[31] Kunita H., *Stochastic flows and stochastic differential equations,* Cambridge Studies in Advanced Mathematics, **24**(1990), Cambridge University Press.

[32] Kusuoka S., *Degree theorem in certain Weiner Riemannian manifolds,* In: Stochastic Analysis; Proceedings, Paris 1987, eds. M. Metivier and S. Watanabe, pp. 93–108. Lecture Notes in Math., **1322**(1988), Springer–Verlag.

[33] Le Jan, Y. and Watanabe S., *Stochastic flows of diffeomorphisms,* In: Stochastic Analysis, Proc. of the Taniguchi Int. Symp. Katata and Kyoto, 1982, ed. K. Itô§, (1984) pp. 307–332, North-Holland.

[34] Malliavin P., *Champ de Jacobi stochastiques,* C. R. Acad. Sc. Paris, **285**(1977b), série A, 789–792.

[35] Meyer P. A., *Geometrie differentielle stochastique (bis),* In: Séminaire de Pobabilités XVI. 1980/81, Supplément: Géometrie Différentielle Stochastique, ed. J. Azéma and M. Yor, 165–207. Lecture Notes in Math., **921**(1981b), Springer–Verlag: Berlin, Heidelberg, New York.

[36] Norris J. R., *Covariant stochastic calculus and applications to heat kernels,* preprint: Statistical Laboratory, University of Cambridge, 16 Mill Lane, Cambridge CB2 1SB, (1990).

[37] Lyons T. J. and Zheng W., *A crossing estimate for the canonical process on a Dirichlet space and a tightness result,* Soc. Math. de France. Astérique, 157–158(1988), 249–271.

K. D. Elworthy
Department of Mathematics
University of Warwick
COVENTRY CV5 7FR
U. K.

PART II

Special Flows and Multipoint Motions

Isotropic Stochastic Flows: A Survey

R.W.R. Darling

Contents:

0. Introduction

This article is intended to be a pedagogical survey of some of the results on isotropic stochastic flows obtained during the last decade by Baxendale, Harris, Le Jan, Matsumoto, and the author; the omission of any other work on this subject is solely due to ignorance, and information from readers is much appreciated. It is neither intended to be balanced nor complete, and proofs which appear elsewhere will be omitted; however outlines of proofs will be given for new results.

In broad terms, a stochastic flow is a probability model for the simultaneous random motion of an ensemble of particles in "space"; think of the diffusion of aerosol particles in a gas, for example. Let us use the notation $X_{st}(x)$ to denote the position at time $t \geq s$ of a particle which is at x at time s. To be a little more precise, a stochastic flow on $\mathbf{R}^d$ means a family of random mappings $\{X_{st}, 0 \leq s \leq t < \infty\}$ from $\mathbf{R}^d$ to $\mathbf{R}^d$ such that with probability 1,

$$X_{tu} \circ X_{st} = X_{su}, \ \forall \ 0 \leq s \leq t \leq u < \infty. \tag{0.1}$$

All random variables are defined on a filtered probability space $(\Omega, \mathcal{F}, \{\mathcal{F}_t\}_{t\geq 0}, P)$ satisfying the usual conditions, and X_{st} is $\mathcal{F}_t$-measurable. Some authors require only that $X_{tu} \circ X_{st} = X_{su}$ a.s. for each fixed $0 \leq s \leq t \leq u < \infty$, but this seems to make some of the other definitions troublesome in the coalescing cases discussed below.

The stochastic flows arising from stochastic differential equations with a finite number of driving Brownian motions have received much attention in the literature: see the texts of Elworthy [1983], Kunita [1984,1990], and Ikeda and Watanabe [1981]. For example, Kunita [1984] shows that given a $C^{k,\alpha}$ (i.e. k^{th} partial derivatives locally Hölder continuous of order α) vector field V on $\mathbf{R}^d$, a $C^{k+1,\alpha}$ random mapping a: $\mathbf{R}^d \to \mathbf{M}(d,k)$ (the real $d \times k$ matrices), and a k-dimensional Brownian motion $\{B_t\}$, there exists for any $\beta < \alpha$ a stochastic flow $\{\Phi_{st}\}$ consisting of $C^{k,\beta}$ diffeomorphims, such that

$$\Phi_{st}(x) = x + \int_s^t \{V(\Phi_{su}(x))ds + a(\Phi_{su}(x)) \circ dB_u\}(\text{Stratonovich})$$

(Generally the domain of $\Phi_{st}(.)$ is not the whole of $\mathbf{R}^d$, but the random set of x such that the explosion time of the trajectory starting at x at time s is greater than t). The Lyapounov exponents and mixing properties of such flows have been calculated by Carverhill [1985] and Baxendale [1990] (and references therein), among others.

The topic of this survey article does not fall within the framework of the previous paragraph. We treat here a very special class of stochastic flows, namely the homogeneous isotropic Brownian ones; here the motion of each point under the flow is a standard Brownian motion in $\mathbf{R}^d$, and so in particular there is no drift; also the motion of every pair of points relative to their initial positions has a law which is invariant under translations and rotations of $\mathbf{R}^d$. The reasons for studying this class are that

(a) It isolates the "correlation" of the motions (as opposed to drift), and presents this correlation in a translation- and rotation- invariant way, and

(b) It is presumed to be the model most relevant to physical phenomena such as diffusion of aerosol particles etc..

Except in the trivial case of rigid motion of the whole space, such a flow cannot however be constructed using a finite number of driving Brownian motions.

As Harris [1981] points out, the best way to understand such a stochastic flow is in terms of an isotropic mean zero Gaussian field U on $\mathbf{R}^d$: the mutual variation for the motion of two points under the flow is to be given by

$$d < X^i_{0t}(y), X^j_{0t}(z) > = b^{ij}(X_{0t}(y) - X_{0t}(z))dt \tag{0.2}$$

where b: $\mathbf{R}^d \to$ {Symmetric $d \times d$ non-negative definite real matrices} is the covariance tensor of U; in other words

$$b(x) \equiv \text{Cov}(U(x),\ U(0)) \equiv \mathbf{E}[U(y+x)U(y)'],\ \ y \in \mathbf{R}^d. \tag{0.3}$$

The (vector) isotropy condition on the Gaussian random field U is precisely the condition that

$$b(x) = G'b(Gx)G,\ \ \forall\ G \in O(d) \tag{0.4}$$

where $O(d)$ means the group of real orthogonal $d \times d$ matrices. This covariance tensor b(.) completely specifies the law of the stochastic flow, as we shall see. Therefore the next section will be devoted to a review of some classical material about covariance tensors of isotropic Gaussian random fields.

1. Covariances of d-Dimensional Isotropic Random Fields on $\mathbf{R}^d$

The material presented here appeared in Yaglom [1957] and is also summarized by Baxendale and Harris [1986]; the book of Yaglom [1987] is highly recommended.

The covariance tensor b(.) appearing in (0.2)-(0.4) will always be normalized so that b(0) is the identity (because the quadratic variation of the process $\{X_{0t}(x),\ t \geq 0\}$ must be t times the identity), and we always assume that all entries in the matrix $b(x)$ converge to 0 as $|\ x\ |$ goes to ∞ (correlation between distant points is zero in the limit). More restrictions will be placed later. In order to qualify as an isotropic covariance function, b(.) has to be of the following form (see Baxendale and Harris [1986], and Yaglom [1987], Sections 7 and 22.4):

$$b^{pq}(x) = \int_{\mathbf{R}^d} e^{ix.\lambda} F^{pq}(d\lambda) \tag{1.1}$$

where for a Borel subset A of $\mathbf{R}^d, F^{pq}(A)$ is a real symmetric nonnegative definite matrix satisfying $F(A) = G'F(GA)G$, in particular $F(A) = F(-A)$. This has the following consequences:

d=1:

$$b(x) = \int_{-\infty}^{\infty} e^{ix\lambda} F(d\lambda) \tag{1.2}$$

where F(.) is a probability measure with $F(d\lambda) = F(-d\lambda)$, and with no atom at 0.

d $\geq$ 2:

$$b^{pq}(x) = (B_L(|x|) - B_N(|x|))x^p x^q/|x|^2 + B_N(|x|)\delta^{pq},\ x \neq 0 \tag{1.3}$$

where

$$B_L(r) = A_d\{\int_0^\infty \left[\frac{J_{d/2}(rs)}{(rs)^{d/2}} - \frac{J_{(d+2)/2}(rs)}{(rs)^{(d-2)/2}}\right]\Phi_1(ds) + (d-1)\int_0^\infty \frac{J_{d/2}(rs)}{(rs)^{d/2}}\Phi_2(ds)\} \tag{1.4}$$

$$B_N(r) = A_d\{\int_0^\infty \frac{J_{d/2}(rs)}{(rs)^{d/2}}\Phi_1(ds) + \int_0^\infty \left[\frac{J_{(d-2)/2}(rs)}{(rs)^{(d-2)/2}} - \frac{J_{d/2}(rs)}{(rs)^{d/2}}\right]\Phi_2(ds)\} \tag{1.5}$$

where $A_d \equiv 2^{(d-2)/2}\Gamma(d/2)$, J_m denotes the Bessel function of the first kind of order m, and Φ_1 and Φ_2 are any positive finite measures on $(0,\infty)$ such that:

$$\frac{1}{d}[\Phi_1(0,\infty) + (d-1)\Phi_2(0,\infty)] = 1 \tag{1.6}$$

They are related to the spectral matrix appearing in (1.1) by the relations (see Yaglom [1987], Section 22.4)

$$\Phi_1(0,t) = \int_{|y|<t} \sum \frac{\lambda_p\lambda_q}{|\lambda|^2} F^{pq}(d\lambda)$$

$$\Phi_2(0,t) = \frac{1}{d-1}\{\int_{|\lambda|<t} \sum F^{pp}(d\lambda) - \Phi_1(0,t)\}$$

Note that the intergrals (1.4) and (1.5) are always well-defined, because the integrands are bounded (Lebedev [1972]); the last fact and Lebesgue's bounded convergence theorem show also that $B_L(r), B_N(r)$, and $b^{pq}(x)$ are all continuous functions. Representations (1.4) and (1.5) are so important that we also give them explicitly for the cases $d = 2$ and $d = 3$:

d=2

$$B_L(r) = \int_0^\infty J_1'(rs)\Phi_1(ds) + \int_0^\infty J_1(rs)\Phi_2(ds) \tag{1.7}$$

$$B_n(r) = \int_0^\infty \frac{J_1(rs)}{rs}\Phi_1(ds) + \int_0^\infty J_1'(rs)\Phi_2(ds)$$

d=3

$$B_L(r) = \int_0^\infty \left[\frac{\sin(rs)}{rs} + \frac{2\cos(rs)}{(rs)^2} - \frac{2\sin(rs)}{(rs)^3}\right]\Phi_1(ds) + \int_0^\infty \left[-\frac{2\cos(rs)}{(rs)^2} + \frac{2\sin(rs)}{(rs)^3}\right]\Phi_2(ds)$$

$$B_N(r) = \int_0^\infty \left[-\frac{\cos(rs)}{(rs)^2} + \frac{\sin(rs)}{(rs)^3}\right]\Phi_1(ds) + \int_0^\infty \left[\frac{\sin(rs)}{rs} + \frac{\cos(rs)}{(rs)^2} - \frac{\sin(rs)}{(rs)^3}\right]\Phi_2(ds) \tag{1.8}$$

The function B_L (.) is called the **longitudinal correlation function**, and $B_L(|x|)$ gives the correlation between the components of the field in the direction of the vector x at the points y and $y+x$. The function B_N (.) is called the **lateral correlation function**, and $B_N(|x|)$ gives the correlation between the components of the field in some direction perpendicular to the direction of the vector x at the points y and $y+x$. A computation using the asymptotics of Bessel functions shows that $B_L(0) = B_N(0) = 1$, which verifies that $b(0)$ is the identity. The **longitudinal and lateral spectral measures** Φ_1 and Φ_2 have physical significance: Φ_1 gives rise to the part of U which is "potential", since when Φ_2 is zero, U is the gradient of a scalar random field; Φ_2 gives rise to the part of U which is "solenoidal", since when Φ_1 is zero, U has zero divergence.

The degree of differentiability of $B_L(.)$ and $B_N(.)$ depends on existence of moments of the spectral measures Φ_1 and Φ_2, as noted by Baxendale and Harris [1986]:

PROPOSITION 1.1 *If k is an integer ≥ 1, and $\int_0^\infty s^k \Phi_j(ds) < \infty$ for $j = 1, 2$, then $b(x)$ has a continuous derivative of order k on $\mathbf{R}^d$.*

Proof. By (1.1) and an analoque of a well-known theorem on characteristic functions (see Chung [1974], 6.4.1), $b(x)$ has a continuous derivative of order k on $\mathbf{R}^d$ if

$$\int |\lambda|^k F^{pq}(d\lambda) < \infty, \quad 1 \leq p,\ q \leq d. \tag{1.9}$$

According to Yaglom [1957], p. 302

$$F^{pq}(d\lambda) = \frac{\sigma_s(d\ell)}{\Omega_{d-1}(s)} \{\ell^p \ell^q \Phi_1(ds) + (\delta^{pq} - \ell^p \ell^q)\Phi_2(ds)\} \tag{1.10}$$

where $\Omega^{d-1}(s)$ denotes the area of the sphere centered at 0 in $\mathbf{R}^d$ of radius s, σ_s is the usual area measure of that sphere, $s = |\lambda|$, and $\ell \equiv (\ell^1, \ldots, \ell^d) = \lambda/|\lambda|$. Evidently (1.9) follows immediately from (1.10) and the assumption stated. ■

2. How to Construct the Corresponding Stochastic Flows

There are essentially two fundamentally different approaches to the construction of isotropic stochastic flows.

I. The Reproducing Kernel Hilbert Space Method. This method will not be used subsequently in this survey, but is outlined for the sake of completeness. It is described clearly in Baxendale and Harris [1986], but its antecedents are Le Jan [1982] and Baxendale [1984]. It assumes that the measures Φ_1 and Φ_2 have finite second moments, so that by Proposition 1.1 the covariance tensor b(.) has two continuous derivatives. Define a reproducing kernel b^{pq} (.,.) by $b^{pq}(x,y) = b^{pq}(x-y)$, and obtain a real separable reproducing kernel Hilbert space $\mathcal{H}$ consisting of vector fields on $\mathbf{R}^d$. Since $b^{pq}(x,y)$ is C^2 in x and y, there is a continuous inclusion of $\mathcal{H}$ into $C_b^2(\mathbf{R}^d, \mathbf{R}^d)$. Take an orthonormal basis $\{V_\alpha\}$ for $\mathcal{H}$, which will consist of bounded twice differentiable vector fields on $\mathbf{R}^d$ such that

$$\sum_\alpha V_\alpha^p(x)V_\alpha^q(y) = b^{pq}(x-y)$$

$$\Rightarrow \sum_\alpha (V_\alpha^p(x))^2 = b^{pp}(0) = 1 \tag{2.1}$$

It is possible to solve the Itô s.d.e

$$dX_t(x) = \sum_\alpha V_\alpha^p(X_t(x))dW_t^\alpha \tag{2.2}$$

with $X_0(x) = x$, where $\{W_t^1\}, \{W_t^2\}, \ldots$ are independent standard Brownian motions, in such a way that X_t and X_t^{-1} are diffeomorphisms of $\mathbf{R}^d$ onto itself for each $t \geq 0$; then set $X_{st} = X_t \circ X_s^{-1}$ for $0 \leq s \leq t < \infty$ to obtain a stochastic flow satisfying (0.2).

II. The Martingale Problem Approach. This method is due to Harris [1981,1984]; it is the only method available when $b''(0)$ does not exist. The version presented here is slightly better than the one in Darling [1987]. Consider the motion of an arbitrary set of k points under a stochastic flow satisfying (0.2), considered as a vector $(X_{0t}(y_1), \ldots, X_{0t}(y_k))$ in $(\mathbf{R}^d)^k$. This must be a diffusion process with generator $\mathcal{A}_k$, where for any smooth function f on $(\mathbf{R}^d)^k$,

$$\mathcal{A}_k f(z_1, \ldots, z_k) \equiv \sum_{i,j=1}^{k} \sum_{p,q=1}^{d} b^{pq}(z_j - z_i)\frac{\partial^2}{\partial z_i^p \partial z_j^q} f(z_1, \ldots, z_k) \tag{2.3}$$

An important object associated with this generator is the $dk \times dk$ symmetric matrix function $B^{(k)}$ on $(\mathbf{R}^d)^k$, called the **k-point covariance function**, defined by

$$B^{(k)}(z_1, \ldots, z_k) \equiv \begin{bmatrix} B_{11} & \cdots & B_{1k} \\ \cdots & \cdots & \cdots \\ B_{k1} & \cdots & B_{kk} \end{bmatrix} \tag{2.4}$$

where $B_{ij} \equiv b(z_j - z_i)$. The next three results are close to those of Harris [1981, 1984], but we are able here to dispense with Lipschitz conditions. The proofs will be found at the end of this section.

Lemma 2.1 *For $d \geq 2$, and for every $k \geq 2$, $B^{(k)}(z_1, \ldots, z_k)$ is continuous in $(z_1, \ldots, z_k)$, and strictly positive definite for all $(z_1, \ldots, z_k) \in D_k$, where D_k denotes the set of k-tuples of distinct points in $(\mathbf{R}^d)^k$; moreover the largest eigenvalue is bounded above by kd. The same conclusions hold when $d = 1$ provided the spectral measure F in (1.2) is not an atomic measure with a finite support.*

Corollary 2.2 *D_k can be expressed as a union of open sets $\bigcup_{m \geq 1} V_{k,m}$, where*

$$V_{k,m} \equiv \{(z_1, \ldots, z_k) \in D_k : ||c||^2/m \leq c^* B^{(k)}(z_1, \ldots, z_k)c \leq kd||c||^2, \ \forall \ c \in \mathbf{C}^{kd}\}. \tag{2.5}$$

Care is needed when solving the martingale problem with generator $\mathcal{A}_k$ as in (2.3): for the k-point process to behave like the k-point motion of a stochastic flow, particles which meet at some time have to stay together. Therefore the path space cannot be $C([0, \infty) : (\mathbf{R}^d)^k)$, but must be

$$\Omega_k \equiv \{\omega = (\omega^1, \ldots, \omega^k) \in C([0, \infty) : (\mathbf{R}^d)^k) : \tag{2.6}$$

$$\omega^i(t) = \omega^j(t) \text{ for some } t \geq 0 \text{ and } i \neq j \Rightarrow \omega^i(s) = \omega^j(s) \ \forall \ s > t\}$$

We say that the martingale problem for $\mathcal{A}_k$ on Ω_k is well-posed when for every $s \geq 0$ and $\mathbf{z} \equiv (z_1, \ldots, z_k) \in D_k$, there is a unique probability measure $P_{s,\mathbf{z}}$ on the path space Ω_k (i.e. paths in $\mathbf{R}^d$ which meet stay together), under which the canonical process $\{(Y_{st}(z_1), \ldots, Y_{st}(z_k)), \ t \geq s\}$ is a diffusion process on $(\mathbf{R}^d)^k)$ with zero drift, such that the mutual quadratic variations satisfy:

$$d < Y^p_{st}(z_i), Y^q_{st}(z_j) > = b^{pq}(Y_{st}(z_j) - Y_{st}(z_i))dt, i, j \in \{1, 2, \ldots, k\}. \tag{2.7}$$

Theorem 2.3 *For $d \geq 2$, for every choice of spectral measures Φ_1 and Φ_2 on $(0, \infty)$, the martingale problem for $\mathcal{A}_k$ on Ω_k is well-posed. Moreover the map $(s, z) \to P_s, z$ is measurable and the family $\{P_{s,z}\}$ is strong Markov. The same conclusions hold when $d = 1$ provided the spectral measure F is not atomic with a finite number of atoms.*

Remark. The probabilty measures constructed here are referred to as "the law of the k-point motion".

Completion of the construction. We have now completed the construction of the "laws of the k-point motions" for the desired stochastic flow, for every $k \geq 1$; a routine calculation shows that they have the obvious consistency property, that the law of any k components of the $(k + r)$-point motion has the law of the k-point motion, for $r \geq 1$. Given strong conditions on the moments of $Y_{st}(x) - Y_{st}(y)$ for fixed $t > s$ as $y \to x$ (Totoki's theorem, proved in Elworthy [1982]), it is not too hard to construct a stochastic flow of continuous mappings. However since in general the stochastic flow will not be spatially continuous, there is a lot more to be done on order to arrive at a family of random mappings satisfying (0.1). Things are not too bad in $\mathbf{R}^1$, where Harris [1984] uses the well-ordered property of the line to construct a stochastic flow of monotone (hence spatially measurable) mappings. In $\mathbf{R}^d$ for $d \geq 2$, the only method known for constructing a stochastic flow from a consistent set of laws for k-point motions is a tricky 20-page projective limit argument, given in Darling [1987], resulting in a stochastic flow of mappings which are not shown to be spatially measurable. Here are two unsolved problems:

Problem 2.4. Given the generators $\{\mathcal{A}_k, k = 1, 2, \ldots\}$ for all the k-point motions, which are consistent, construct a stochastic flow consisting of spatially measurable mappings such that for each k, the k-point motion under the flow has the given generator (with the rule that paths in $\mathbf{R}^d$ which meet must coalesce). You may assume that the 2-point motion satisfies $P(|X_{0t}(x) - X_{0t}(y)| > \epsilon) \to 0$ as $y \to x$, for fixed $\epsilon > 0$ and $t \geq 0$. It would be interesting to see whether an efficient proof could be given by the methods of non-standard analysis.

The following problem arises from the point of view that the real object of interest in stochastic flows is not the family of mappings $\{X_{st}\}$, but the way that an initial measure on $\mathbf{R}^d$ is transported under the flow.

Problem 2.5 Given the generators $\{\mathcal{A}_k, k = 1, 2, \ldots\}$ for all the k-point motions, which are consistent, construct a Markov process $\{\mu_t\}$ in the space of finite measures on $\mathbf{R}^d$ such that for each k, if the initial value of the process is an atomic measure distributing its mass equally among k points, then the support of μ_t evolves according to the diffusion with generator $\mathcal{A}_k$ (with the rule that paths in $\mathbf{R}^d$ which meet must coalesce). It may be that this problem can be treated within the framework of measure-valued diffusions: see Dawson and Kurtz [1982].

Proof of Lemma 2.1. Continuity of $B^{(k)}(.)$ follows from that of b(.); see the remarks preceding (1.7). We give the proof of positive definiteness for $d \geq 2$; the proof for $d = 1$ is similar but easier. Let $c = ((c^1)', \ldots, (c^k)')'$ be an arbitrary element of $(\mathbf{C}^d)^k$, whose conjugate transpose is denoted c^*. Using the spectral representation

(1.1) and formula (1.10), we obtain

$$c^*B^{(k)}c = \sum_{h,j=1}^{k}\sum_{p,q=1}^{d} c_p^h b^{pq}(z_j - z_h)c_q^j$$

$$= \int_{\mathbf{R}^d}\sum_{p,q=1}^{d}\sum_{h,j=1}^{k}(c_p^h \exp(-i\lambda.z_h))F^{pq}(d\lambda)(c_q^j \exp(i\lambda.z_j))$$

$$= \int_0^\infty \int_{S^{d-1}} \frac{\sigma_s(d\ell)}{\Omega_{d-1}(s)} \sum_{p,q=1}^{d} \{\bar{a}_p(s,\ell)\ell^p\ell^q a_q(s,\ell)\Phi_1(ds) + a_p(s,\ell)(\delta^{pq}-\ell^p\ell^q)a_q(s,\ell)\Phi_2(ds)\}$$

$$\Rightarrow c^*B^{(k)}c = \int_0^\infty \int_{S^{d-1}} \frac{\sigma^s(d\ell)}{\Omega_{d-1}(s)}\{|\sum_{p=1}^{d} a_p(s,\ell)\ell^p|^2\Phi_1(ds) + \tag{2.8}$$

$$[\sum_{p=1}^{d}|a_p(s,\ell)|^2 - |\sum_{p=1}^{d} a_p(s,\ell)\ell^p|^2]\Phi_2(ds)\}$$

where $s = |\lambda|, \ell = \lambda/|\lambda|$, and

$$a_q(s,\ell) \equiv \sum_{j=1}^{k} c_q^j \exp(is\ell.z_j). \tag{2.9}$$

Assume that $(z_1,\dots,z_k) \in D_k$ and $c \neq 0$. By the Cauchy-Schwarz inequality,

$$|\sum_{p=1}^{d} a_p(s,\ell)\ell^p|^2 \leq \sum_{p=1}^{d}|a_p(s,\ell)|^2 \sum_{p=1}^{d}(\ell^p)^2 = \sum_{p=1}^{d}|a_p(s,\ell)|^2 \tag{2.10}$$

with strict inequality for ℓ in a set of positive σ_s-measure, for every $s > 0$. Therefore

$$\int_{S^{d-1}} \frac{\sigma_s(d\ell)}{\Omega_{d-1}(s)}|\sum_{p=1}^{d} a_p(s,\ell)\ell^p|^2 > 0, \text{ and}$$

$$\int_{S^{d-1}} \frac{\sigma_s(d\ell)}{\Omega_{d-1}(s)}[\sum_{p=1}^{d}|a_p(s,\ell)|^2 - |\sum_{p=1}^{d} a_p(s,\ell)\ell^p|^2] > 0,$$

for all $s > 0$, and this proves that, for every choice of Φ_1 and Φ_2, $c^*B^{(k)}c > 0$ whenever $c \neq 0$.

It follows easily from (2.9) that

$$\sum_{p=1}^{d}|a_p(s,\ell)|^2 \leq k||c||^2$$

and so (2.8), (2.10) and (1.6) imply that

$$c^*B^{(k)}c \leq k(\Phi_1(0,\infty) + \Phi_2(0,\infty))||c||^2 \leq kd||c||^2$$

which proves the upper bound assertion. ■

Proof of Theorem 2.3 (Outline only). Result 7.1.7 of Stroock andVaradhan [1979] shows that the martingale problem for the generator

$$Lf(y) \equiv \sum_{i,j} a_{ij}(y) \frac{\partial^2 f(y)}{\partial y_i \partial y_j}, \; f \in C_0^\infty(\mathbf{R}^n) \tag{2.11}$$

is well-posed provided a: $\mathbf{R}^n \to$ {Symmetric $n \times n$ matrices} is measurable, and for some $0 < \lambda \leq \Lambda$,

$$\lambda|\theta|^2 \leq < \theta, a(y)\theta > \leq \Lambda|\theta|^2, \text{ for all } y. \tag{2.12}$$

Also their localization principle 6.6.1 shows that if, for each $m \geq 1$, there is an open set G_m in $\mathbf{R}^n$, and $a^{(m)} : \mathbf{R}^n \to$ { Symmetric $n \times n$ matrices} which agrees with a on G_m, such that (2.12) holds with $a^{(m)}$ in place of a, then the martingale problem for L is well-posed for trajectories starting inside $G \equiv \cup_{m \geq 1} G_m$ and stopped at the first exit from G. Moreover the resulting probability measures $P_{s,y}$ (for a diffusion starting at y at time s) are measurable in (s, y), and form a strong Markov family (Stroock and Varadhan [1979], 6.2.2). Applying these ideas to $\mathcal{A}_k$, and using the results of Corollary 2.2, we see that the martingale problem for $\mathcal{A}_k$ is well-posed up to the first exit from D_k. Since for more than two trajectories to collide at once is an event with probability zero, the process is now a.s. in D_{k-1}, and has a unique law (the solution of the martingale problem for $\mathcal{A}_{k-1}$) until the first exit from D_{k-1}, and so on. To conclude that the martingale problem for $\mathcal{A}_k$ on Ω_k is well-posed, use the technique of Ethier and Kurtz [1986], Theorem 4.6.2. The measurability and strong Markov assertions follow easily from Stroock and Varadhan [1979], 6.6.2. The identity (2.7) follows from the form of the generator $\mathcal{A}_k$ in (2.3). ■

3. Coalescing and Non-Coalescing Stochastic Flows on the Line

This section is based on the work of Harris [1984], Matsumoto [1989], and a little on Darling [1988]. The first point to notice when $d = 1$ is that, for $y < x$, the "distance process"

$$\eta_t \equiv X_{0t}(x) - X_{0t}(y), \;\; t \geq 0 \tag{3.1}$$

is a diffusion on $[0, \infty)$, with 0 as an absorbing boundary, and with generator

$$Lf(z) = (1 - b(z))\frac{\partial^2 f(z)}{\partial z^2} \tag{3.2}$$

The probability that trajectories from x and y coalesce in finite time (i.e. the probability of absorption of $\{\eta_t\}$ in finite time) is 1 if

$$\int_{0+} \frac{z\,dz}{1-b(z)} < \infty \tag{3.3}$$

and is 0 otherwise, by Feller's criterion for accessibility. A more difficult problem is to show, in the former case, that the cardinality of the image of a finite interval under the map X_{0t} (t positive) is almost surely finite.

Theorem 3.1 (Harris [1984], Matsumoto [1989]). *Suppose there exists a continuous function $\beta : (0,1] \to (0,\infty)$ such that (i) $1-b(z) \geq \beta(z)$, (ii) $z/\beta(z)$ is an integrable function on (0,1), and (iii) $z^{-2}\beta(z)$ is monotone decreasing on $(0,\alpha)$, for some $\alpha > 0$. (Of course (i) and (ii) together imply (3.3).) Then*

(a)*For every $t_0 > 0$ and every compact interval $K, X_{0t}(k)$ is a finite set for all $t \geq t_0$, a.s..*

(b)*For every $t_0 > 0$ and every compact interval $I, X_{0t}(\mathbf{R}) \cap I$ is a finite set for all $t \geq t_0$, a.s..*

Theorem 3.2 (Matsumoto [1989]). *If the integral in (3.3) is infinite, then X_{0t} is a homeomorphism of $\mathbf{R}$ with probability 1, for any $t \geq 0$.*

The proof of Theorem 3.1 involves analysis of the asymptotics of eigenfunctions of the operator L in (3.2), and Harris' result that, if J_n denotes the set $\{k2^{-n} : k = 0,1,\ldots,2^n\}$, then

$$\mathbf{E}[\text{card}(X_t(J_n))] = 1 + 2^n Q_{2^{-n}}(\eta_t > 0) \tag{3.4}$$

where Q_x is the probability law for the diffusion $\{\eta_t\}$ starting at x, and the expectation is taken with regard to the law P of the stochastic flow. Theorem 3.2 uses the result that, for any $t > 0$ and $\epsilon > 0$,

$$\lim_{x\to 0} \frac{1}{x} Q_x(\eta_t \geq \epsilon) = 0 \tag{3.5}$$

provided the integral in (3.3) is infinite.

In the coalescing case, how big is the interval which has coalesced by time t with the point which started at 0? To answer this, let J_t denote the (random) set of points x such that $X_{0t}(x) = X_{0t}(0)$; of course, by monotonicity J_t has to be a random interval.

Proposition 3.3 (Darling [1988]). *The Lebesgue measure $m(J_t)$ of the coalescent set J_t satisfies*

$$\lim_{t\to\infty} \frac{\mathbf{E}[m(J_t)]}{\sqrt{t}} = \frac{4}{\sqrt{\pi}} \tag{3.6}$$

Here the value of the limit is the same as in the case of a system of independent coalescing Brownian motions, as studied by Arratia (1979), and does not depend at all on the precise form of the covariance function.

Example 3.4 Let $0 < m \leq 1, c > 0$, and $b(x) = 1 - c|x|^m$ for $|x| \leq c$, and $=$ 0 otherwise. The function b(.) is then convex on $[0,\infty)$, and so by a theorem of Polya (see Yaglom [1987], p.136), b(.) is indeed a covariance function, i.e. it has an expression of the form (1.2). Then the conditions of Theorem 3.1 hold. On the other hand the covariance function $b(x) = e^{-cx^2}$ (see Yaglom [1987], p. 133) satisfies the conditions of Theorem 3.2.

4. Isotropic Stochastic Flows in Dimension $d \geq 2$ With C^4 Covariance Tensor

The material of this section is taken from various papers of Baxendale, Harris and Le Jan. Suppose the spectral measures Φ_1 and Φ_2 in (1.4) - (1.6) have finite fourth moments, so that $b(x)$ is C^4, and $B_L(r)$ and $B_N(r)$ are C^4 on $[0,\infty)$ (right derivatives at 0). Consequently there are expansions

$$B_L(r) = 1 - \beta_L r^2/2 + O(r^4), \text{ as } r \to 0 \tag{4.1}$$

$$B_N(r) = 1 - \beta_N t^2/2 + O(r^4), \text{ as } r \to 0$$

Using the asymptotics of Bessel functions, it may be shown that for potential flows (i.e. $\Phi_1 = 0$), $3\beta_N - \beta_L = 0$, and for solenoidal flows (i.e. $\Phi_1 = 0$), $(d-1)\beta_N - (d+1)\beta_L = 0$. These are the extreme cases, and in general

$$\frac{d-1}{d+1} \leq \frac{\beta_L}{\beta_N} \leq 3 \tag{4.2}$$

Baxendale and Harris [1986] show that the requirement that $b^{pq}(x)$ converges to 0 as $|x|$ goes to ∞ implies that β_N and β_L are both strictly positive.

Henceforward abbreviate X_{0t} to X_t. With or without the differentiability assumptions made in this section, it is true that, for fixed x and y, each component of $V_t \equiv X_t(x) - X_t(y)$ is a martingale and a diffusion, and its absolute value $\rho_t \equiv |X_t(x) - X_t(y)|$ is a diffusion on $[0,\infty)$, with absorption at 0, and with generator

$$Ag(\rho) = (1 - B_L(\rho))g''(\rho) + (d-1)\frac{1 - B_N(\rho)}{\rho}g'(\rho), \quad g \in C_b^2(\mathbf{R}). \tag{4.3}$$

Of course, 0 is inaccessible in the cases considered in this section because of an integral condition discussed in Theorem 5.1. By studying the scale function for this diffusion, Le Jan [1985] and Baxendale and Harris [1986] reach the following conclusions:

$$\mathbf{d} = \mathbf{2}: P(\rho_t \to \infty) = 0, \text{ and } P(\rho_t \to 0) = 1 (\text{resp. } 0) \text{ if } \frac{\beta_L}{\beta_N} > 1 (\text{resp. } \le 1) \tag{4.4}$$

$$\mathbf{d} = \mathbf{3}: P(\rho_t \to \infty) > 0; P(\rho_t \to 0) = 1 - P(\rho_t \to \infty), \text{ and } > 0 \Longleftrightarrow \frac{\beta_L}{\beta_N} > 2.$$

$$\mathbf{d} \ge \mathbf{4}: P(\rho_t \to \infty) = 1.$$

In particular, $\{\rho_t\}$ is transient unless $d = 2$ and $\frac{\beta_L}{\beta_N} \le 1$. Comparison with (4.2) shows $P(\rho_t \to 0) = 0$ for solenoidal flows, but is > 0 in dimensions 2 and 3 provided the potential component is large enough.

To understand the asymptotics of $\{\rho_t\}$ more precisely, and for other reasons, it is desirable to calculate the Lyapunov exponents of the stochastic flow. As explained in Section 2 above, the differentiability properties of the covariance tensor are sufficient for the existence of a stochastic flow consisting of diffeomorphisms, and so for x in $\mathbf{R}^d$ there is a well-defined linear mapping $DX_t(x)$ from $\mathbf{R}^d$ to $\mathbf{R}^d$, and DX_t is called for the tangent flow. Isotropy implies that the law of $DX_t(x)$ is the same for all x. Let $\lambda_{1t} \ge \lambda_{2t} \ge \dots \ge \lambda_{dt} > 0$ be the characteristic values of the positive definite symmetric matrix $(DX_t(x))'DX_t(x)$. Then the Lyapunov exponents μ_i may be defined as the limits

$$\mu_i \equiv \lim_{t\to\infty} \frac{\log(\lambda_{it})}{2t} \tag{4.5}$$

which exist, are constant a.s., and satisfy $\mu_1 > \dots > \mu_d$ (for isotropic flows); see Carverhill [1985], Le Jan [1985], and Baxendale [1986]. Here they are given by the formula

$$\mu_i = \frac{(d-i)\beta_N - i\beta_L}{2}. \tag{4.6}$$

For example, the case $\mu_1 < 0$ corresponds to the situation where length of tangent vectors (and hence, by linearization, small interpoint distances) go to zero exponentially fast under the stochastic flow; evidently

$$\mu_1 < 0 \Longleftrightarrow \frac{\beta_L}{\beta_N} > d - 1 \tag{4.7}$$

which cannot occur for $d \ge 4$, by (4.2); it can occur for stochastic flows when $d = 2$ or 3 and the potential component is large enough, however.

A quantity related to the Lyapunov exponents is the change in volume of an infinitesimal solid under the stochastic flow. Let $v_1, \dots, v_n$ denote "tangent vectors" in $\mathbf{R}^d$, and let $v_i(t) = DX_t(x)v_i$. The volume in $\mathbf{R}^n$ of the parallellopiped with edges parallel to $v_1(t),\ldots,v(t)$ is denoted $||v_1(t) \wedge \ldots \wedge v_n(t)||$. Le Jan [1985] shows that, provided $||v_1(0) \wedge \ldots \wedge v_n(0)|| = 1$,

$$\log||v_1(t) \wedge \ldots \wedge v_n(t)|| = \tag{4.8}$$

$$\sqrt{n[(n+1)\beta_L - (n-1)\beta_N]}W_t - \frac{n}{2}[(n+1)\beta_L - (2d-1-n)\beta_N]t$$

where $\{W_t\}$ is a standard Brownian motion. In particular, when $n = d$,

$$\lim_{t\to\infty}\frac{\log||v_1(t) \wedge \ldots \wedge v_d(t)||}{2t} = -\frac{d}{4}[(d+1)\beta_L - (d-1)\beta_N] \tag{4.9}$$

$$= \sum_{i=1}^{d}\mu_i$$

In other words, the sum of the Lyapunov exponents is the exponent in the long term behavior of infinitesimal volumes in $\mathbf{R}^d$.

Let v be a tangent vector at any point, and let $v_t \equiv DX_t(x)v$, i.e. the image of v under the tangent flow. Baxendale and Harris [1986] show that, if v_t is expressed as $(|v_t|, u_t) \in (0,\infty) \times S^{d-1}$, where $u_t = v_t/|v_t|$, then $|v_t|$ and u_t are independent diffusion processes, such that

$$|v_t| = |v_0|\exp\{\sqrt{\beta_L}W_t + \frac{1}{2}[(d-1)\beta_N - \beta_L]t\} \tag{4.10}$$

(compare (4.7) and (4.8)), and $u_t = \tilde{W}(\beta_N t)$, where $\tilde{W}$ denotes Brownian motion on S_{d-1}.

Suppose tangent vectors v_i and v_2 are noncollinear. Baxendale and Harris [1986] show that, if

$$\gamma_t \equiv \frac{< v_1(t), v_2(t) >}{|v_1(t)||v_2(t)|}, \quad \theta_t \equiv \cos^{-1}\gamma_t \in (0,\pi) \tag{4.11}$$

then $\{\gamma_t\}$ is a diffusion on (-1,1) with operator

$$\frac{1}{2}(\beta_L + \beta_N)(1-\gamma^2)\frac{\partial^2}{\partial\gamma^2} - \frac{1}{2}(\beta_L + \beta_N)(1-\gamma^2)\frac{\partial}{\partial\gamma} \tag{4.12}$$

while $\{\theta_t\}$ is a diffusion on $(0,\pi)$ with operator

$$\frac{1}{2}(\beta_L + \beta_N)\sin^2(\theta)\frac{\partial^2}{\partial\theta^2} \tag{4.13}$$

Evidently $\{\theta_t\}$ is a bounded martingale, and hence converges a.s.; the limit is π (with probability θ_0/π), or 0 (with probability $1 - \theta_0/\pi$). In other words, any two initial tangent vectors in the same tangent space line up in the same or directly opposite directions, under the tangent flow. It can also be proved that the rotation of the mean of the two vectors in the plane spanned by both of them is a martingale; possible further results in this vein are suggested by the theorem of Arnold and San Martin [1988] in the deterministic case.

The preceding discussion depends on an analysis of the first derivative of the flow. Le Jan [1989] also studies the second derivative, in the context:

$$X_t^{(2)}(x,u,w) \equiv (X_t(x), DX_t(x)u, D^2X_t(x)(w,w))$$

If Γ is a parametrized curve and if $\Gamma_t \equiv X_t(\Gamma)$, then

$$X_t^{(2)}(\Gamma(0),\Gamma'(0),\Gamma''(0)) = (\Gamma_t(0),\Gamma_t'(0),\Gamma_t''(0))$$

Let $\tau_t \equiv \Gamma_t'/||\Gamma_t'||, \rho_t \equiv \Gamma_t''/||\Gamma_t'||^2$; the **curvature vector** of the curve Γ_t at 0 is given by

$$\rho_t^\perp \equiv \frac{d\tau_t(s)}{ds}|_{s=0} = \rho_t - < \rho_t, \tau_t > \tau_t \tag{4.14}$$

Assume that the stochastic flow is solenoidal (i.e. $\Phi_1 = 0$), and let m_i denote the ith moment of the spectral measure Φ_2. Le Jan [1989] states that $||\rho_t^\perp||^2$ is a positive recurrent diffusion, whose asymptotic law is of the form $cx^v(x+k)^{u-v}$, where

$$v = \frac{d-3}{2}, \quad u = -\frac{d^2+11d-2}{10d-2}, \quad k = \frac{3m_4(d-3)}{m_2(d+4)(5d-1)}$$

Problem 4.1. Show positive recurrence of sectional curvatures, and other quantities derived from the full curvature tensor.

A final topic in this section is the statistical equilibrium of the flow. For this it is desirable to extend the definition of the stochastic flow to $\{X_{st}, -\infty < s \le t < \infty\}$ (possible since the maps are diffeomorphic), and then define random measures $\{\nu_{s,t},\ s \le t\}$ on $\mathbf{R}^d$ by duality as follows:

$$< \phi, \nu_{s,t}(\omega) > = \int \phi(X_{st}(\omega))dx, \ \phi \text{ continuous with compact support.} \tag{4.15}$$

As discussed in Le Jan [1985], the left side of (4.15) is a backwards martingale in s for fixed t, and in fact the difference of two positive backwards martingales, and therefore converges almost surely as $s \to -\infty$. Since the convergence occurs for all ϕ, this proves that for each fixed t, $\{\nu_{s,t},\ s \le t\}$ converges a.s. in the vague topology to a random measure $\nu_{-\infty,t}$ as $s \to -\infty$. By time-homogeneity of the flow, $\nu_{-\infty,t}$ has the same law for each t. We abbreviate $\nu_{-\infty,0}$ to $\nu_{-\infty}$, called the **statistical** equilibruim. If the longitudinal spectral measure Φ_1 is zero, then the stochastic flow is incompressible, and $\nu_{-\infty}$ is equal to Lebesgue measure almost surely; on the other hand if Φ_1 is not zero, then $\nu_{-\infty}$ is a.s. singular.

Darling and Le Jan [1988] show that, if the top Lyapunov exponent $\mu_1 < 0$, then $\nu_{-\infty}$ is a.s. the zero measure. On the other hand Le Jan [1985] shows that if $\mu_1 > 0$ ($\Longleftrightarrow (d-1)\beta_N/\beta_L > 1$, by (4.7)), then $\mathbf{E}[\nu_{-\infty}(dx)] = dx$, and

$$\mathbf{E}[\nu_{-\infty}(dx)\nu_{-\infty}(dy)] = |x-y|^{1-d}h(|x-y|)dxdy \tag{4.16}$$

where h is the invariant measure of the interpoint distance process, and may be computed from (4.3). In particular, the Hausdorff dimension of $\nu_{-\infty}$ is a.s. at least $[(d-1)\beta_N/\beta_L] - 1$.

Problem 4.2. Describe $\nu_{-\infty}$ when $\mu_1 = 0$.

5. Coalescing Isotropic Stochastic Flows in Dimensions 2 and 3.

Since two independent Brownian motions in $\mathbf{R}^d$ almost surely never meet for $d \geq 2$, it seems at first hard to imagine that there could exist coalescing stochastic flows in dimensions higher than 1. To see how this is possible, at least in dimensions 2 and 3, notice that the generator of the interpoint distance process $\rho_t \equiv |X_t(x) - X_t(y)|$ is still given by (4.3). Any function $S : (0,\infty) \to (0,\infty)$ such that

$$S'(r) = \exp\{-(d-1)\int_1^r \frac{1-B_N(s)}{s[1-B_L(s)]}ds\} \tag{5.1}$$

is a scale function for $\{\rho_t\}$. Let $\tau(x,y)$ denote $\inf\{t : |X_t(x) - X_t(y)| = 0\}$. The following result, which appears in Darling [1987], was essentially discovered by Harris.

Theorem 5.1 *Suppose there is a solution S to the differential equation (5.1) such that $S(0) = 0$. If*

$$\int_{0+} \frac{S(r)dr}{S'(r)[1-B_L(r)]} < \infty \tag{5.2}$$

then $P(\tau(x,y) < \infty) = 1 - S(|y-x|)/S(\infty)$, where $S(\infty) \equiv \lim_{r\to\infty} S(r)$. If (5.2) is false, then $P(\tau(x,y) < \infty) = 0$.

Example 5.2 (d=2). When $d = 2$, choose the longitudinal and lateral spectral measures Φ_1 and Φ_2 appearing in (1.7) as follows: let $\Phi_2 = 0$ (i.e. no solenoidal component), and for any $\delta \in (1,3)$, let

$$\Phi_1(ds) = 2(\delta-1)s^{-\delta}1_{[1,\infty)}(s)ds \tag{5.3}$$

Thus $\Phi_1(0,\infty)=2$ as required by (1.6), and the longitudinal and lateral correlation functions (see (1.7)) are given by

$$B_L(r)=\int_1^\infty J_1'(rs)2(\delta-1)s^{-\delta}ds \tag{5.4}$$

$$B_N(r)=\int_1^\infty \frac{J_1(rs)}{rs}2(\delta-1)s^{-\delta}ds$$

Study of the asymptotics of these expressions shows that there is a positive constant γ such that

$$B_L(r)=1-\delta\gamma r^{\delta-1}+O(r^2),\ B_N(r)=1-\gamma r^{\delta-1}+O(r^2),\ \text{as } r\downarrow 0. \tag{5.5}$$

Compare with Example 3.4. It can be checked (Darling [1987]) that there is a solution S to the differential equation (5.1) such that $S(0)=0$, and (5.2) holds; for this example, $S(\infty)=\infty$, so $P(\tau(x,y)<\infty)=1$ for all $x,y\in\mathbf{R}^d$.

Example 5.3 (d=3). Follow the same procedure as in the previous example, exept that now $\Phi_1(ds)=3(\delta-1)s^{-\delta}1_{[1,\infty)}(s)ds$, for some $\delta\in(2,3)$. This time the longitudinal and lateral correlation functions (see (1.8)) are given by

$$B_L(r)=\int_1^\infty\left[\frac{\sin(rs)}{rs}+\frac{2\cos(rs)}{(rs)^2}-\frac{2\sin(rs)}{(rs)^3}\right]3(\delta-1)s^{-\delta}ds \tag{5.6}$$

$$B_N(r)=\int_1^\infty\left[-\frac{\cos(rs)}{(rs)^2}+\frac{\sin(rs)}{(rs)^3}\right]3(\delta-1)s^{-\delta}ds$$

The asymptotic formula is exactly the same as (5.5), with possibly a different values of γ. There is a solution S to the differential equation (5.1) such that $S(0)=0$, and (5.2) holds (see Darling [1988]); for this example, $S(\infty)<\infty$, so $P(\tau(x,y)<\infty)$ decreases continuously from 1 to 0 as $|x-y|$ increases to ∞. It is interesting to note that, in this example, the covariance tensor b(.) is C^1 at 0 but not C^2.

Problem 5.4 Derive an analog to (4.4) in the general case, where no moments of Φ_1 or Φ_2 are assumed to be finite; in particular, show that there are no coalescing isotropic flows in dimension $d\geq 4$. (Work on this problem is currently in progress.)

An interesting object of study in these coalescing isotropic flows in dimensions 2 and 3 is the coalescent set

$$J_t=\{x\in\mathbf{R}^d : X_t(x)=X_t(0)\}. \tag{5.7}$$

There is no reason to expect this random set to be connected, open or closed, and since the construction (see Section 2 above) of the stochastic flow $\{X_{st}: 0\leq$

$s \leq t \leq \infty\}$ does not necessarily yield measurable mappings from $\mathbf{R}^d$ to $\mathbf{R}^d$, J_t is not known to be a Lebesgue measurable set a.s.. Until Problem 2.4 is solved, the easiest way around this problem is to note that, for examples such as 5.2 and 5.3 above, the random mappings $\{X_{st} : 0 \leq s \leq t \leq \infty\}$ at least satisfy $P(|X_{st}(x) - X_{st}(y)| > \epsilon) \to 0$ as $y \to x$, for fixed $\epsilon > 0$ and $t \geq s$, and so one can construct versions of $\{X_{st}\}$ for s and t in a countable set of times (such as $\{j/n : j = 0, 1, 2, \ldots\}$ for some fixed n) which are measurable mappings such that the flow property (0.1) still holds. Then Fubini's theorem shows that the Lebesgue measure of J_t satisfies

$$\mathbf{E}[m(J_t)] = \int P(\tau(0, y) \leq t) m(dy) \tag{5.8}$$

The law of $\tau(0, y)$ is the same of that of the first passage time to 0 of the diffusion with generator (4.3), started at $|y|$. It is shown in Darling [1988] that if $B_L(r)$ and $B_N(r)$ are both $O(r^{-\alpha})$ as $r \to \infty$, for some $\alpha > 0$, then given $h > 0$ there exist constants c_i such that for t sufficiently large

$$\frac{c_1 t}{(\log t)^{1+h}} \leq \mathbf{E}[m(J_t)] \leq \frac{c_2 t}{\log t} \text{ when } d = 2, \tag{5.9}$$

$$\frac{c_3 t}{(\log t)^h} \leq \mathbf{E}[m(J_t)] \leq c_4 t(\log t) \text{ when } d = 3.$$

Here c_2 is universal, c_4 depends on b(.) only, and c_3 and c_4 depend on both h and b(.).

Problem 5.5 Find conditions in dimensions 2 and 3 under which an analog of Theorem 3.1 holds; when is it true that for every $t_0 > 0$ and every compact $K \subset \mathbf{R}^d$, $X_t(K)$ is a finite set for all $t \geq t_0$, a.s.? Can this occur at all in dimension 3?

Problem 5.6 If Problem 2.4 can be solved, then what can be said about the topological characteristics of J_t? Possibly the law of J_t as a random set can be studied by defining a suitable measure-valued process in backwards time.

Problem 5.7 Improve (5.9) by showing that $\lim_{t \to \infty} (\log t) \mathbf{E}[m(J_t)]/t$ exists in dimension 2, and $\lim_{t \to \infty} \mathbf{E}[m(J_t)/t$ exists in dimension 3, and find the limits.

Acknowledgements: The author thanks T.E. Harris and Y. Le Jan for their comments and corrections to this survey.

References

[1] ARNOLD, L. and SAN MARTIN, L. (1988). A multiplicative ergodic theorem for rotation numbers. J. Dynamics and Differential Equations 1, 95-119.

[2] ARRATIA, R. (1979). Coalescing Brownian motions on the line. Ph.D. thesis, Univ. of Wisconsin, Madison.

[3] BAXENDALE, P. (1984). Brownian motions in the diffeomorphism group, I.Compositio Math 53, 19-50.

[4] BAXENDALE, P. (1986). The Lyapunov spectrum of a stochastic flow of diffeomorphisms. Lecture Notes in Math. 1186, 322-337.

[5] BAXENDALE, P. (1990)(companion paper to appear in this volume).

[6] BAXENDALE, P. and HARRIS, T.E. (1986). Isotropic stochastic flows. Annals of Probability 14, 1155-1179.

[7] CARVERHILL, A. (1985). Flows of stochastic dynamical systems: Ergodic theory. Stochastics 14, 273-318.

[8] CHUNG, K.L. (1974).*A course in Probabiltiy Theory*, 2nd Ed. Academic Press, New York.

[9] DARLING, R.W.R. (1987).*Constructing nonhomeomorphic stochastic flows.* Memoirs of the American Math.Soc. 376.

[10] DARLING, R.W.R. (1988). Rate of growth of the coalescent set in a coalescing stochastic flow. Stochastics 23, 465-508.

[11] DARLING, R.W.R. and LE JAN, Y. (1988). The statistical equilibrium of an isotropic stochastic flow with negative Lyapunov exponents is trivial. Lecture Notes in Mathematics 1321, 175-185.

[12] DAWSON, D.A. and KURTZ, T.G. (1982). Applications of duality to measure-valued diffusion processes. Springer Lecture Notes in Control and Info. Sci. 42, 91-105.

[13] ETHIER, S.N. and KURTZ, T.G. (1986).*Markov Processes: Characterization and Convergence.* John Wiley, New York.

[14] ELWORTHY, K.D. (1982).*Stochastic Differential Equations on Manifolds.* Cambridge University Press.

[15] HARRIS, T.E. (1981). Brownian motions on the homeomorphisms of the plane. Annals of Probability 9, 232-254.

[16] HARRIS, T.E. (1984). Coalescing and noncoalescing stochastic flows in $\mathbf{R}_1$. Stochastic Processes and their Applications 17, 187-210.

[17] IKEDA, N. and WATANABE, S. (1981). *Stochastic Differential Equations and Diffusion Processes.* North Holland, New York.

[18] KUNITA, H. (1984). *Stochastic Differential Equations and Stochastic Flows of Diffeomorphisms.* Lecture Notes in Math 1097, 143-303.

[19] KUNITA, H. (1990). *Stochastic Flows and Stochastic Differential Equations.* Cambridge University Press.

[20] LE JAN, Y. (1982). Flots de diffusion dans $\mathbf{R}^d$. Comptes Rendus Acad.Sci.Paris 294, 697-699.

[21] LE JAN, Y. (1985). On isotropic Brownian motions. Z.Wahrsch. verw. Gebiete 70, 609-620.

[22] LE JAN, Y. (1989). Propriétés asymptotiques des flots browniens isotropes. Comptes Rendus Acad.Sci.Paris 309, 63-65.

[23] LEBEDEV, N.N. (1972). *Special functions and their applications.* Dover, New York.

[24] MATSUMOTO, H. (1989). Coalescing stochastic flows on the real line. Osaka J. Math. 26, 139-158.

[25] STROOCK, D.W. and VARADHAN, S.R.S. (1979). *Multidimensional Diffusion Processes.* Springer, New York.

[26] YAGLOM, A. (1957). Some classes of random fields in n-dimensional space, related to stationary random processes. Theory of Probabiltiy and its Applications 2, 273-320.

[27] YAGLOM, A. (1987). *Correlation theory of stationary and related random functions.* Springer, New York.

Mathematics Department
University of South Florida
Tampa, Fl 33620-5700

The Existence of Isometric Stochastic Flows for Riemannian Brownian Motions

Ming Liao [1]

Abstract Let A be a G-invariant differential operator on a homogeneous space $M = G/H$, which is the generator of some diffusion process. We study the existence of a G-valued stochastic flow whose one point motion is an A- diffusion in terms of the Lie algebra of G. When M is a Riemannian symmetric space, we show that there exists an isometric stochastic flow whose one point motion is a Brownian motion if M is a symmetric space of compact type and such a flow does not exist if M is of non-compact type. The uniqueness of such a flow is also discussed.

1. Introduction

Let M be a smooth d-dimensional manifold and G be a Lie group of diffeomorphisms: $M \to M$. Suppose ϕ_t is a diffusion on G with $\phi_0 = e$, the identity of G. ϕ_t will be called a G-valued stochastic flow on M. For any $x \in M$, $\phi_t(x)$ is a stochastic process on M with $\phi_0(x) = x$ and will be called the one point motion of ϕ_t starting from x. In this paper, we will be interested in the following question. Let A be a differential operator on M. Does there exist a G-valued flow ϕ_t whose one point motion is an A-diffusion? This means that $\forall x \in M$, $\phi_t(x)$ is a diffusion process on M with generator A. When M is a Riemannian manifold and G is the group of isometries, a G-valued flow will be called an isometric flow. We would like to know whether there exists an isometric flow whose one point motion is a Brownian motion on M, i.e., a diffusion process generated by Δ, the Laplace-Betrami operator of M.

1980 Mathematics Subject Classification (1985 Revision). Primary 58G32; Secondary 53C05.

Key words and phrases. Stochastic flows, diffusion processes, homogeneous spaces, symmetric spaces of compact type and of non-compact type.

[1] Research supported by the Huo Ying Dong Educational Fundation and NSF of P.R. China.

Suppose there are vector fields $U_0, U_1, \ldots, U_n$ on M such that

$$\Delta = \frac{1}{2}\sum_{i=1}^{n} U_i U_i + U_0. \tag{1}$$

Then a Brownian motion process X_t on M can be obtained as a solution to the following stochastic differential equation.

$$dX_t = \sum_{i=1}^{n} U_i(X_t) \circ dW_t^i + U_0(X_t)dt, \tag{2}$$

where $W_t = (W_t^1, \ldots, W_t^n)$ is an n-dimensional Wiener process and $\circ dW_t$ denotes the Stratonovich stochastic differential. This stochastic differential equation generates a stochastic flow ϕ_t on M, which is contained in the isometry group G of M if and only if all the vector fields U_i, $i = 0, 1, \ldots, n$, are killing vector fields, i.e., they generate isometric dynamic flows. See [K]. Hence, if there are killing vector fields U_i, $i = 0, 1, \ldots, n$, on M such that they satisfy (1), then there is an isometric stochastic flow whose one point motion is a Brownian motion on M. On the other hand, if this is the case, by Remark 3 in Section 3, such killing vector fields must exist. Therefore, our question can be re-formulated as follows. Can we find killing vector fields $U_0, U_1, \ldots, U_n$, for some integer n, on M such that they satisfy (1)?

We give two examples. First assume that M is the Euclidean space R^d. We can choose $U_0 = 0$ and $U_1, \ldots, U_d$ to be the vector fields on R^d of unit speed translations along coordinate axes. For the next example, assume $M = S^d$, the unit d-dimensional sphere (embedded in R^{d+1}). We can choose $U_0 = 0$ and $U_1, \ldots, U_n$ to be the vector fields on S^d of unit speed rotations determined by 2-dimensional coordiate planes in R^{d+1}, where n is the dimension of the orthogonal group $O(d+1)$ of R^{d+1}. These are killing vector fields on S^d and satisfy (1). See Section 4 for more details. Hence, if $M = R^d$ or S^d, there is an isometric stochastic flow whose one point motion is a Brownian motion.

Since Brownian motions are non-degenerate, if there is an isometric flow whose one point motion is a Brownian motion, the isometry group G must be transitive on M and M is a Riemannian homogeneous space. On the other hand, the Laplace-Betrami operator is G-invariant. This leads us to consider the existence of a G-valued flow on a homogeneous space G/H, whose one point motion is an A-diffusion, where A is a G-invariant differential operator on G/H.

As we will see, not all Riemannian homogeneous spaces have isometric stochastic flows whose one point motions are Brownian motions. It seems to be an interesting problem to characterize all Riemannian manifolds having this property. We are unable to solve this problem completely. However, for Riemannian symmetric spaces, which is an important class of Riemannian homogeneous spaces, we obtain a satisfactory classification. We will show in Section 4 that a Riemannian symmetric space of compact type, of which a sphere is a special case, has an isometric stochastic flow whose one point motion is a Brownian motion, whereas a Riemannian symmetric space of non-compact type, of which a hyperbolic space is a special case, does not. In general, a simply connected Riemannian symmetric space is a direct product of a Euclidean space, a symmetric space of compact type and a symmetric space of non-compact type, with the corresponding product structure for their isometry groups. See Proposition 4.2 in [H1, Ch.V]. From this, we can show that a simply connected Riemannian symmetric space has an isometric stochastic flow whose one point motion is a Brownian motion if and only if its component of non-compact type symmetric space is trivial.

In the next section, we will start with a general manifold M, a Lie group G of diffeomorphisms and a diffusion generator A on M, and we will show that the existence of a G-valued stochastic flow whose one point motion is an A-diffusion corresponds to the existence of a naturally lifted differential operator on G of A. Section 3 deals with a homogeneous space $M = G/H$ with a G-invariant differential operator A. We will identify right invariant differential operators on G with polynomial functions on the dual space of the Lie algebra and will obtain a necessary and sufficient condition for the existence of a G-valued stochastic flow whose one point motion is an A-diffusion using these polynomial functions. Section 4 contains our main results above Riemannian symmetric spaces mentioned above. In the last section, Section 5, we will prove the uniqueness of isometric stochastic flows for some special cases, including Euclidean spaces and spheres.

The standard reference on diffusion processes is [IW]. The reader is referred to [H1], [H2] and [KN] for differential geometry, Lie groups and symmetric spaces. The symbol $\|$ marks the end of a proof.

2. G-valued Stochastic Flows

Let M be a smooth d-dimensional manifold. A differential operator A on M is said to be of generator type if it is the generator of some diffusion process on M. This means that, under local coordinates $x^1, \ldots, x^d$, A has the following expression.

$$A = \sum_{j,k=1}^{d} a_{jk}(x) \frac{\partial}{\partial x^j} \frac{\partial}{\partial x^k} + \sum_{i=1}^{d} b_i(x) \frac{\partial}{\partial x^i},$$

where a_{jk} and b_i are smooth functions and the matrix a_{jk} is symmetric and non-negative definite.

Let G be a Lie group of diffeomorphisms: $M \to M$ and for any $x \in M$, let $E_x : G \to M$ be the evaluation map defined by $E_x\phi = \phi(x)$. If ϕ_t is a diffusion process on G with generator D and $\phi_0 = e$, the identity of G, and whose one point motion is an A-diffusion, then $\forall f \in \mathcal{D}(M)$, the space of smooth functions on M with compact supports,

$$f \circ E_x(\phi_t) - f \circ E_x(\phi_0) - \int_0^t D(f \circ E_x)(\phi_s) ds$$

is a martingale. On the other hand,

$$\begin{aligned} & f \circ E_x(\phi_t) - f \circ E_x(\phi_0) - \int_0^t (Af) \circ E_x(\phi_s) ds \\ = \ & f(\phi_t(x)) - f(\phi_0(x)) - \int_0^t Af(\phi_s(x)) ds \end{aligned}$$

is also a martingale. This implies: $D(f\circ E_x)(\phi_s) = (Af)\circ E_x(\phi_s)$ *a.s.*. Hence,

$$\forall x \in M \text{ and } f \in \mathcal{D}(M),\quad D(f\circ E_x)(e) = (Af)\circ E_x(e). \tag{3}$$

For $\psi \in G$, let $L_\psi : G \to G$ and $R_\psi : G \to G$ be defined by $L_\psi\phi = \psi\phi$ and $R_\psi\phi = \phi\psi$. A differential operator $\tilde{D}$ is said to be right invariant on G if

$$\forall \psi \in G \text{ and } u \in \mathcal{D}(G),\quad \tilde{D}(u\circ R_\psi) = (\tilde{D}u)\circ R_\psi.$$

Similarly, we can define left invariant differential operators on G.

Let $\tilde{D}$ be a right invariant differential operator on G defined by $\tilde{D}u(e) = Du(e)$. Since D is of generator type, so is $\tilde{D}$. This can be easily checked. To simplify our notation, we will write D for $\tilde{D}$. Now (3) is still true. The right invariance of D and the fact that $E_x\circ R_\phi = E_{\phi(x)}$ yield

$$\forall x \in M,\ \phi \in G \text{ and } f \in \mathcal{D}(M),\quad D(f\circ E_x)(\phi) = (Af)\circ E_x(\phi). \tag{4}$$

Now it is easy to check that if ϕ_t is a D-diffusion on G with $\phi_0 = e$, then its one point motion is an A-diffusion on M. Conversely, any right invariant differential operator D on G defines a differential operator A on M via (3), which satisfies (4). To summarize, we obtain the following result.

Proposition 1 *If there is a G-valued flow whose one point motion is an A-diffusion, then there is a G-valued flow with right invariant generator on G, whose one point motion is an A-diffusion. On the other hand, if ϕ_t is a diffusion process on G with right invariant generator D and $\phi_0 = e$, then its one point motion is an A diffusion on M, where A is defined by (3).*

A right invariant differential operator D on G will be called a natural G-lift of a differential operator A on M if it satisfies (3). It is clear that if D is a natural G-lift of A, then (4) holds and, when D is of generator type, the one point motion of the flow generated by D is an A-diffusion on M.

Corollary 1 *Let A be a generator type differential operator on M. Then there exists a G-valued flow whose one point motion is an A-diffusion if and only if A has a generator type natural G-lift.*

Remark 1 If ϕ_t is a D-diffusion on G with right invariant D and $\phi_0 = e$, then $\forall\phi \in G$, $\phi_t\phi$ is a D-diffusion starting from ϕ. The collection of all diffusion processes ϕ_t on G with right invariant generators is precisely the family of continuous stochastic processes on G with stationary independent increments on the left. Such processes have been completely characterized under more general settings. See [B] and references therein.

A differential operator A on M is said to be G-invariant if $\forall\psi \in G$ and $f \in \mathcal{D}(M)$,

$$A(f\circ\psi) = (Af)\circ\psi.$$

Remark 2 A differential operator on G is said to be bi-invariant if it is both right and left invariant. It is clear that if D is bi-invariant, then it is a natural G-lift of some G-invariant differential operator on M. On the other hand, assume that G is compact and A is a G-invariant differential operator on M. If A has a natural G-lift, then A has a bi-invariant natural G-lift. In fact,let D be a natural G-lift of A and define a right invariant differential operator $\tilde{D}$ by

$$\tilde{D}u(e) = \int_G D(u \circ L_\psi \circ R_{\psi^{-1}})(e)d\psi,$$

where $d\psi$ is the normalized Haar measure on G. Then $\tilde{D}$ is a bi-invariant natural G-lift of A.

3. Homogeneous Spaces

In this section, we will assume that G is transitive on M.

Proposition 2 *Let D be a natural G-lift of some differential operator A on M and let p be a fixed point of M. The following four statements are equivalent:*
(i) A is G-invariant.
(ii) $\forall\psi \in G$, $x \in M$ and $f \in \mathcal{D}(M)$,

$$D(f \circ E_x \circ L_\psi) = D(f \circ E_x) \circ L_\psi.$$

(iii) $\forall\psi \in G$ and $\forall f \in \mathcal{D}(M)$,

$$D(f \circ E_p \circ L_\psi)(e) = D(f \circ E_p) \circ L_\psi(e).$$

(iv) $\forall\psi \in G$ and $\forall f \in \mathcal{D}(M)$,

$$D(f \circ E_p \circ L_\psi \circ R_{\psi^{-1}})(e) = D(f \circ E_p)(e).$$

Proof The equivalence of (i) and (ii) follows from the following computation.

$$\begin{aligned} D(f \circ E_x \circ L_\psi)(\phi) &= D(f \circ \psi \circ E_x)(\phi) &&= A(f \circ \psi) \circ E_x(\phi) \\ &= (Af) \circ \psi \circ E_x(\phi) &&= Af(\psi\phi(x)) \\ &= D(f \circ E_x)(\psi\phi) &&= D(f \circ E_x)(L_\psi\phi). \end{aligned}$$

Since (ii) clearly implies (iii) and (iv), so it remains to show that (iii) $\Rightarrow$ (ii) and (iv) $\Rightarrow$ (iii). Fix $x \in M$ and choose $\xi \in G$ such that $\xi(x) = p$. Assume (iii).

$$\begin{aligned} & D(f \circ E_x \circ L_\psi)(\phi) = D(f \circ E_x \circ L_\psi)(R_\xi\phi\xi^{-1}) = D(f \circ E_x \circ L_\psi \circ R_\xi)(\phi\xi^{-1}) \\ = \ & D(f \circ \psi \circ E_p)(\phi\xi^{-1}) = D(f \circ \psi \circ E_p \circ L_{\phi\xi^{-1}})(e) = D(f \circ E_p \circ L_{\psi\phi\xi^{-1}})(e) \\ = \ & D(f \circ E_x \circ R_\xi)(\psi\phi\xi^{-1}) = D(f \circ E_x)(\psi\phi) \end{aligned}$$

This proves (ii). Now assume (iv). We have

$$\begin{aligned} D(f \circ E_p)(\psi) &= D(f \circ E_p \circ R_\psi)(e) &&= D(f \circ \psi \circ E_p \circ L_{\psi^{-1}} \circ R_\psi)(e) \\ &= D(f \circ \psi \circ E_p)(e) &&= D(f \circ E_p \circ L_\psi)(e) \end{aligned}$$

This proves (iii). ||

Let $\mathcal{G}$ be the Lie algebra of G, $\{X_1, X_2, \ldots, X_n\}$ be a basis of $\mathcal{G}$ and $S(\mathcal{G})$ be the space of linear combinations of the following objects:

$$X_1^{k_1} X_2^{k_2} \cdots X_n^{k_n},$$

where $k_1, k_2, \ldots, k_n$ are non-negative integers. $S(\mathcal{G})$ is an algebra under the obvious addition and multiplication. Note that the multiplication is commutative. Since each $X \in \mathcal{G}$ is a linear functional on $\mathcal{G}^*$, the dual space of $\mathcal{G}$, so $S(\mathcal{G})$ can be considered as the space of polynomial functions on $\mathcal{G}^*$. The definition of $S(\mathcal{G})$ is independent of the choice of basis of $\mathcal{G}$.

Let $P \in S(\mathcal{G})$. If we replace each X_i by $\frac{\partial}{\partial t_i}$ in the expression of P, we obtain a linear combination of

$$(\frac{\partial}{\partial t_1})^{k_1}(\frac{\partial}{\partial t_2})^{k_2} \cdots (\frac{\partial}{\partial t_n})^{k_n},$$

which will be denoted $P(\frac{\partial}{\partial t})$. We define a right invariant differential operator P' on G as follows. For $u \in \mathcal{D}(G)$,

$$P'u(\phi) = P(\frac{\partial}{\partial t})u(\exp(\sum_{i=1}^{n} t_i X_i)\phi) \mid_{t=0} .$$

The definition of P' is independent of the choice of basis of $\mathcal{G}$. Our discussion of right invariant differential operators is modelled after that in [H2, II.2] for left invariant differential operators, to which the reader is referred for details. We have

$$(XY)' = \frac{1}{2}(X'Y' + Y'X') = X'Y' + \frac{1}{2}[X,Y]'. \tag{5}$$

Any right invariant differential operator D on G is P' for some $P \in S(\mathcal{G})$. For example, if $\forall u \in \mathcal{D}(G)$,

$$Du(e) = (\sum_{j,k=1}^{n} c_{jk}\frac{\partial}{\partial t_j}\frac{\partial}{\partial t_k} + \sum_{i=1}^{n} d_i \frac{\partial}{\partial t_i})u(\exp(\sum_{i=1}^{n} t_i X_i)) \mid_{t=0},$$

for some constants c_{jk} and d_i with $c_{jk} = c_{kj}$, then $D = P'$, where $P \in S(\mathcal{G})$ is given by

$$P = \sum_{j,k=1}^{n} c_{jk}X_jX_k + \sum_{i=1}^{n} d_iX_i. \tag{6}$$

P' is of generator type if and only if the symmetric matrix c_{jk} is non-negative definite. Such a $P \in S(\mathcal{G})$ will be said to be of generator type. We will let $S_g(\mathcal{G})$ be the space of generator type $P \in S(\mathcal{G})$.

Remark 3 Let A be a generator type differential operator on M. Then there is a G-valued stochastic flow ϕ_t whose one point motion is an A-diffusion if and only if there are vector fields $U_0, U_1, U_2, \ldots, U_n$ on M, whose dynamic flows are contained in G, such that

$$A = \frac{1}{2}\sum_{i=1}^{n} U_iU_i + U_0. \tag{7}$$

Assume that such a flow ϕ_t exists. We may assume that its generator D is right invariant on G. Then $D = P'$ for some polynomial function P expressed by (6). Let $\{\sigma_{ih}\}$ be the square root of the matrix $2\{c_{jk}\}$, $Y_i = \sum_h \sigma_{ih} X_h$ and let $\overline{U}_i = Y_i'$, for $i = 1, 2, \ldots, n$. It is clear that there is $Y_0 \in \mathcal{G}$ such that, with $\overline{U}_0 = Y_0'$,

$$P = \frac{1}{2}\sum_{i=1}^{n} Y_i Y_i + Y_0$$

and

$$D = \frac{1}{2}\sum_{i=1}^{n} \overline{U}_i \overline{U}_i + \overline{U}_0.$$

Hence, ϕ_t can be obtained as a solution to the following stochastic differential equation.

$$d\phi_t = \sum_{i=1}^{n} \overline{U}_i(\phi_t) \circ dW_t^i + \overline{U}_0(\phi_t)dt.$$

As a right invariant differential operator on G, $\overline{U}_i$ is the natural G-lift of some vector field U_i on M. It follows from Itô's formula for stochastic integrals and the fact that $\phi_t(x) = F(\phi_t, x)$ for some smooth function F on $G \times M$ that $\forall x \in M$, $\phi_t(x)$ is a solution to the following stochastic differential equation.

$$dX_t = \sum_{i=1}^{n} U_i(X_t) \circ dW_t^i + U_0(X_t)dt.$$

This implies (7).

Conversely, assume that A has the expression (7), where $U_0, U_1, \ldots, U_n$ are vector fields on M whose dynamic flows are contained in G. By [K], the stochastic flow generated by the above stochastic differntial equation is contained in G and its one point motion is an A-diffusion.

Let $\phi \in G$ and $Ad(\phi)$: $\mathcal{G} \to \mathcal{G}$ be the Lie algebra automorphism induced by the Lie group automorphism

$$L_\phi \circ R_{\phi^{-1}} : G \to G.$$

Let $P \in S(\mathcal{G})$. If we replace X_i by $Ad(\phi)X_i$ in the linear combination which expresses P, we obtain an element in $S(\mathcal{G})$, which is denoted by $Ad(\phi)P$. We have,

$$\forall \phi \in G,\ P \in S(\mathcal{G}) \text{ and } u \in \mathcal{D}(G),\quad P'(u \circ L_\phi) = ([Ad(\phi)P]'u) \circ L_\phi. \tag{8}$$

Fix $p \in M$ and let H be the isotropy group at p, i.e., $H = \{\phi \in G;\ \phi(p) = p\}$, and let $\mathcal{H}$ be its Lie algebra. M is identified with the homogeneous space $G/H = \{\phi H;\ \phi \in G\}$. We will also assume that G/H is reductive, i.e.,

$$\mathcal{G} = \mathcal{H} \oplus \mathcal{P},$$

for some subspace $\mathcal{P}$ of $\mathcal{G}$, which is $Ad(H)$-invariant in the sense that $\forall h \in H$, $Ad(h)\mathcal{P} \subset \mathcal{P}$. G/H is reductive when H is compact, or more generally, when $Ad(H)$ is a compact subgroup of the linear transformation group of $\mathcal{G}$.

Let $\pi = E_p$: $G \to M$ and $\{X_1, X_2, \ldots, X_n\}$ be a basis of $\mathcal{G}$ such that $\{X_1, \ldots, X_d\}$ is a basis of $\mathcal{P}$ and $\{X_{d+1}, \ldots, X_n\}$ is a basis of $\mathcal{H}$. For $P \in S(\mathcal{G})$, there is a unique element in $S(\mathcal{P})$, which is a polynomial involving only $X_1, \ldots, X_d$ and is denoted by P_π, such that

$$\forall f \in \mathcal{D}(M), \quad P'(f \circ \pi)(e) = P'_\pi(f \circ \pi)(e).$$

When P has the expression (6), P_π is given by

$$P_\pi = \sum_{j,k=1}^{d} c_{jk} X_j X_k + \sum_{j=i}^{d} \sum_{k=d+1}^{n} c_{jk} [X_j, X_k]_{\mathcal{P}} + \sum_{i=1}^{d} d_i X_i, \tag{9}$$

where $[X_j, X_k]_{\mathcal{P}}$ denotes the $\mathcal{P}$-component of $[X_j, X_k]$. This follows from (5) and the fact that $X_i^l(f \circ \pi)(e) = 0$ for $i = d+1, \ldots, n$.

Proposition 3 *Let $P \in S(\mathcal{G})$ and let A be the differential operator on M having P' as its natural G-lift. Then A is G-invariant if and only if P satisfies*

$$\forall \phi \in G, \quad [Ad(\phi)P]_\pi = P_\pi. \tag{10}$$

Proof By Proposition 2, A is G-invariant if and only if

$$P'(f \circ \pi \circ L_\phi \circ R_{\phi^{-1}})(e) = P'(f \circ \pi)(e).$$

By (8),

$$[Ad(\phi)P]'(f \circ \pi)(e) = P'(f \circ \pi)(e).$$

This is equivalent to (10). ||

Let A be a G-invariant generator type differential operator on M. There is a unique $P \in S_g(\mathcal{P})$ such that

$$\forall f \in \mathcal{D}(M), \quad Af(p) = P(\frac{\partial}{\partial t}) f \circ \pi(\exp(\sum_{i=1}^{d} t_i X_i)) \mid_{t=0}.$$

We will write P^A for this P. The following result is a direct consequence of Proposition 3.

Corollary 2 *Let A be a G-invariant generator type differential operator on M. Then A has a generator type natural G-lift if and only if there exists $P \in S_g(\mathcal{G})$ such that $P^A = P_\pi$ and (10) holds.*

Remark 4 Assume that M_1, M_2 are two smooth manifolds and $M = M_1 \times M_2$. Let $p_1 \in M_1$, $p_2 \in M_2$ and $p = (p_1, p_2) \in M$. For $i = 1, 2$, we will use the subscript i to denote anything associated to M_i. For example, G_i is a transitive Lie group of diffeomorphisms on M_i and H_i is the isotropy group at p_i. Assume that each $M_i \cong G_i/H_i$ is reductive with $\mathcal{G}_i = \mathcal{H}_i \oplus \mathcal{P}_i$. Let $G = G_1 \times G_2$, $H = H_1 \times H_2$ and $\mathcal{P} = \mathcal{P}_1 \oplus \mathcal{P}_2$. Then $M \cong G/H$ is reductive with $\mathcal{G} = \mathcal{H} \oplus \mathcal{P}$. Let A_i be a differential operator on M_i. A_i will also be regarded as a differential operator on M by applying it along M_i. Let $A = A_1 + A_2$ and assume that A is G-invariant and of generator type. Then A has a generator type natural G-lift if and only if each A_i has a generator type natural G_i-lift. To see this, let S' be the subspace of $S(\mathcal{G})$ spanned by $\{XY;\ X \in \mathcal{G}_1$ and $Y \in \mathcal{G}_2\}$. For any $\phi \in G$, $Ad(\phi)$: $S(\mathcal{G}_i) \to S(\mathcal{G}_i)$ and $Ad(\phi)$: $S' \to S'$. Now the conclusion follows from the observation that any $P \in S(\mathcal{G})$ can be expressed as $P = P_1 + P_2 + \tilde{P}$, where $P_i \in S(\mathcal{G}_i)$ and $\tilde{P} \in S'$.

4. Symmetric Spaces

Let M be a d-dimensional Riemannian manifold. For $x \in M$, there is a uniquely defined smooth map S_x from a neighborhood of x into M such that $S_x(x) = x$ and $S_{x*}(X) = -X$ for any $X \in T_xM$. M is said to be a Riemannian locally symmetric space if S_x is a local isometry for any $x \in M$. This is equivalent to assuming that the covariant derivative of curvature tensor vanishes. If all these S_x can be extended to global isometries: $M \to M$, then M is said to be a Riemannian globally symmetric space, or for short in this paper, a Riemannian symmetric space.

In this section, we will assume that M is a Riemannian symmetric space. We will summarize some facts about symmetric spaces. The reader is referred to [H1] and [KN] for their proofs.

Let G be the identity component of the isometry group of M. G is transitive on M. Fix $p \in M$ and let H be the isotropy group at p, which is a compact subgroup of G. We will identify M with the homogeneous space G/H. As before, $\mathcal{G}$ and $\mathcal{H}$ are the corresponding Lie algebras. Let θ: $G \to G$ be the group automorphism defined by

$$\theta(\phi) = S_p \circ \phi \circ S_p^{-1}$$

and use θ also to denote the induced Lie algebra automorphism θ: $\mathcal{G} \to \mathcal{G}$. For $X \in \mathcal{H}$, $\theta(X) = X$. Let $\mathcal{P} = \{X \in \mathcal{G};\ \theta(X) = -X\}$. Then $\mathcal{G} = \mathcal{H} \oplus \mathcal{P}$ and $\mathcal{P}$ is $Ad(H)$- invariant. We have $[\mathcal{H},\mathcal{H}] \subset \mathcal{H}$, $[\mathcal{H},\mathcal{P}] \subset \mathcal{P}$ and $[\mathcal{P},\mathcal{P}] \subset \mathcal{H}$. For $X \in \mathcal{G}$, define $ad(X)$: $\mathcal{G} \to \mathcal{G}$ by $ad(X)Y = [X,Y]$. Then

$$\forall X,Y \in \mathcal{G},\quad \frac{d}{dt}Ad(e^{tX})Y\mid_{t=0} = ad(X)Y.$$

The killing form of $\mathcal{G}$ is defined to be $B(X,Y) = Trace(ad(X)ad(Y))$ for $X,Y \in \mathcal{G}$. $B(X,Y) = 0$ for $X \in \mathcal{H}$ and $Y \in \mathcal{P}$, and $B(\cdot,\cdot)$ is negative definite on $\mathcal{H}\times\mathcal{H}$. G/H is said to be a symmetric space of compact type if $B(\cdot,\cdot)$ is negative definite on $\mathcal{P}\times\mathcal{P}$, it is said to be of non-compact type if $B(\cdot,\cdot)$ is positive definite on $\mathcal{P}\times\mathcal{P}$. M and G are compact if G/H is of compact type and they are non-compact if G/H is of non-compact type. A sphere is a symmetric space of compact type and a hyperbolic space is a symmetric space of non-compact type. A Euclidean space is a symmetric space but it is of neither compact type nor non-compact type.

Let Δ be the Laplace-Betrami operator on M. Δ is G-invariant.

Proposition 4 *If $M = G/H$ is a Riemannian symmetric space of compact type, then Δ has a generator type natural G-lift.*

Proof By Corollary 2, it suffices to show the existence of $P \in S_g(\mathcal{G})$ such that $P_\pi = P^\Delta$ and P satisfies (10). Let $< \cdot,\cdot >$ be the symmetric bilinear form on $\mathcal{P}$ induced by the Riemannian metric using the identification of T_pM and $\mathcal{P}$. $< \cdot,\cdot >$

is $Ad(H)$-invariant and so by the lemma on Page 257 of [KN, vol.2], it extends to be an $Ad(G)$-invariant inner product on $\mathcal{G}$. Let $X_1, \ldots, X_n$ be an orthonormal basis of $\mathcal{G}$ with respect to $< \cdot, \cdot >$. Then $P = \sum_{i=1}^n (X_i)^2$ satisfies the requirement. In fact, P' is a bi-invariant generator type natural G-lift of Δ. ||

Example Let $M = S^d$, the unit sphere in the Euclidean space R^{d+1} with the induced metric. See [KN, XI.10] for the details of the following discussion. G is the rotation group $SO(d+1)$ and H can be identified with $SO(d)$. The Lie algebra of $SO(d+1)$ is $o(d+1)$, the space of all skew symmetric $(d+1)\times(d+1)$ matices. $o(d+1) = \mathcal{H} \oplus \mathcal{P}$ with $\mathcal{H} \cong o(d)$ and $\mathcal{P} \cong R^d$. $Ad(H)$ acts on $\mathcal{P}$ as the rotation group $SO(d)$ on R^d. The Riemannian metric induces an inner product on $\mathcal{P}$, which extends to $\mathcal{G} = o(d+1)$, given by $< X, Y > = -\frac{1}{2}Trace(X, Y)$ for $X, Y \in \mathcal{G}$. Let e_{jk} be a $(d+1)\times(d+1)$ matrix with 1 at entry (j,k), -1 at entry (k,j) and 0 elsewhere. Then $\{e_{jk};\ j<k\}$ is an orthonormal basis of $\mathcal{G}$ with respect to $< \cdot, \cdot >$. Let $P = \sum_{j<k} e_{jk}^2$. P' is a generator type natural G-lift of Δ. Let U_{jk} be the vector fields on S^d such that e'_{jk} are their natural G-lifts. They are vector fields corresponding to unit speed rotations determined by 2-dimensional coordinate planes. The stochastic flow generated by the following stochastic differential equation on S^d is an isometric flow whose one point motion is a Brownian motion.

$$dx_t = \sum_{j<k} U_{jk}(x_t) \circ dW_t^{jk}. \tag{11}$$

Now assume that $M = G/H$ is of non-compact type. Define $B_\theta(X,Y) = -B(X, \theta Y)$ for $X, Y \in \mathcal{G}$. B_θ is an inner product on $\mathcal{G}$. Let $\mathcal{A}$ be a maximal abelian subspace of $\mathcal{P}$. For $\lambda \in \mathcal{A}^*$, the dual space of $\mathcal{A}$, let

$$\mathcal{G}_\lambda = \{X \in \mathcal{G};\ \ \forall W \in \mathcal{A},\ ad(W)X = \lambda(W)X\}$$

We have $\theta(\mathcal{G}_\lambda) = \mathcal{G}_{-\lambda}$ and $[\mathcal{G}_\lambda, \mathcal{G}_\mu] \subset \mathcal{G}_{\lambda+\mu}$. For $W \in \mathcal{P}$, $ad(W)$: $\mathcal{G} \to \mathcal{G}$ is symmetric with respect to B_θ. The family of the linear maps, $\{ad(W),\ W \in \mathcal{A}\}$, is commutative, hence, by simultaneous diagonalization, we have the following orthogonal decomposition of $\mathcal{G}$ with respect to B_θ.

$$\mathcal{G} = \sum_{\lambda \in T} \mathcal{G}_\lambda,$$

where $T = \{\lambda \in \mathcal{A}^*;\ \ \mathcal{G}_\lambda \neq 0\}$. For each $\lambda \in T$, let $m(\lambda) = \dim \mathcal{G}_\lambda$ and let $X_1^\lambda, \ldots, X_{m(\lambda)}^\lambda$ be a basis of $\mathcal{G}_\lambda$ consisting mutually orthogonal elements. When $\lambda \not\equiv 0$, $X_i^\lambda = Y_i^\lambda + Z_i^\lambda$, where $Y_i^\lambda \in \mathcal{P}$ and $Z_i^\lambda \in \mathcal{H}$. Then $-Y_1^\lambda + Z_1^\lambda, \ldots, -Y_{m(\lambda)}^\lambda + Z_{m(\lambda)}^\lambda$ is a basis of $\mathcal{G}_{-\lambda}$. We may assume $X_i^{-\lambda} = -Y_i^\lambda + Z_i^\lambda$. When $\lambda \equiv 0$, $\mathcal{G}_0 = \mathcal{A} \oplus \mathcal{M}$, where $\mathcal{M} = \{X \in \mathcal{H};\ \ \forall W \in \mathcal{A},\ [W, X] = 0\}$. So we can take $X_1^0 = Y_1^0, \ldots, X_k^0 = Y_k^0$ to be a basis of $\mathcal{A}$ and $X_{k+1}^0 = Z_1^0, \ldots, X_{m(0)}^0 = Z_{m(0)-k}^0$ to be a basis of $\mathcal{M}$. Choose $U \subset T$ with the property that $T = U \cup (-U) \cup \{0\}$ and $U \cap (-U)$ is empty, where $-U = \{-\lambda;\ \lambda \in U\}$. Then $\{Y_i^\lambda, Z_i^\lambda;\ \ \lambda \in \{0\} \cup U$ and $i = 1, 2, \ldots\}$ is a basis of $\mathcal{G}$ consisting mutually orthogonal elements, $\{Y_i^\lambda\}$ is a basis of $\mathcal{P}$ and $\{Z_i^\lambda\}$ is a basis of $\mathcal{H}$. We may assume that $\{Y_i^\lambda\}$ is an orthonormal basis of $\mathcal{P}$ with respect to B.

Proposition 5 *If $M = G/H$ is a Riemannian symmetric space of non-compact type, then Δ does not have a generator type natural G-lift.*

Proof It is clear that

$$P^{\Delta} = \sum_{\lambda,i}(Y_i^{\lambda})^2,$$

where λ ranges over the set $\{0\} \cup U$. If Δ has a generator type natural G-lift, then there exists $P \in S_g(\mathcal{G})$ such that $P_\pi = P^\Delta$ and P satisfies (10). Under the basis $\{X_i^\lambda\}$ of $\mathcal{G}$, P can be written as

$$P = \sum_{\lambda \in U}\sum_j [a_j^\lambda (X_j^\lambda)^2 + b_j^\lambda X_j^\lambda X_j^{-\lambda} + c_j^\lambda (X_j^{-\lambda})^2] + Q,$$

where Q is a polynomial containing terms different from those shown above. For any $W \in \mathcal{A}$, $Ad(e^{tW})Q$ is also such a polynomial, because $ad(W)X_j^\lambda = \lambda(W)X_j^\lambda$ and $Ad(e^{tW})X_j^\lambda = e^{\lambda(W)t}X_j^\lambda$. The term $(Y_j^\lambda)^2$ in $P_\pi = P^\Delta$ comes from

$$a_j^\lambda (X_j^\lambda)^2 + b_j^\lambda X_j^\lambda X_j^{-\lambda} + c_j^\lambda (X_j^{-\lambda})^2$$

in P. Let $W \in \mathcal{A}$. Since $[Ad(e^{tW})P]_\pi = P_\pi$, we have,

$$a_j^\lambda e^{2t\lambda(W)} - b_j^\lambda + c_j^\lambda e^{-2t\lambda(W)} = 1.$$

This implies $a_j^\lambda = c_j^\lambda = 0$ and $b_j^\lambda = -1$, and

$$P = \sum_{\lambda \in U}\sum_j (-X_j^\lambda X_j^{-\lambda}) + Q.$$

If we express P under the basis $\{Y_j^\lambda, Z_j^\lambda\}$, the coefficient of $(Z_j^\lambda)^2$ in P is -1. This means that P cannot be of generator type and contradicts $P \in S_g(\mathcal{G})$. ||

Remark 5 If M is a simply connected Riemannian symmetric space, then $M = M_0 \times M_c \times M_n$, where M_0 is a Euclidean space, M_c is a symmetric space of compact type and M_n is a symmetric space of non-compact type, and any of components may be trivial. G is a direct product of the corresponding Lie groups on M_0, M_c and M_n. By Remark 4, Δ on M has a natural G-lift if and only if the component of non-compact type symmetric space is trivial.

The stochastic versions of our results are summarized in the following corollary.

Corollary 3 *Let M be a Riemannian symmetric space. If M is of compact type, then there exists an isometric flow on M whose one point motion is a Brownian motion. If M is of non-compact type, such flow does not exist. When M is simply connected, such a flow exists if and only if it does not have the component of non-compact type symmetric space.*

5. The Uniqueness

In this last section, we will prove the uniqueness of an isometric stochastic flow whose one point motion is a Brownian motion for some special cases, including Euclidean spaces and spheres. The general uniqueness problem remains to be investigated.

Proposition 6 *Assume that $M = G/H$ is a symmetric space of compact type and the adjoint action $Ad(G)$ is transitive on $\mathcal{G}$. This means that for any two non-zero vectors X and Y in $\mathcal{G}$, there is $\phi \in G$ such that $Ad(\phi)X = cY$ for some constant c. Then any differential operator A on M has at most one natural G-lift. In particular, there is a unique isometric stochastic flow on a sphere whose one point motion is a Brownian motion.*

Proof Let $\mathcal{N}$ be the space of second degree polynomial functions P such that P' is a natural G-lift of the zero operator. Our conclusion will follow if we can show $\mathcal{N} = \{0\}$. Let $P \in \mathcal{N}$ with

$$P = \sum_{j,k=1}^{n} a_{jk} X_j X_k + Y,$$

for some $Y \in \mathcal{G}$. We may assume that $X_1, \ldots, X_n$ are orthonormal with respect to $-B(\cdot,\cdot)$, where B is the killing form of $\mathcal{G}$. Define a symmetric bilinear form $L(\cdot,\cdot)$ on $\mathcal{G}$ by setting $L(X_j, X_k) = a_{jk}$. For $\phi \in G$, let $Ad(\phi)X_j = \sum_{p=1}^{n} c_{jp} X_p$. Since $Ad(\phi)$: $\mathcal{G} \to \mathcal{G}$ is isometric with respect to $-B$, $\{c_{jk}\}$ is an orthogonal matrix. Therefore,

$$Ad(\phi^{-1}) = \sum_{j,k,p,q} a_{pq} c_{jp} c_{kq} X_j X_k + Y'$$

for some $Y' \in \mathcal{G}$. Since $[Ad(\phi^{-1})P]_\pi = 0$, we have

$$L(Ad(\phi)X_j, Ad(\phi)X_k) = \sum_{p,q} a_{pq} c_{jp} c_{kq} = 0$$

for $j, k \le d$. Hence,

$$\forall \phi \in G, \ X, Y \in \mathcal{P}, \ L(Ad(\phi)X, Ad(\phi)Y) = 0.$$

Since for any $X \in \mathcal{G}$, there is $\phi \in G$ such that $Ad(\phi)X \in \mathcal{P}$, so $L(X,X) = 0$. It follows that $L = 0$ and so all $a_{jk} = 0$. Now, $[Ad(\phi)Y]_\pi = [Ad(\phi)P]_\pi = 0$, for any $\phi \in G$, which clearly implies $Y = 0$ and proves $\mathcal{N} = \{0\}$. ||

Remark 6 Let $M = G/H$ be a Riemannian symmetric space of compact type which is a direct product of two symmetric spaces of compact type, i.e., $G = G_1 \times G_2$, $H = H_1 \times H_2$ and $G/H = G_1/H_1 \times G_2/H_2$. So it has a product structure as in Remark 4. Assume that the adjoint actions of $Ad(G_1)$ on $\mathcal{G}_1$ and $Ad(G_2)$ on $\mathcal{G}_2$ are transitive. Then $\mathcal{N} = \{0\}$, where $\mathcal{N}$ as before is the space of second degree polynomial functions $P \in S(\mathcal{G})$ such that P' is a natural G-lift of the zero operator on M.

By Proposition 6, it suffices to prove that if $P = \sum_{j,k} a_{jk} X_j Y_k \in \mathcal{N}$, then $P = 0$, where $X_1, X_2, \ldots, X_n$ form a basis of $\mathcal{G}_1$ and $Y_1, Y_2, \ldots, Y_m$ form a basis of $\mathcal{G}_2$. We may assume that $X_1, \ldots, X_d$ form a basis of $\mathcal{P}_1$, $X_{d+1}, \ldots, X_n$ form a basis of $\mathcal{H}_1$, $Y_1, \ldots, Y_c$ form a basis of $\mathcal{P}_2$ and $Y_{c+1}, \ldots, Y_m$ form a basis of $\mathcal{H}_2$. Since $P \in \mathcal{N}$, $a_{jk} = 0$ for $j \le d$ and $k \le c$. Now we show that $a_{jk} = 0$ for $j \le d$ and any k. Fix $j \le d$. Choose $\phi \in G_2$ such that $Ad(\phi)Y_j' \in \mathcal{P}_2$, where $Y_j' = \sum_q a_{jq} Y_q$. Then

$$Ad(\phi)P = \sum_{p,q} a_{pq} X_p Ad(\phi) Y_q = \sum_p X_p Ad(\phi) Y_p'.$$

If a_{jq} were not zero for some q, then $Ad(\phi)Y_j'$ would be a non-zero element of $\mathcal{P}_2$. This would imply that the coefficient of X_jY_r in $Ad(\phi)P$ is not zero for some $r \leq c$ and, therefore, would contradict the assumption that $P \in \mathcal{N}$, since we should have $[Ad(\phi)P]_\pi = 0$. Now fix $j > d$ and assume $a_{jk} \neq 0$ for some k. Choose $\phi \in G_1$ such that $Ad(\phi)X_k' \in \mathcal{P}_1$, where $X_k' = \sum_p a_{pk}X_p$. Since $a_{jk} \neq 0$, $Ad(\phi)X_k'$ is a non-zero element of $\mathcal{P}_1$. This implies that

$$Ad(\phi)P = \sum_{p,q} a_{pq}Ad(\phi)X_pY_q = \sum_q Ad(\phi)X_q'Y_q$$

contains a term X_rY_k with non-zero coefficient for some $r \leq d$. By the above proof, this is impossible if $Ad(\phi)P \in \mathcal{N}$. Hence, all $a_{jk} = 0$ and $\mathcal{N} = \{0\}$.

The above remark may lead to a general uniqueness result for compact type symmetric spaces if we can check the condition in Proposition 6 for any irreducible compact type symmetric space.

The rest of this section is devoted to a proof for the uniqueness of an isometric stochastic flow whose one point motion is a Brownian motion on a Euclidean space.

Let G be the isometry group of R^d and let $\{X_1, X_2, \ldots, X_n\}$ be a basis of $\mathcal{G}$ such that $X_1, \ldots, X_d$ correspond to unit speed translations along d coordinate axes of R^d and $X_{d+1}, \ldots, X_n$ correspond to unit speed rotations of R^d determined by 2-dimensional coordinate planes. Let $Q \in \mathcal{S}(\mathcal{G})$ be such that Q' is a natural G-lift of Δ. It suffices to show that $Q = \sum_{i=1}^d X_iX_i$. Let $P = Q - \sum_{i=1}^d X_iX_i$. Then P' is a natural G-lift of the zero operator. Assume

$$P = \sum_{j,k=1}^n a_{jk}X_jX_k + Y,$$

for constants a_{jk} and some $Y \in \mathcal{G}$. It is clear that the coefficients a_{jk} form the following matrix

$$\begin{pmatrix} 0 & C \\ C* & D \end{pmatrix}$$

where $C = \{a_{jk}\}_{j=1,\ldots,d;\, k=d+1,\ldots,n}$, $C*$ is the transpose of C and $D = \{a_{jk}\}_{j,k=d+1,\ldots,n}$. We will show that D is not non-negative definite unless all its entries are zero. Assume this for the moment. The coefficient matrix of Q has the following form.

$$\begin{pmatrix} I & C \\ C* & D \end{pmatrix}$$

This matrix is non-negative definite, since Q is of generator type. It then follows that D has to be non-negative definite and, by our claim, D is zero. Now the non-negaive definiteness of the coefficient matrix of Q will also imply that C is zero, hence, all $a_{jk} = 0$ and $Q = \sum_{i=1}^n X_iX_i + Y$. It is easy to show that $Y = 0$.

It remains to prove that D is not non-negative definite unless it is zero. The following relations are easily checked. For $h, i = 1, 2, \ldots, d$ and $k = d+1, \ldots, n$, we have

$$Ad(e^{tX_h})X_i = X_i \quad \text{and} \quad Ad(e^{tX_h})X_k = X_k + t\sum_{r=1}^d c_{hk}^r X_r,$$

where the coefficients c_{hk}^r are determined by

$$[X_h, X_k] = \sum_{r=1}^{d} c_{hk}^r X_r.$$

Our P has the form

$$P = \sum_{j,k=1}^{n} a_{jk} X_j X_k + \sum_{i=1}^{n} a_i X_i.$$

$$\begin{aligned}
Ad(e^{tX_h})P &= P + 2t\sum_{j=1}^{n}\sum_{k=d+1}^{n} a_{jk}X_j \sum_{r=1}^{d} c_{hk}^r X_r + t\sum_{i=d+1}^{n} a_i \sum_{r=1}^{d} c_{hi}^r X_r \\
&\quad + t^2 \sum_{j,k=d+1}^{n} a_{jk} \sum_{p,q=1}^{d} c_{hj}^p c_{hk}^q X_p X_q \\
&= P + 2t\sum_{j=1}^{n}\sum_{k=1}^{d}[\sum_{r=d+1}^{n} a_{jr} c_{hr}^k] X_j X_k + t\sum_{i=1}^{d}[\sum_{r=d+1}^{n} a_r c_{hr}^i] X_i \\
&\quad + t^2 \sum_{j,k=1}^{d}[\sum_{p,q=d+1}^{n} a_{pq} c_{hp}^j c_{hq}^k] X_j X_k.
\end{aligned}$$

It follows from the above that the coefficient of $t^2 X_j X_k$, for $j, k \leq d$, in $[Ad(e^{tX_h})P]_\pi$ is

$$\sum_{p,q=d+1}^{n} a_{pq} c_{hp}^j c_{hq}^k,$$

which should be zero since $P \in \mathcal{N}$. If D is non-negative definite, then D has a square root $\{c_{pq}\}_{p,q=d+1,\ldots,n}$, so $a_{pq} = \sum_{r=d+1}^{n} c_{pr} c_{qr}$. It follows that

$$\sum_{r>d}(\sum_{p>d} c_{pr} c_{hp}^j)^2 = \sum_{p,q>d} a_{pq} c_{hp}^j c_{hq}^j = 0$$

and, hence, $\sum_{p>d} c_{pr} c_{hp}^j = 0$ for any $h, j \leq d$ and $r > d$. Let $Y = \sum_{p>d} c_{pr} Y_p$. Then $Y \in \mathcal{H}$ and

$$[X_h, Y] = \sum_{p>d} c_{pr}[X_h, X_p] = \sum_{i=1}^{d}\sum_{p=d+1}^{n} c_{pr} c_{hp}^i X_i = 0$$

for any $h \leq d$. This implies $Y = 0$. Therefore, c_{pq} and a_{pq} are all zero for $p, q > d$. Our claim is proved.

The author wishes to thank the referee for his helpful commends and for pointing out an error in a previous version of the above proof.

REFERENCES

[B] Baxendale, P.H., "Brownian motions in the diffeomorphism group I", Compositio Math. 53 (1984), pp 19-50.

[E] Elworthy, K.D., "Stochastic differential equations on manifolds", Cambridge Univ. Press, 1982.

[H1] Helgason, S., "Differntial geometry, Lie groups, and symmetric spaces", Academic Press, 1978.

[H2] Helgason, S., "Groups and geometric analysis", Academic Press, 1984.

[IW] Ikeda, N. & Watanabe, S., "Stochastic differential equations and diffusion processes", North-Holland, 1981.

[K] Kunita, H., "Stochastic differential equations and stochastic flows of diffeomorphisms", École d'Été de Probabilités de Saint-Flour XII, Springer Lecture Notes in Math. 1079, 1984, pp 143-303.

[KN] Kobayashi, S. & Nomizu, K., "Foundations of differential geometry", vols I & II, Interscience Publishers, 1963 & 1969.

Department of Mathematics
Nankai University
Tianjin, P.R. China.

TIME REVERSAL OF SOLUTIONS OF EQUATIONS DRIVEN BY LÉVY PROCESSES

P.Sundar*

Abstract

Time reversal of solutions of stochastic differential equations driven by Lévy processes is shown under suitable hypotheses. Possible weakening of the sufficient conditions for time reversal is briefly discussed.

1.INTRODUCTION

The aim of the present paper is to show that the solutions of stochastic differential equations driven by Lévy processes are time reversible semimartingales. Once this is accomplished, it is easy to write down a backward stochastic differential equation whose solution is the time reversed process. A result on stochastic flows plays an essential role in the proof of the main result. The corresponding problem in the context of equations driven by a Brownian motion was first studied by Pardoux [10] and later by a number of authors.

The importance of the problem stems from the fact that semimartingales are the most general of stochastic integrators and therefore, it is desirable to get conditions under which they are reversible. It is well-known that every semimartingale Hunt process is obtained by a random time change from a Markov process that satisfies a stochastic differential equation driven by a Brownian motion and a Poisson random measure. Markov property is preserved when Hunt processes are reversed. The question as to whether the semimartingale property of Hunt processes is preserved under time reversal will therefore be answered by the problem posed above.

The present paper owes a lot to the beautiful results of Jacod [5] on enlargement of filtrations as well as the paper of Jacod and Protter [6] . The sufficient conditions given in this paper for the time reversal result are strong and yield the C^∞ - diffeomorphism property of the stochastic flows.

---

* Research supported by the Air Force Office of Scientific Research Contract No. F49620 85C 0144.

In section 2, we give three examples of semimartingales that are time reversible. The main result of the paper is given after a brief look at the background results. Section 3 is meant to give the reader a flavour of the kind of results that one could expect on stochastic flows by using the Markov property of solutions of such stochastic differential equations. In fact, the Markov property is the additional structure we gain by having Lévy processes instead of general semimartingales as integrators.

2. TIME REVERSAL

Let $(\Omega, \mathcal{F}, P)$ be a given complete probability space with at least two filtrations $\underline{\underline{F}} = (F_t : 0 \leq t \leq 1)$ and $\underline{\underline{\widetilde{H}}} = (\widetilde{H}_t : 0 \leq t \leq 1)$.Let X be a process with paths that are right continuous with left limits (hereafter referred to as cadlag), defined on [0,1] as follows:

$$\widetilde{X}_t = \begin{cases} 0 & \text{if } t = 0 \\ X_{(1-t)-} - X_{1-} & \text{if } 0 < t < 1 \\ X_0 - X_{1-} & \text{if } t = 1 \end{cases} \tag{2.1}$$

where X_{t-} denotes the left limit at t.

Definition 2.2: X is called an $(\underline{\underline{F}}, \underline{\underline{\widetilde{H}}})$ reversible semimartingale if:

(1) X is an $\underline{\underline{F}}$-semimartingale on [0,1]

(2) $\widetilde{X}$ is a $\underline{\underline{\widetilde{H}}}$-semimartingale on [0,1[.

It is customary to take $\underline{\underline{F}}$ as the minimal filtration with respect to which the process X is adapted. In addition, if $\underline{\underline{\widetilde{H}}}$ is taken to be the minimal filtration generated by $\underline{\underline{\widetilde{X}}}$, then we shall say that X is a reversible semimartingale.

Example 1: Let Z be a Lévy process. Clearly, $\widetilde{Z}$ is also a Lévy process since it has the same law as $-Z$. It is well known that Lévy processes are semimartingales, so that Z is a reversible semimartingale.

Example 2: Let B be a Brownian motion and set $\widetilde{H}_t = \sigma(\widetilde{B}_s : 0 \leq s \leq t) \vee \sigma(B_1)$. Since $\widetilde{B}_1 = B_0 - B_{1-} = B_1, \widetilde{H}_t = \sigma(\widetilde{B}_s : 0 \leq s \leq t) \vee \sigma(\widetilde{B}_1)$.

Clearly, $\widetilde{B}$ is also a Brownian motion, so that it suffices to show that B is a $\underline{\underline{H}}$-semimartingale where $\underline{\underline{H}} = (H_t : 0 \leq t \leq 1)$ with $H_t = \sigma(B_s : 0 \leq s \leq t) \vee \sigma(B_1)$.

Towards this, note that

$$E(B_t - B_s|H_s) = \frac{(t-s)}{(1-s)}(B_1 - B_s)$$

by a result in Karlin and Taylor ([7]; see Theorem 2.1 in Ch. 7), for each $o \leq s \leq t \leq 1$. Let $M_t = B_t - \int_0^t (B_1 - B_s)/(1-s)ds$ Then,

$$\begin{aligned} E(M_t - M_s|H_s) &= E(B_t - B_s|H_s) - \int_s^t \frac{1}{1-u} E(B_1 - B_u|H_s)du \\ &= \frac{t-s}{1-s}(B_1 - B_s) - \int_s^t \frac{1}{1-u}\frac{1-u}{1-s}(B_1 - B_s)du \\ &= 0 \end{aligned}$$

so that (M_t) is a $\underline{\underline{H}}$ martingale. Therefore, B is a $\underline{\underline{H}}$- semimartingale admitting the decomposition

$$B_t = M_t + A_t \text{ where } A_t = \int_0^t \frac{B_1 - B_s}{1-s} ds.$$

Remark: The above decomposition was first used by Itô [3]. Example 2, with B replaced by an integrable Lévy process, follows along similar lines, and a nice proof of it is due to Kurtz. The reader is referred to Jacod and Protter [6] for a continuation of Example 2 when B is replaced by any Lévy process (not necessarily integrable).

Example 3: Let $(X_t : o \leq t \leq 1)$ be the solution of the SDE:

$$dX_t = b(t, X_t)dt + c(t, X_t)dB_t.$$

Let $\widetilde{H}_t = \sigma(\widetilde{B}_s : 0 \leq s \leq t) \vee \sigma(X_1)$. Then under suitable conditions on the coefficients b, c and on the initial r.v. X_0, Pardoux [10] has shown that (B_t) is a $(\underline{\underline{F}}, \underline{\underline{H}})$-reversible semimartingale. Let p(s,x) denote the probability density of X_s. Then the martingale part of the semimartingale decomposition of $\widetilde{B}_t$ is

$$M_t = \widetilde{B}_t - \int_{1-t}^1 \frac{1}{p(s, X_s)} \left(c(s, X_s)\frac{dp}{dx}(s, X_s) + p(s, X_s)\frac{dc}{dx}(s, X_s)\right) ds \qquad (2.3)$$

where the integrand in the above expression is taken to be zero whenever p is zero. In fact, M is a $\underline{\underline{\widetilde{H}}}$- Brownian motion.

Remark: Given the knowledge of Example 3, it is an easy task to show that $\widetilde{X}_t$ solves a backward stochastic differential equation driven by $\widetilde{B}_t$ and t with an appropriate modification of the coefficients. Example 3 can also be written for a stochastic system of equations driven by t and an R^l-valued Brownian motion, in which case, (2.3) becomes

$$M_t^i = \widetilde{B}_t^i - \int_{1-t}^{1} \frac{1}{p(s,X_s)} \operatorname{div}(c^i p)(s,X_s)ds$$

where c^i is the i-th column of c. The appearance of $\frac{d}{dx}\ln p(s,X_s)$ in (2.3) is very natural (see Emery [2], pages 215-216), though not obvious at first sight.

The stochastic differential equation that we shall consider in the rest of this section is one driven a Lévy process and is given by

$$X_t = X_0 + \int_0^t b(X_{s-})ds + \int_0^t \sigma(X_{s-})dB_s + \int_0^t \int_{|u|\leq J} F(X_{s-})u(\mu - \nu)(ds,du) \tag{2.4}$$

where X_0 is a bounded random variable, measurable with respect to F_0. μ is a Poisson random measure with the compensator ν of the form $dt \times G(du)$ where G is a positive, σ-finite measure.

Let $\underline{\underline{F}} = (F_t : 0 \leq t \leq 1)$ where F_t = the minimal filtration generated by B_t and $\mu(t,.)$.

Let $\underline{\underline{\widetilde{H}}} = (\widetilde{H}_t : 0 \leq t \leq 1)$ be the smallest right continuous filtration relative to which $\widetilde{B}$ and $\widetilde{\mu}$ are adapted and X_1 is $\widetilde{H}_0$-measurable.

Our aim is to show that X is an $(\underline{\underline{F}}, \underline{\underline{\widetilde{H}}})$- reversible semimartingale. Towards this, we need the following theorems.

Theorem 2.5: (Jacod [4]) *If the functions b, σ, and F are continuously differentiable with bounded derivatives, then there exists a unique strong solution X of (2.4) such that $X \in L^p$ for any $p > 0$.*

Theorem 2.6: (Jacod and Protter [6]) *Let K be an $(\underline{\underline{F}}, \underline{\underline{\widetilde{J}}})$-reversible semimartingale. Let Y be a process with cadlag paths such that:*

(a) For all $0 \leq t \leq 1, Y_t$ is F_t and $\widetilde{J}_{1-t}$ measurable.

(b) The quadratic variation [Y, K] exists as the limit in probability of discrete approximations and is of finite variation. Then, the processes [Y, K] and $X_t = \int_0^t Y_{s-}dK_s$ are $(\underline{\underline{F}}, \underline{\underline{\widetilde{J}}})$-reversible semimartingales.

Note: Theorem 2.5 is a particular case of a general result of Jacod (Also see

Bichteler et al.[1]). Theorem 2.6 is a general result in the sense that it holds for any given pair of filtrations $(\underline{\underline{F}}, \underline{\underline{\tilde{J}}})$.

The following is a result due to Jacod [5] on the enlargement of filtrations.

Theorem 2.7: *Suppose we are given a filtration* $\underline{\underline{\tilde{G}}} = (\tilde{G}_t : t \geq 0)$ *and a random variable L. Let* $\underline{\underline{\tilde{J}}} = (\tilde{J}_t : t \geq 0)$ *with* $\tilde{J}_t = \tilde{G}_t \vee \sigma(L)$. *Let* $Q_t(\omega, dx)$ *designate a regular version of the conditional law of L with respect to* $\tilde{G}_t$. *If* $Q_t(\omega, \cdot)$ *is absolutely continuous with respect to a positive σ-finite measure η a.s. in ω, then* $\underline{\underline{\tilde{G}}}$ *semimartingales will also be* $\underline{\underline{\tilde{J}}}$ *semimartingales.*

To suit our needs, we set in theorems 2.6 and 2.7, $\underline{\underline{\tilde{J}}} = \underline{\underline{\tilde{H}}}; \underline{\underline{\tilde{G}}} = (\tilde{G}_t : 0 \leq t \leq 1)$ to be the smallest right continuous filtration relative to which $\tilde{B}$ and $\tilde{\mu}$ are adapted. Besides, $L = X_1$ and $\eta =$ the Lebesgue measure.

Theorem 2.8: *Assume that*
(1) σ is a non-vanishing function.
(2) $F \in C^\infty$ *and the map* $a \to a + F(a)u$ *is a diffeomorphism for each u with* $|U| \leq J$. *Then, the solution of (2.4) is an* $(\underline{\underline{F}}, \underline{\underline{\tilde{H}}})$*-reversible semimartingale.*

Proof: By Theorem 2.6, it suffices to show that
(a) X_t is $\tilde{H}_{1-t}$ measurable, and
(b) Z is an $(\underline{\underline{F}}, \underline{\underline{\tilde{H}}})$ reversible semimartingale, where

$$Z_s = s + B_s + \int_0^s \int_{|u| \leq J} u(\mu - \nu)(ds, du).$$

The proof of (a) is as follows:

Let $\phi(x; s, t)$ denote the stochastic flow for the equation (2.4), so that it is equal to

$$x + \int_s^t b(\phi(x; t, a))da + \int_s^t \sigma(\phi(x; t, a))dB_a + \int_s^t \int_{|u| \leq J} F(\phi(x; t, a-))u(\mu - \nu)(da, du)$$

The assumption (2) made in this theorem implies that the stochastic flow ϕ is a C^∞ diffeomorphism (see Leandre [9]). In particular ϕ is invertible. Note that

$\phi(X_t : t, 1) = X_1$. Call $\phi(x : t, 1)$ as $\phi_t(x)$ for convenience. ϕ_t is easily seen to be in $\sigma(Z_r - Z_t : t \leq r \leq 1) \subseteq \widetilde{H}_{1-t}$. Therefore, ϕ_t^{-1} is $\widetilde{H}_{1-t}$ -measurable. $X_t = \phi_t^{-1}(X_1)$ so that X_t is $\widetilde{H}_{1-t}$ -measurable. To prove part (b), note that if σ is non-vanishing, then the law of X_t has a density with respect to the Lebesgue measure (see Bichteler et al. [1]). Let the density of X_t be denoted by f_t. By the regular conditional distribution of X_1 given $\widetilde{G}_{1-t}$ is meant a random measure $Q_{1-t}(\omega, dx)$ such that for every $g \in b\mathcal{B}$,

$$E(g(X_1)|\widetilde{G}_{1-t})(\omega) = \int g(x) Q_{1-t}(\omega, dx).$$

We want to show that $Q_{1-t}(\omega, dx) = \alpha_t(\omega, x)dx$.

Note that $g(X_1) = g(\phi_t(X_t))$, so that it is enough to consider functions of the form $h(\omega, x) = h_1(\omega)h_2(x)$, we have

$$\begin{aligned} Eh_1(\omega)h_2(X_t)|\widetilde{G}_{1-t}(\omega) &= h_1(\omega)Eh_2(X_t)|\widetilde{G}_{1-t} \\ &= h_1(\omega)Eh_2(X_t) \\ &= h_1(\omega)\int h_2(x)f_t(x)dx \end{aligned}$$

A monotone class argument extends this to $h(\omega, x)$ so that

$$\begin{aligned} Eh(\omega, X_t)|\widetilde{G}_{1-t} &= \int h(\omega, x)f_t(x)dx \\ Eg(X_1)|\widetilde{G}_{1-t} &= \int g(\phi_t(x))f_t(x)dx \\ &= \int g(y)f_t(\phi_t^{-1}(y))\frac{d\phi_t^{-1}}{dy}(y)dy \end{aligned}$$

by putting $y = \phi_t(x)$. Thus $Q_{1-t}(\omega, dx) = \alpha_t(\omega, y)dy$ where

$$\alpha_t(\omega, y) = f_t(\phi_t^{-1}(y))\frac{d\phi_t^{-1}}{dy}(y)$$

An application of Theorem 2.7 finishes the proof of (b). ∎

It is easy to see from the above proof that our sufficient conditions are stronger than what we need. In fact condition (2) would have given us the C^∞ property of the flows even for equations with general integrators. However, an effective use of the strong Markov property of solutions should yield weaker sufficient conditions. A hint to this is given in the next section.

3. A RESULT ON STOCHASTIC FLOWS

In this section we will consider the following SDE:

$$X_t^x = x + \int_0^t F(X_{s-}^x)dZ_s \qquad (3.1)$$

where F is Lipschitz continuous with c as the Lipschitz constant, and Z is a Lévy process (possibly with unbounded jumps). By use of the Ito lemma,

$$(X_t^x - X_t^y)^2 = (x-y)^2\mathcal{E}(S)_t \qquad (3.2)$$

where $\mathcal{E}$ denotes the stochastic exponential and

$$S_t = 2\int_0^t \frac{F(X_{s-}^x) - F(X_{s-}^y)}{(X_{s-}^x - X_{s-}^y)} I_{[(X_{s-}^x - X_{s-}^y)\neq 0]} dZ_s + \int_0^t \frac{(F(X_{s-}^x) - F(X_{s-}^y))^2}{(X_{s-}^x - X_{s-}^y)^2} I_{[(X_{s-}^x - X_{s-}^y)\neq 0]} d[Z]_s$$

The form of the stochastic exponential shows that if $T = inf(t : X_t^x = X_t^y)$, then $(T<\infty)$ if and only if $\Delta S_T = -1$. Therefore

$$X_{t-}^x \neq X_{t-}^y$$

In other words confluence can occur only because of a jump at time T in Z.

$$X_T^x - X_T^y = 0 = (X_{T-}^x - X_{T-}^y) + (F(X_{T-}^x) - F(X_{T-}^y))\Delta Z_T$$

$$\Delta Z_T = \frac{-(X_{T-}^x - X_{T-}^y)}{F(X_{T-}^x) - F(X_{T-}^y)} \text{ on } (T<\infty) \qquad (3.3)$$

The following result can be found in [11] and is given here for the reader's convenience.

Theorem 3.4: *Let Z be a Lévy process whose Lévy measure G(du) is non-singular. Then the solution X_T^x of the SDE (3.1) has non- confluent paths a.s.. That is $P(\omega$: there exists t such that $X_t^x(\omega) = X_t^y(\omega)) = 0$.*

Proof: On $(T<\infty), |\Delta Z_T| \geq \frac{1}{c}$. Let $T_1 = inf(t : |\Delta Z_t| \geq \frac{1}{c})$ and

$$T_n = inf(t > T_{n-1} : |\Delta Z_t| \geq \frac{1}{c})$$

By (3.3),

$$(T = T_n) = (\Delta Z_{T_n} = \frac{-(X^x_{T_n-} - X^y_{T_n-})}{F(X^x_{T_n-}) - F(X^y_{T_n-})} I_{(F(X^x_{T_n-}) - F(X^y_{T_n-}) \neq 0)}) \qquad (3.5)$$

ΔZ_{T_n} however is independent of F_{T_n-} by a result of Jacod (4, Page 96) whereas

$$\frac{-(X^x_{T_n-} - X^y_{T_n-})}{F(X^x_{T_n-}) - F(X^y_{T_n-})} I_{(F(X^x_{T_n-}) - F(X^y_{T_n-}) \neq 0)}$$

is in F_{T_n-}.

Thus (3.5) implies that $P(T = T_n) = 0$ since the law of ΔZ_{T_n} has a density with respect to the Lebesgue measure. Since this is true for all n,

$$P(T < \infty) = 0$$

∎

In fact, invertibility of stochastic flows is proved in [11] under milder conditions than those in Leandre [9], though the latter covers a more general context that includes non-markov situations as well. The above reference also provides the form of the backward stochastic differential equation for which the time reversed process namely $\widetilde{X}$ acts as the solution.

REFERENCES

[1] K.Bichteler, J-B. Gravereaux and J. Jacod (1987): Malliavin Calculus for processes with jumps. Stochastics Monographs. Gordon and Breach Science Publishers.

[2] M.Emery (1982): En marge de l'exposé de Meyer: "Geométrie differéntielle stochastique". Springer-Verlag Lecture Notes in Math. 921, 208-216.

[3] K.Itô (1978): Extension of stochastic integrals. Proc. Int. Symp. SDE's. Wiley, New York.

[4] J. Jacod (1979): Calcul stochastique et Problemes de martingales. Springer-Verlag Lecture Notes in Math. 714.

[5] J.Jacod (1985): Grossissement initial, hypothese (H'), et théorème de Girsanov. Springer-Verlag Lecture Notes in Math., 1118, 15-35.

[6] J.Jacod and Ph.E. Protter (1988): Time reversal on Lévy processes. Ann. Probab. 16, 620-641.

[8] P.A. Meyer (1981): Flot d'une equation differentielle stochastique . Springer-Verlag Lecture Notes in Math. 850, 103-117.

[9] R. Leandre (1984): Flot d'une equation differentielle stochastique avec semi-martingales disectrice discontinue. Springer-Verlag Lecture Notes in Math. 1123, 271-275.

[10] E.Pardoux (1986): Grossissement d'une filtration et retournement du temps d'une diffusion. Springer-Verlag Lecture Notes in Math.1204, 48-56.

[11] P. Sundar (1989): Invertibility of stochastic flows and time reversal of solutions of SDE's driven by Lévy processes with bounded jumps. Soochow J.Math. 15, 193-204.

P. Sundar
Department of Mathematics
Louisiana State University
Baton Rouge
LA 70803.

BIRTH AND DEATH ON A FLOW

by

Erhan Çinlar and John S. Kao

Princeton University

1. INTRODUCTION

In recent years there has been much interest in the equilibrium behavior of stochastic flows; see for instance BAXENDALE [1], CARVERHILL [2], LE JAN [6],[7] and [8]. Most of the work seems to be concentrated on the limiting distribution, as $t \to \infty$, of the random measure

$$\mu_t(\omega, A) = \mu_0\{x: F^{\omega}_{0,t}x \in A\}$$

for a stochastic flow $F = \{F_{s,t} : 0 \le s \le t \le \infty\}$ and a given mass distribution μ_0 with total mass 1. In some applications, notably in transport of pollutant particles by groundwater flows, one is interested in similar questions but with creation and annihilation of mass over time and space.

To that end, we start with a random configuration of a countable collection of particles at time 0, allow additional particles to enter the space at random times and locations, and assume that each particle is carried by the flow until it is killed according to a killing rule with position dependent killing rate. With this setup, we are interested in the measure-valued process $\{M_t: t \ge 0\}$ where $M_t(A)$ is the (random) number of live particles in the set A at time t. This is basically a birth-death process where the live particles move inertly on the flow. Assuming that births (entrances of particles) over time and space are regulated by a Poisson random measure, we are looking for the limiting law of the random measure M_t as $t \to \infty$.

Although our results lack generality, they illustrate a new approach to questions of equilibrium. To better contrast the issues involved, in the next section, we shall give an account of the limiting behaviors of μ_t and M_t in the case of an Ornstein-Uhlenbeck flow F, which flow is essentially the real line version of the North-South flow of CARVERHILL[2]. We shall see that, in that case, μ_t converges in distribution to δ_Z, a Dirac measure sitting at the random point Z, whereas M_t converges in distribution to a Poisson random measure M whose mean is the Lebesgue measure. This special example illustrates the interplay between the contractive action of the flow (which constantly brings particles closer) and the thinning action of the killing mechanism. Thus, in the general case as well, the drift and diffusion coefficients of the flow and the birth and death rates of the particles will all have to satisfy certain balance conditions.

2. PARTICLE SYSTEM

In this section we describe the stochastic process of interest and mention a few preliminary facts to be used later.

Throughout, positive means ≥ 0, all measures are positive, μf denotes the integral of the function f with respect to the the measure μ, and $\exp_{-}x$ is another notation for e^{-x}.

The space E will always be a fixed locally compact space with a countable base. We will write $\mathbb{C}_K^+$ for the set of all positive contimuous functions with compact supports in E.

Let $(\Omega, \mathcal{H}, \mathbb{P})$ be a probability space. Let F be a right-continuous stochastic flow on E. We write $F_{st}x$ or $F(s, x, t)$ for the position, at time t, of the particle that was at x at time s. Consider a countable system of particles indexed in some manner. To the particle i, we associate three random variables: a positive random variable S_i denoting its time of "birth," and a E-valued random variable X_i denoting its position at birth, and a positive random variable $T_i > S_i$ denoting its time of "death." Assuming that it is alive at time t, that is, on the event $\{S_i \le t \le T_i\}$, its position then is $F(S_i, X_i, t)$. Thus,

2.1 $$M_t(A) = \sum_i 1_A \circ F(S_i, X_i, t)\ 1_{[0,t]} \circ S_i\ \ 1_{(t,\infty)} \circ T_i$$

is the number of live particles in the Borel set $A \subset E$ at time t. We are interested in the measure-valued process $M = \{M_t :\ t \in \mathbb{R}_+\}$ and its limiting behavior.

We specify next the probabilistic assumptions. As mentioned already, F is a right-continuous flow on E. We shall assume that its probability law is invariant under time shifts, that is, the probability law of the process $t \to F(s, x, s+t)$ is free of s. We leave the probability law of M_0 unspecified, but we assume that M_0 and F are independent. We assume that the birth process

2.2 $$N = \sum_i 1_{(0,\infty)} \circ S_i\ \ \delta_{(S_i, X_i)}$$

is independent of M_0 and F and is a Poisson random measure with mean measure ν given by

2.3 $$\nu(ds, dx) = ds\ \pi(dx), \quad s > 0,\ x \in E,$$

where π is some Σ-finite measure on E (that is, $\pi = \pi_1 + \pi_2 + \ldots$ for some finite measures $\pi_1,\ \pi_2, \ldots$ on E). Finally, the killing mechanism that determines the lifetimes is regulated by a positive Borel function k on E as follows: heuristically, each particle's lifetime is independent of all things except its path in life, and its probability of death within dt has the form $k(x)\ dt + o(dt)$ if its position is x at time t. More precisely, we assume that

2.4 $$T_i = \inf\ \{t > S_i :\ \int_{S_i}^{t} dr\ k \circ F(S_i, X_i, r) \ge U_i\}$$

where the collection of random variables U_i is independent of M_0, F, N, and the U_i are independent of each other, are strictly positive, and have the exponential distribution with mean 1. Note that, then,

2.5 $$\lim_{u \to 0} \frac{1}{u}\ \mathbb{P}\ \{T_i \le t + u \mid M_0,\ F,\ N,\ T_i > t\} = k \circ F(S_i, X_i, t),$$

which is the usual meaning of being killed at rate $k(x)$ when at x. This completes the probabilistic specification of the model. It is clear that M is a time-homogeneous Markov process.

Life Process

The following auxiliary random measure L indicates the birth times, birth process, and the intrinsic lifetimes of all the particles:

2.6
$$L = \sum_i \delta_{(S_i, X_i, U_i)}.$$

The following master formula for M_t is immediate from 2.1 and 2.4:

2.7
$$M_t f = \int L(ds, dx, du)\ f(F_{st}x)\ 1_{[0,t]}(s)\ 1_{[0,u)}\Big(\int_s^t dr\ k(F_{sr}x)\Big)$$

for every t in $\mathbb{R}_+$ and every positive Borel function f on E.

Preliminaries

A sequence (μ_n) of measure on E is said to converge vaguely to a measure μ if $\mu_n f \to \mu f$ for every f in $\mathbb{C}_K$. This convergence induces the so-called vague topology on the space $\mathcal{M}$ of all measures on E. A sequence (M_n) of random measures on E is said to converge in distribution in the vague topology to a random measure M_∞, and then we write $M_n \xrightarrow{vd} M_\infty$, if $\mathbb{E}\ g(M_n) \to \mathbb{E}\ g(M_\infty)$ for every real-valued function g with compact support in $\mathcal{M}$. According to a theorem in KALLENBERG [4],

2.8
$$\begin{aligned} M_n \xrightarrow{vd} M_\infty &\Leftrightarrow M_n f \xrightarrow{d} M_\infty f && \forall f \in \mathbb{C}_K^+ \\ &\Leftrightarrow \mathbb{E} \exp_- M_n f \to \mathbb{E} \exp_- M_\infty f && \forall f \in \mathbb{C}_K^+. \end{aligned}$$

Let D be an arbitrary space equipped with a σ-algebra (of measurable sets). Given a random measure H on D, its probability law is determined by its Laplace functional

2.9
$$f \mapsto \mathbb{E}\ e^{-Hf}$$

from the set of all positive measurable functions into $[0,1]$. In particular, H is a Poisson random measure on D with mean (measure) η if and only if its Laplace functional has the form

2.10
$$\mathbb{E}\ e^{-Hf} = \exp_-\ \eta(1 - e^{-f}).$$

The following folk theorem has some uses.

2.11 THEOREM. Suppose that $H = \sum_i \delta_{Y_i}$ is a Poisson random measure on D with some mean η. Suppose that the Z_i are random variables taking values in some measurable space D' and are such that

$$\mathbb{P}\{Z_i \in B \mid H,\ Z_j,\ j \neq i\} = Q(Y_i, B)$$

for some transition probability kernel Q from D into D'. Then, $\tilde{H} = \sum_i \delta_{(Y_i, X_i)}$ is a Poisson random measure on the product space $D \times D'$, and its mean $\tilde{\eta}$ is given by

$$\tilde{\eta}(dy, dz) = \eta(dy)\ Q(y, dz).$$

The next two propositions apply the preceding to our particle system.

2.12 PROPOSITION. a) The restriction L_+ of L to $(0, \infty) \times E \times \mathbb{R}_+$ is a Poisson random measure with mean λ_+, where

$$\lambda_+(ds, dx, du) = ds\ \pi(dx)\ e^u du.$$

b) Suppose that M_0 is a Poisson random measure on E with mean μ_0. Then, the restriction L_0 of L to $\{0\} \times E \times \mathbb{R}_+$ is Poisson with mean λ_0, where

$$\lambda_0(ds, dx, du) = \delta_0(ds)\ \mu_0(dx)\ e^{-u} du.$$

Moreover, L_0 and L_+ are independent, and L is Poisson with mean $\lambda = \lambda_0 + \lambda$.

PROOF. By re-labeling the particles if necessary, we may assume that for each i the birth time S_i is either 0 everywhere or strictly positive everywhere. Now, take $Y_i = (S_i, X_i)$ and $Z_i = U_i$ for those i with $S_i > 0$ everywhere. Then, $H = N$ and $\tilde{H} = L_+$ in the notation of Theorem 2.11. In view of the assumption that N is Poisson with mean given by 2.3, and in view of assumptions regarding U_i, the condition of 2.11 is satisfied with $Q(y, dz) = e^{-z} dz$. So, $\tilde{H} = L_+$ is Poisson with mean λ_+ as claimed. Proof of the part (b) is similar.

In the next proposition, and later, we write $\mathbb{P}_F$ (and $\mathbb{E}_F$) for the conditional probability (and the conditional expectation) given the σ-algebra generated by the stochastic flow F.

2.13 PROPOSITION. Suppose that M_0 is Poisson with mean μ_0. Then, for each time t, the probability law of M_t under $\mathbb{P}_F$ is that of a Poisson random measure with mean μ_t where

$$\mu_t f = \mathbb{E}_F\ M_t f = \int_E \mu_0(dx)\ f(F_{0t}x)\ \exp_- \int_0^t dr\ k(F_{0t}x)$$
$$+ \int_0^t ds \int_E \pi(dx)\ f(F_{st}x)\ \exp_- \int_s^t dr\ k(F_{st}x).$$

PROOF. By the preceding proposition, L is Poisson with mean $\lambda = \lambda_0 + \lambda_+$ specified there. Next, in preparation for applying Theorem 2.11, let $Y_i = (S_i, X_i, U_i)$ and define

$$Z_i = \begin{cases} F(S_i, X_i, t) & \text{on } \{S_i \le t, \int_{S_i}^t dr\ k \circ F(S_i, X_i, r) > U_i\} \\ \Delta & \text{elsewhere,} \end{cases}$$

where Δ is some point outside E. Under $\mathbf{P}_F$, the random measure L is still Poisson with mean λ in view of the independence of L and F.

Moreover, under $\mathbf{P}_F$ again, the Z_i satisfy the condition of Theorem 2.11 (since they are "deterministic" transformations of the Y_i) with

$$Q(s, x, u; B) = 1_B(F_{st}x)\ 1_{[0,t)}(s)\ 1_{[0,u)}\left(\int_s^t\ dr\ k(F_{sr}x)\right)$$
$$+ 1_B(\Delta)\ \left[1_{(t,\infty)}(s) + 1_{[0,t)}(s)\ 1_{[u,\infty)}\left(\int_s^t\ dr\ k(F_{sr}x)\right)\right]$$

for all measurable $B \subset E \cup \{\Delta\}$. It follows from Theorem 2.11 that, under $\mathbb{P}_F$, the random measure $\tilde{H} = \sum \delta_{(Y_i, Z_i)}$ is Poisson with mean

$$\lambda(ds, dx, du)\ Q(s, x, u, dz)$$

on the space $\mathbb{R}_+ \times E \times \mathbb{R}_+ \times (E \cup \{\Delta\})$. It follows that $M_t(B) = \tilde{H}(\mathbb{R}_+ \times E \times \mathbb{R}_+ \times B)$, with Borel $B \subset E$, is a Poisson random measure under $\mathbf{P}_F$ with mean

$$\mu_t(B) = \int\ \lambda(ds, dx, du)\ Q(s, x, u; B).$$

This completes the proof modulo an easy computation.

3. ORNSTEIN-UHLENBECK FLOWS

Our aim is to contrast the limiting behavior of the random measure

3.1 $$\bar{M}_t(A) = \mu_0\{x \in E\colon\ F_{0t}x \in A\}$$

with the limiting behavior of M_t in a special case where the computations are easy to do explicitly. We shall see that the limiting behavior of M_t is closer to what a physicist would call equilibrium, whereas that of $\bar{M}_t$ is in fact degenerate.

To this end, we take $E = \mathbf{R}$ and let F be the flow induced by the Ornstein-Uhlenbeck process, that is, $F_{st}x$ is the solution of the stochastic differential equation

3.2 $$dV_t = -aV_t\, dt + b\, dW_t, \quad t > s,\ V_s = x,$$

where a and b are strictly positive constants and W is the Wiener process. Solution is well-known:

3.3 $$F_{st}x = (x - Y_s)\, e^{-a(t-s)} + Y_t, \quad t \geq s,$$

where

3.4 $$Y_t = b \int_0^t e^{-a(t-s)}\, dW_s, \quad t \geq 0.$$

This flow is equivalent, on the real-line, of CARVERHILL's North-South flow on the circle. Therefore, the following is the version, for our flow, of his limiting result:

3.5 PROPOSITION. Let μ_0 have total mass 1. Then, as $t \to \infty$,

$$\bar{M}_t \xrightarrow{vd} \delta_Z$$

where Z is a Gaussian random variable with mean 0 and variance $b^2/2a$.

PROOF. We use 2.8 with 3.1-3.4. For $f \in \mathbb{C}_K^+(\mathbb{R})$,

$$\begin{aligned} \mathbb{E}\exp_-\tilde{M}_t f &= \mathbb{E}\exp_- \int_{\mathbb{R}} \mu_0(dx)\, f(xe^{-at} + Y_t) \\ &= \mathbb{E}\exp_- \int_{\mathbb{R}} \mu_0(dx)\, f(xe^{-at} + v_t Z) \end{aligned}$$

where $v_t^2 = 1 - e^{-2at}$ and Z is Gaussian with mean 0 and variance $b^2/2a$. This is immediate from 3.4, since Y_t has the same distribution as $v_t Z$. Letting $t \to \infty$, using the fact that f is continuous and bounded, and applying the bounded convergence theorem once to the integral and once to the expectation, we get

$$\lim_{t\to\infty} \mathbb{E}\exp_-\tilde{M}_t f = \mathbb{E}\exp_- f(Z) = \mathbb{E}\exp_-\delta_Z f,$$

which completes the proof via 2.8.

An entirely different picture emerges when one considers our particle system. In the following proposition, the conditions listed are in addition to the standing assumptions imbedded in the preceding section and the special flow we are considering now.

3.6 PROPOSITION. Suppose that M_0 is a Poisson random measure on $\mathbb{R}$ with mean measure $\mu_0(dx) = c_0\, dx$. Suppose that the mean measure of N is $ds\, \pi(dx) = c\, ds\, dx$. Suppose that the killing rate function k is constant. Then, for each t, the random measure M_t is Poisson with mean measure $\mu_t(dx) = c_t\, dx$, where

$$c_t = \begin{cases} c_0 + ct & \text{if } k = a\,, \\ \dfrac{c}{k-a} + \left(c_0 - \dfrac{c}{k-a}\right)e^{-(k-a)t} & \text{if } k \neq a\,. \end{cases}$$

If $k = a$ and $c = 0$, then M_t is Poisson with mean measure $c_0\, dx$ for all t. If $k > a$, then $M_t \overset{vd}{\to} M_\infty$ where M_∞ is Poisson with mean $\frac{c}{k-a}\, dx$. Otherwise, $M_t \overset{vd}{\to} M_\infty$ where $M_\infty(A) = +\infty$ for every open interval A.

PROOF. We prove the first assertion; the others are easy consequences of the first via 2.8.

Under the hypotheses here, Proposition 2.13 implies that

3.7 $$\mathbb{E}_F \exp\_M_t f = \exp\_\mu_t(1 - e^{-f}),$$

where

$$\mu_t f = \mathbb{E}_F\, M_t f = \int_{\mathbf{R}} c_0\, dx\, f(F_{0t}x)\, e^{-kt} + \int_0^t ds \int_{\mathbf{R}} c\, dx\, f(F_{st}x)\, e^{-k(t-s)} .$$

Replacing $F_{st}x$ by 3.3 and doing the integration over $\mathbf{R}$ we get

$$\begin{aligned}\mu_t f &= c_0\, e^{-(k-a)t} \int_{\mathbf{R}} dx\, f(x) + c \int_0^t ds\, e^{-(k-a)(t-s)} \int_{\mathbf{R}} dx\, f(x) \\ &= c_t \int_{\mathbf{R}} dx\, f(x) ,\end{aligned}$$

which shows that $\mu_t(dx) = c_t\, dx$, independent of F. Putting this into 3.7, we see that the Laplace functional of M_t is that of a Poisson (see 2.10) random measure with mean measure $\mu_t(dx) = c_t\, dx$. ∎

Note that this flow F is a contraction:

$$|F_{st}x - F_{st}y| = |x - y|\, e^{-a(t-s)} .$$

Therefore, its effect on a finite mass distribution μ_0 will be to contract all the mass onto the path (Y_t), so that the limiting behavior of $\bar{M}_t = M_0(F_{0t}^{-1})$ will be the same as that of $\mu_0(\mathbf{R})\, \delta_{Y_t}$. This is the content of Proposition 3.5. By contrast, in the particle model, the contraction action of the flow is compensated by the killing (assuming $k > a$) so that we obtain a "homogeneous" distribution of mass in the limit (in the physicists' language).

4. EQUILIBRIUM IN GENERAL

Consider the flow F and the particle system M as described in Section 2. In addition to the standing assumptions of Section 2, we assume the following.

a) For each s in $\mathbb{R}_+$ and x in E, the path $t \mapsto F_{st}x$ is right-continuous.

b) F is strong Markov in the following sense: For each stopping time T of the filtration $\mathcal{F}_t = \sigma\{F_{su}: 0 \le s \le u \le t\}$, $F_{T,T+t}$ is independent of $\mathcal{F}_T$ and has the same law as F_{0t}. If F is the flow associated with a stochastic differential equation with homogeneous coefficients, these conditions are fulfilled automatically.

Our aim is to show that M_t converges in distribution in the vague topology. To this end, we will put three conditions: killing rate should be bounded below by a strictly positive constant, the numbers of particles over compacts should remain within bounds, and the effect of the initial particle distribution should wear off. Of course, all these imply certain relationships between the killing function k, birth rate measure π, and the infinitesimal characteristics of the flow F. We plan to present such more explicit results elsewhere for flows with more explicit specifications.

In preparation for the statement of the main result, we start by decomposing M as

4.1
$$M_t = M_t^0 + M_t^+$$

where M^0 and M^+ are defined by 2.7 with L there replaced by L_0 and L_+ respectively, where L_0 and L_+ are as described in Proposition 2.12. Thus, for instance

4.2
$$M_t^+ f = \int L_+(ds, dx, du)\ f(F_{st}x)\ 1_{(0,t]}(s)\ 1_{[0,u)}\Big(\int_s^t dr\ k(F_{sr}x)\Big).$$

Note that $M_0 = M_0^0$ and that $M_t^0(A)$ is the number of particles that were present at time 0 and are in A at time t.

4.3 THEOREM. Suppose that

a) $k \ge a$ for some constant $a > 0$,

b) $M_t^0 f \xrightarrow{d} 0$ for every f in $\mathbb{C}_K^+$,

c) $\limsup \mathbf{E}\, M_t^+ f < \infty$ for every f in $\mathbb{C}_K^+$.

Then, $M_t \xrightarrow{vd} M_\infty$, where M_∞ is a random measure on E with $\mathbf{E}\, M_\infty f < \infty$ for every f in $\mathbb{C}_K^+$.

4.4 REMARK. The process $t \mapsto F_{0t}x$ is a temporally homogeneous Markov process with initial state x. Let (P_t) be its transition semigroup and (Q_t) the transition semigroup obtained from it by killing at rate k:

4.5 $$P_t f(x) = \mathbf{E}\, f(F_{0t}x), \quad Q_t f(x) = \mathbf{E}\, f(F_{0t}x) \exp_- \int_0^t dr\ k(F_{0r}x).$$

It follows from Proposition 2.12 and 4.2 that

4.6 $$\begin{aligned}\mathbf{E}\, M_t^+ f &= \mathbf{E} \int_0^t ds \int_E \pi(dx)\ f(F_{st}x) \exp_- \int_s^t dr\ k(F_{sr}x) \\ &= \int_0^t ds \int_E \pi(dx)\ Q_{t-s} f(x) = \int_0^t ds\ \pi Q_s f.\end{aligned}$$

Thus, the limit superior appearing in 4.3c is in fact a supremum:

4.7 $$\limsup\ \mathbf{E}\, M_t^+ f = \sup_t\ \mathbf{E}\, M_t^+ f = \int_0^\infty ds\ \pi Q_s f = \pi V f$$

where V is the potential kernel for (Q_t). In other words, condition 4.3c is equivalent to assuming that $\pi V f < \infty$ for every $f \in \mathbb{C}_K^+$.

Moreover, in the presence of condition 4.3a, we see from 4.5 that

4.8 $$Q_t f \leq e^{-at} P_t f,$$

which implies that, in order for 4.3c to hold, it is sufficient that π be a sub-invariant Radon measure for (P_t), for, then,

$$\pi V f = \int_0^\infty ds\ \pi Q_s f \leq \int_0^\infty ds\ e^{-as}\ \pi P_t f \leq \tfrac{1}{a}\ \pi f < \infty$$

for f in $\mathbb{C}_K^+$. So, the condition 4.3c does not eliminate recurrence for F as a possibility (of course, transience is better).

4.9 REMARK. Convergence in distribution to a constant is equivalent to convergence in probability. Thus, using the Markov inequality $\mathbf{P}\{M_t^0 f > \epsilon\} \leq \mathbf{E}\, M_t^0 f/\epsilon$, we see that condition 4.3b is implied by the condition that

$$\mathbf{E}\, M_t^0 f \to 0 \quad \text{for every} \quad f \in \mathbb{C}_K^+.$$

Now, by the independence of M_0 and F,

$$\mathbb{E}\ M_t^0 f = \mathbb{E} \int M_0(dx)\ f(F_{0t}x)\ \exp_- \int_0^t dr\ k(F_{0r}x) = \mathbb{E}\ M_0 Q_t f = \mu_0 Q_t f,$$

where μ_0 is the mean measure for M_0. Thus, since 4.3a implies 4.8, it is sufficient to have μ_0 be a sub-invariant Radon measure for (P_t). Again in the presence of 4.3a, a very simple condition that ensures 4.3b is that M_0 be almost surely finite. For, then, conditioning on $Z = M_0(E)$,

$$\mathbb{P}\{M_t^0(E) > 0\} \le 1 - \mathbb{E}(1 - e^{-at})^Z \to 0$$

as $t \to \infty$, and therefore $M_t^0(E) \to 0$ in probability.

The remainder of this note is devoted to proving Theorem 4.3. We start with the following lemma of independent interest.

4.10 LEMMA. Suppose that $M_0 = 0$, π is finite, and $k \ge a$ for some constant $a > 0$. Then, $M_t \overset{vd}{\to} M_\infty$ where M_∞ is an almost surely finite random measure (in fact, $\mathbb{E}\ M_0(E) < \infty$).

PROOF. Let $b = \pi 1$. Let $\bar{M}$ be defined by 4.2 but with k replaced by the constant a. Then, the process $(\bar{M}_t 1)$ is a birth death process, on the positive integers, with birth rate b and death rate na when n particles are alive. It is well-known that this process is positive recurrent and its limiting distribution is Poisson with mean b/a. Moreover, setting (recall that $\bar{M}_0 = M_0 = 0$)

$$\bar{S} = \inf\ \{t: \bar{M}_t 1 > 0\}, \quad \bar{T} = \inf\ \{t > S: \bar{M}_t 1 = 0\}, \tag{4.11}$$

a renewal theoretic argument shows that the probability $e^{-b/a}$ that $\bar{M}_t 1$ is 0 in the limit is equal to $\mathbb{E}\bar{S}/\mathbb{E}\bar{T}$. Since $\bar{S}$ is the time of first birth, which is exponential with mean $1/b$, it follows that $\mathbb{E}\bar{T} = e^{b/a}/b$.

b) Let S and T be defined by 4.11 but for $M = M^+$. Clearly, since $M_0 = 0$, we have $S = \bar{S}$. Since $k \ge a$, we have $M_t \le \bar{M}_t$ for all t, which implies that $T \le \bar{T}$. Thus,

$$\mathbb{E}T < \infty. \tag{4.12}$$

But at the stopping time T we have $M_T = 0$. The strong Markov property of F extends to this stopping time (of F and L_+) by the independence of L_+ and F. Thus, conditioning at time T and using the invariance of the laws under time shifts, we obtain that

4.13 $$\mathbb{E}\,\exp\_M_t f = g(t) + \int_0^t \mathbb{P}\{T \in ds\}\,\mathbb{E}\,\exp\_M_{t-s} f,$$

where $f \in \mathbb{C}_K^+$ and

4.14 $$g(t) = \mathbb{E}\,(\exp\_M_t f)\,I_{\{T>t\}}, \quad t \in \mathbb{R}_+.$$

We now appeal to renewal theory (see ÇINLAR [3] for an account). By the various independence properties of the Poisson random measures, the random variables S and $T-S$ are independent, and S has an exponential distribution. It follows that the distribution of T is not arithmetic. Thus, by the key renewal theorem, the solution of the renewal equation 4.13 has the limit

4.15 $$\lim_{t\to\infty} \mathbb{E}\,\exp\_M_t f = \frac{1}{\mathbb{E}T}\int_0^\infty dt\; g(t)$$

provided that the function g be directly Riemann integrable. We show next that g is indeed so.

The paths $t \mapsto F_{st}x$ are right-continuous, and M_t is almost surely finite (in fact $\mathbb{E}\,M_t 1 \le bt$). Thus, the continuity of f and 4.2 shows that $t \mapsto M_t f$ is right-continuous. Thus, $t \mapsto (\exp\_M_t f)\,I_{\{T>t\}}$ is right-continuous and bounded, which implies that g is right-continuous via the bounded convergence theorem. Moreover, g is dominated by the function $t \mapsto \mathbb{P}\{T > t\}$, which is Riemann integrable since $\mathbb{E}T < \infty$ as was shown above (see 4.12). It follows that g is directly Riemann integrable. So, 4.15 holds.

b) We show that the right side of 4.15 has the form $\mathbb{E}\,\exp\_M_\infty f$ for some random measure M_∞. To this end, we recall that $M_t \le \bar{M}_t$ for all t. Moreover, an easy computation using 4.2 with k replaced by a shows that

4.16 $$\mathbb{P}\{\bar{M}_t 1 = i\} = \frac{e^{-c(t)}\,c(t)^i}{i!},$$

with $c(t) = (1-e^{-at})b/a$. Thus, $\mathbf{E}\, M_t 1 \le b/a$, which shows that (M_n) is relatively compact in the vague topology (see KALLENBERG [4]). Thus, there is a sequence (t_n) increasing to $+\infty$ and a random measure M_∞ such that

$$M_{t_n} \xrightarrow{vd} M_\infty.$$

In view of 2.8, this and 4.15 imply that $M_t \xrightarrow{vd} M_\infty$.

There remains to show that $\mathbf{E}\, M_\infty 1 \le b/a$. Using 4.16, we check that the family $(\bar{M}_t 1)$ is uniformly integrable. Since $M_t f \le \bar{M}_t f$, this shows that $(M_t f)$ is uniformly integrable for every $f \in \mathbb{C}_K^+$. Since $M_t f \xrightarrow{d} M_\infty f$, uniform integrability implies that $M_t f \to M_\infty f$ in L^1. Hence, since $\mathbf{E}\, M_t f \le \mathbf{E}\, \bar{M}_t f \le (b/a)\, \| f \|$,

$$\mathbf{E}\, M_\infty f = \lim_t \mathbf{E}\, \bar{M}_t f \le (b/a)\, \| f \| \, .$$

This completes the proof.

4.17 PROOF OF THEOREM 4.3. a) Recall the decomposition 4.1 and note that $M_t^0 f \xrightarrow{d} 0$ for every f in $\mathbb{C}_K^+$. By standard theorems on weak convergence, these imply that it is enough for us to show that $M_t^+ \xrightarrow{vd} M_\infty$.

b) Recall that birth rate measure π is Σ-finite, that is, there exist finite measures $\pi_1, \pi_2, \ldots,$ on E such that $\pi = \Sigma\, \pi_j$. We start by decomposing the birth process L in accordance with this decomposition of π.

For each integer $j \ge 1$, let L_j be a Poisson random measure on $(0,\infty) \times \mathbf{E} \times \mathbb{R}_+$ with mean

4.18 $$ds\ \pi_j(dx)\ e^{-u}\ du.$$

Choose the L_j so that they are independent of each other and of F. Let

4.19 $$M_t^n f = \sum_{j=1}^n \int L_j(ds, dx, du)\ f(F_{st}x)\ 1_{[0,t]}(s)\ 1_{[0,u)}\Big(\int_s^t dr\ k(F_{sr}x)\Big)$$

and note that the process M defined by

4.20 $$M_t f = \lim_n M_t^n f$$

has the same probability law as M^+, because $\sum_1^\infty L_j$ has the same probability law as L_+ and is independent of F. Thus, it is enough to show that, with M defined here, we have $M_t \overset{vd}{\to} M_\infty$.

c) Fix n. Since $M_0^n = 0$, $\sum_1^n \pi_j$ is finite, and $k \geq a$ for some constant $a > 0$ (see condition 4.3a), Lemma 4.10 applies to M^n:

4.21 $$M_t^n \overset{vd}{\to} M_\infty^n \quad \text{as} \quad t \to \infty.$$

Condition 4.3c holds for the present M and implies, via Markov's inequality, that for each bounded subset B of E

$$\lim_{b\to\infty} \limsup_{t\to\infty} \mathbf{P}\{M_t(B) > b\} \leq \lim_b \limsup_t \mathbf{E}\, M_t(B)/b = 0.$$

Thus, (M_t) is relatively compact in the vague topology, and there is a sequence (t_p) increasing to $+\infty$ such that

4.22 $$M_{t_p} \overset{vd}{\to} M_\infty \quad \text{as} \quad p \to \infty$$

for some random measure M_∞.

Fix f in $\mathbb{C}_K^+$. It follows from 4.18-4.20 that

$$\begin{aligned}\mathbf{E}\,|M_t f - M_t^n f| &= \mathbf{E} \sum_{j=n+1}^{\infty} \int_0^t ds \int_E \pi_j(dx)\, f(F_{st}x) \exp_- \int_s^t dr\; k(F_{sr}x) \\ &= \sum_{j=n+1}^{\infty} \int_0^t ds\; \pi_j Q_s f\end{aligned}$$

by computations of the sort in Remark 4.4 and in the notation of 4.4. Still in the notation of 4.4, this shows via Fubini's theorem and the monotone convergence theorem that

$$\sup_t \mathbf{E}\,|M_t f - M_t^n f| = \sum_{j=n+1}^{\infty} \pi_j V f,$$

and the right side vanishes as $n \to \infty$ since

$$\sum_1^\infty \pi_j V f = \pi V f = \sup_t \ \mathbb{E}\, M_t f < \infty$$

by condition 4.3c. Hence,

4.23
$$\begin{aligned}\limsup_{n\to\infty} \ \sup_t \ |\mathbb{E}\exp\_M_t f - \mathbb{E}\exp\_M_t^n f| \\ \leq \limsup_n \ \sup_t \ \mathbb{E}|M_t f - M_t^n f| \ = \ 0.\end{aligned}$$

d) Finally, we bring 4.21-4.23 together to show that $M_t \overset{vd}{\to} M_\infty$, which will complete the proof. To this end we use the scheme

$$M_t - M_\infty = (M_t - M_t^n) + (M_t^n - M_\infty^n) + (M_\infty^n - M_{t_p}^n) + (M_{t_p}^n - M_{t_p}) + (M_{t_p} - M_\infty)$$

and combine the first and the fourth terms to write, for f in $\mathbb{C}_K^+$,

$$\begin{aligned}|\mathbb{E}\exp\_M_t f - \mathbb{E}\exp\_M_\infty f| \ &= \ 2\sup_u |\mathbb{E}\exp\_M_u f - \mathbb{E}\exp\_M_u^n f| \\ &+ |\mathbb{E}\exp\_M_t^n f - \mathbb{E}\exp\_M_\infty^n f| \\ &+ |\mathbb{E}\exp\_M_\infty^n f - \mathbb{E}\exp\_M_{t_p}^n f| \\ &+ |\mathbb{E}\exp\_M_{t_p} f - \mathbb{E}\exp\_M_\infty f| \,.\end{aligned}$$

On the right side, letting $p \to \infty$ makes the third and fourth terms vanish in view of 4.21 and 4.22. Letting $t \to \infty$ makes the second term vanish in view of 4.21. Finally, letting $n \to \infty$ makes the first term vanish in view of 4.23. Thus, for every f in $\mathbb{C}_K^+$, the left side goes to 0 as $t \to \infty$, which implies via 2.8 that $M_t \overset{vd}{\to} M_\infty$.

REFERENCES

[1] P. BAXENDALE. Asymptotic behavior of stochastic flows of diffeomorphisms: two case studies. *Prob. Theory Rel. Fields* **73** (1986), 51-85.

[2] A. P. CARVERHILL. Flows of stochastic dynamical systems: ergodic theory. *Stochastics* **14** (1985), 209-226.

[3] E. ÇINLAR. *Introduction to Stochastic Processes*. Prentice-Hall, Englewood Cliffs, 1975.

[4] O. KALLENBERG. *Random Measures*. Academie-Verlag, Berlin, 1983.

[5] H. KUNITA. *Lectures on Stochastic Flows and Applications*. Tata Institute of Fundamental Research. Springer-Verlag, Berlin, 1986.

[6] Y. LE JAN. Flots de diffusion dans $\mathbb{R}^d$. *C. R. Acad. Sci. Paris Ser. I* **294** (1982), 697-699.

[7] ———. Equilibre et exposants de Lyapunov de certaines flots Browniens. *C. R. Acad. Sci. Paris Ser. I* **298** (1984), 361-364.

[8] ———. Equilibrium state for a turbulent flow of diffusion. *Research Notes in Math.* **124**, 83-93. Pitman, Boston, 1985.

PART III

Infinite Dimensional Systems

Lyapunov Exponents and Stochastic Flows of Linear and Affine Hereditary Systems*

SALAH-ELDIN A. MOHAMMED

1. The General Problem

In this article we intend to review known results—and also discuss new ones—concerning the existence of flows and the characterization of Lyapunov exponents for trajectories of stochastic linear and affine hereditary systems. Such systems (also called *stochastic functional differential equations*) are stochastic differential equations in which the differential of the state variable x depends on its current value $x(t)$ at time t as well as its previous values $x(s)$, $t-r \leq s < t$. We shall be concerned almost exclusively with the *finite history* case $0 \leq r < \infty$.

More specifically, consider the stochastic affine hereditary system

$$\left.\begin{aligned} dx(t) &= \sum_{i=0}^{m}\left[\int_{-r}^{0} \nu_i(t)\,(ds)x(t+s)\right] dZ_i(t) + dQ(t), \qquad t>0 \\ x(0) &= v, \quad x(s) = \eta(s), \quad -r < s < 0. \end{aligned}\right\}(I)$$

The above system lives on a complete filtered probability space $(\Omega, \mathcal{F}, (\mathcal{F}_t)_{t\geq 0}, P)$ satisfying the "usual conditions" (Métivier and Pellaumail [31], Métivier [30], Dellacherie and Meyer [14]). Vectors in $\mathbb{R}^n$ (or $\mathbb{C}^n$) are column vectors, given the Euclidean norm $|\cdot|$. The *noise* in (I) is provided by $(\mathcal{F}_t)_{t\geq 0}$-semimartingales $Z_i: \mathbb{R}^+ \times \Omega \to \mathbb{R}$, $i = 0, 1, \ldots, m$, $Q: \mathbb{R}^+ \times \Omega \to \mathbb{R}^n$ with jointly stationary increments. The *memory* is prescribed by stationary $(\mathcal{F}_t)_{t\geq 0}$-adapted measure-valued processes ν_i, $i = 0, \ldots, m$, such that each $\nu_i(t,\omega)$ is an $n\times n$-matrix-valued measure on $[-r, 0]$. The *solution* $x: [-r,\infty)\times\Omega \to \mathbb{R}^n$ is a measurable $(\mathcal{F}_t)_{t\geq 0}$-adapted process with $x|(0,\infty)\times\Omega$ having a.a. sample paths cadlag (viz. right-continuous with left limits). The *initial condition* is a (possibly random) pair $(v,\eta) \in \mathbb{R}^n \times \mathcal{X}$, where $\mathcal{X}$ is some Banach space containing all cadlag paths $[-r,0] \to \mathbb{R}^n$, e.g., $\mathcal{X} = C([-r,0],\mathbb{R}^n)$, $D([-r,0],\mathbb{R}^n)$, $\mathbf{L}^2([-r,0],\mathbb{R}^n)$, or a weighted $\mathbf{L}^2$ space $\mathbf{L}^2_\rho((-r,0],\mathbb{R}^n)$ so as to allow for the infinite fading memory case $r=\infty$ (cf. Mizel and Trutzer [32], Coleman and Mizel [9], [10]). In order to observe the dynamics of (I) it is convenient to define the *segment* $x_t \in \mathcal{X}$ by

$$x_t(\cdot,\omega)(s) := x(t+s,\omega), \qquad t \geq 0, \quad -r \leq s \leq 0. \tag{1}$$

---

* Research supported in part by NSF Grant DMS-8907857 and by NATO Collaborative Research Grant No. 0010/88.

This idea goes back to Krasovskii [24](pp. 126–175) in the deterministic case: $Q \equiv 0$, $\nu_i(t,\omega)$ fixed in (t,ω), and $Z_i(t) = t$, $i = 0, 1, \ldots, m$, a.s. In this case the existence of solutions and the asymptotic stability of the trajectories $x_t \in \mathcal{X} = C([-r,0],\mathbb{R}^n)$ were studied extensively by J.K. Hale and his school in the sixties (Hale [19], [20]), Krasovskii [24], El'sgol'tz [17], Bellman and Cooke [6] and others. The corresponding issues in the case $\mathcal{X} = \mathbf{L}^2([-r,0],\mathbb{R}^n)$ were studied by Delfour and Mitter [13] in the finite memory case (cf. also Corduneanu and Lakshmikantham [11] and the references therein for systems with infinite memory).

For the stochastic hereditary white-noise case ($Z_0(t) = t$, $Z_i(t)$, $Q(t)$ independent Brownian motions, $\nu_i(t,\omega)$ fixed), the existence of $(\mathcal{F}_t)_{t\geq 0}$-adapted solutions and their asymptotic stability were treated by several authors, e.g., K. Itô and M. Nisio [21], Kushner [25], Mohammed [33], [34], [36], Mizel and Trutzer [32], Mohammed, Scheutzow and Weizsäcker [40], Scheutzow [45], Kolmanovskii and Nosov [23]. Extensions of the existence results to the case of semimartingale noises Z_i, Q were discussed by Doleans-Dade [15], Métivier and Pellaumail [31], Métivier [30], Protter [42] and others.

Our present discussion will focus on results concerning almost sure asymptotic stability of the *trajectory* $(x(t), x_t) \in \mathcal{E} := \mathbb{R}^n \times \mathcal{X}$ of the stochastic hereditary system (I). In particular the following issues will be discussed:

(i) Existence of *measurable stochastic (semi-)flows* $X: \mathbb{R}^+ \times \Omega \times \mathcal{E} \to \mathcal{E}$ for (I) with the properties
 (a) If x is the solution of (I) with initial data $(v,\eta) \in \mathcal{E}$, then $X(t,\cdot,(v,\eta)) = (x(t), x_t)$ for all $t \geq 0$ a.s.
 (b) Each map $X(t,\omega,\cdot)$, $t \in \mathbb{R}^+$, a.a. $\omega \in \Omega$, is a continuous affine linear operator on $\mathcal{E}$.

(ii) A characterization of the almost sure *Lyapunov exponents*

$$\overline{\lim_{t\to\infty}} \frac{1}{t} \log \|(x(t), x_t)\|_{\mathcal{E}}$$

for a given natural norm on the state space $\mathcal{E}$; for example, if $\mathcal{X} = \mathbf{L}^2([-r,0],\mathbb{R}^n)$, one usually takes the Hilbert norm

$$\|(v,\eta)\|_{M_2}^2 := |v|^2 + \int_{-r}^{0} |\eta(s)|^2\, ds, \quad (v,\eta) \in \mathbb{R}^n \times \mathbf{L}^2([-r,0],\mathbb{R}^n), \tag{2}$$

on the classical Delfour-Mitter space $\mathcal{E} := M_2 := \mathbb{R}^n \times \mathbf{L}^2([-r,0],\mathbb{R}^n)$ (Delfour and Mitter [13]).

(iii) A study of *hyperbolicity* in (I), viz. the case of non-zero Lyapunov exponents. This is of interest for two reasons. In the linear case $(Q \equiv 0)$, hyperbolicity leads to an exponential dichotomy with a flow-invariant saddle-point splitting of $\mathcal{E}$ (see §3 A, B). When $Z_i(t) = t$ a.s.

with $\nu_i(t,\omega)$ fixed, $1 \le i \le m$, Q having stationary increments and (I) hyperbolic, it turns out that the affine hereditary equation admits a unique stationary solution (§3 C, Theorem 13).

In the following section we examine the question of the existence of a robust flow for (I).

2. Classification of Stochastic Hereditary Systems. Existence of Flows.

(A) Linear Equations Driven by White Noise:

Consider first the non-delay case $r = 0$, $\nu_i(t,\omega) = A_i(t,\omega)\delta_{\{0\}}$ with $A_i(t,\omega)$, $i = 0, 1, \ldots, m$, stationary $n \times n$-matrix-valued processes and $\delta_{\{0\}}$ the Dirac measure at 0. In this case the state space $\mathcal{E}$ may be identified with $\mathbb{R}^n$, and it is well-known that the trajectories $\{x(t): t \ge 0, x(0) = v \in \mathbb{R}^n\}$ admit a measurable flow $X: \mathbb{R}^+ \times \Omega \times \mathbb{R}^n \to \mathbb{R}^n$ such that $X(t,\cdot,v) = x(t)$ for all $t \ge 0$ a.s. and $X(t,\omega,\cdot)$ is linear (invertible) on $\mathbb{R}^n$ for a.a. $\omega \in \Omega$ and all $t \ge 0$, (Arnold [1], Leandre [26], Jacod [22]).

However, when $r > 0$ in the hereditary system (I), stochastic flows may no longer exist. To be more specific we introduce the following classification of hereditary systems:

Definition: The hereditary system (I) is said to be *regular* (with respect to the state space $\mathcal{E}$) if the family of its *trajectories*

$$\{(x(t), x_t): (x(0), x_0) = (v,\eta) \in \mathcal{E}, \quad t \ge 0\}$$

admits a (Borel $\mathbb{R}^+ \otimes \mathcal{F} \otimes$ Borel $\mathcal{E}$, Borel $\mathcal{E}$)-measurable version $X: \mathbb{R}^+ \times \Omega \times \mathcal{E} \to \mathcal{E}$ such that, for a.a. $\omega \in \Omega$ and each $t \ge 0$, the map $X(t,\omega,\cdot): \mathcal{E} \to \mathcal{E}$ is affine continuous linear. The system (I) is called *singular* (w.r.t. $\mathcal{E}$) if it is not regular (w.r.t. $\mathcal{E}$).

Unfortunately *singular hereditary linear systems do exist*, e.g., the one-dimensional stochastic linear delay equation

$$\left.\begin{aligned} dx(t) &= x(t-r)\,dW(t), \qquad t > 0 \\ (x(0), x_0) &= (v,\eta) \end{aligned}\right\}(II)$$

driven by a Wiener process W and with a *positive* delay r is singular with respect to either of the state spaces

$$\mathcal{E} := \{(v,\eta): v \in \mathbb{R}, \eta \in C([-r,0],\mathbb{R}), v = \eta(0)\} \cong C([-r,0],\mathbb{R})$$

or $\mathcal{E} := \mathbb{R} \times \mathbf{L}^2([-r,0],\mathbb{R})$. Indeed we have

Theorem 1. (Mohammed [33], [35])
Let $\mathcal{E} \cong C([-r,0],\mathbb{R})$ or $\mathbb{R} \times \mathbf{L}^2([-r,0],\mathbb{R})$ and suppose the delay r in (II) is positive. Suppose $Y: [0,r] \times \Omega \times \mathcal{E} \to \mathbb{R}$ is any (Borel $[0,r] \otimes \mathcal{F} \otimes$ Borel $\mathcal{E}$, Borel $\mathbb{R}$)-measurable version of the solution field $\{x(t): 0 \leq t \leq r, (x(0), x_0) = (v,\eta) \in \mathcal{E}\}$ to (II). Then, for a.a. $\omega \in \Omega$ and each $t \in (0,r]$, the map $Y(t,\omega,\cdot): \mathcal{E} \to \mathbb{R}$ is locally unbounded and (hence) non-linear.

The above pathological phenomenon is peculiar to the delay case $r > 0$. On the other hand when $r = 0$ we have the simple explicit solution

$$x(t,\omega,v) = ve^{\omega(t)-\frac{1}{2}t} \qquad t > 0,\ \omega \in \Omega,\ v \in \mathbb{R}. \tag{3}$$

This version of the solution is (a.s.) continuous linear in the initial state $v \in \mathcal{E} = \mathbb{R}$. The pathology in the delay case $r > 0$ is attributed to the *Gaussian nature* of the Wiener process W coupled with the *infinite-dimensionality* of the state space $\mathcal{E}$. A proof of Theorem 1 may be found in (Mohammed [33], pp. 144–147) for the case $\mathcal{E} \cong C([-r,0],\mathbb{R})$. Essentially the same proof also covers the case $\mathcal{E} = \mathbb{R} \times \mathbf{L}^2([-r,0],\mathbb{R})$.

Remarks:
(i) The conclusion of Theorem 1 imposes non-trivial limitations on the applicability of the general existence theorems for stochastic differential equations given in (Métivier and Pellaumail [31], Métivier [30], Protter [42], et al.) The hypotheses in these theorems (e.g., Theorem 6.10, pp. 74–75 in [31]) require that the random coefficients in the s.d.e. admit a.s. Lipschitz versions

$$\Omega \times \mathcal{E} \to \mathbb{R}^n$$

rather than just being random fields

$$\mathbf{L}^2(\Omega,\mathcal{E}) \to \mathbf{L}^2(\Omega,\mathbb{R}^n)$$

(e.g., as in Mohammed [33], Theorem (2.1), p. 36; cf. also Berger and Mizel [7], Weizsäcker and Winkler [49], Remark (a), p. 274). As an example the one-dimensional stochastic linear hereditary equation

$$\left.\begin{aligned} dx(t) &= \left\{\int_{-r}^{0} x(t+s)\,dW(s)\right\} dW(t), \qquad t > 0 \\ (x(0),x_0) &\in \mathcal{E} \end{aligned}\right\} (III)$$

with a Wiener process $\{W(t): t \geq -r\}$ does not appear to be covered by the existence theorems in [31], [30], [42]. This is because the coefficent in (III) is a random field

$$\begin{aligned} \mathcal{E} &\to \mathbf{L}^2(\Omega,\mathbb{R}) \\ (v,\eta) &\mapsto \int_{-r}^{0} \eta(s)\,dW(s) \end{aligned}$$

which *does not admit* measurable a.s. locally bounded or linear versions

$$\Omega \times \mathcal{E} \to \mathbb{R}$$

let alone Lipschitz ones! (For the cases $r > 0$, $\mathcal{E} \cong C([-r,0],\mathbb{R})$ or $\mathbb{R} \times \mathbf{L}^2([-r,0],\mathbb{R})$, see Theorem (8.6), p. 28 in Mohammed [33]).
(ii) The erratic behavior in Theorem 1 above suggests similar difficulties in certain types of stochastic linear partial differential equations driven by *multi-dimensional* white noise (cf. Flandoli and Schaumlöffel [18]).

Recent work by V.J. Mizel and the author shows that the conclusion of Theorem 1 also holds for the one-dimensional hereditary equation

$$\left.\begin{aligned} dx(t) &= \int_{-r}^{0} x(t+s)\,d\nu(s)\,dW(t), \qquad t>0 \\ &(x(0),x_0) \in \mathbb{R} \times \mathbf{L}^2([-r,0],\mathbb{R}) \end{aligned}\right\}(IV)$$

where W is a Wiener process and ν is a fixed finite real-valued Borel measure on $[-r,0]$ satisfying the hypotheses

$$\overline{\operatorname{supp}\nu} \subset [-r,0), \quad \liminf_{n\to\infty} \left|\int_{-r}^{0} e^{2\pi i n(s/r)}\,d\nu(s)\right| > 0.$$

On the other hand, (IV) is regular if ν has a C^1 (or even $\mathbf{L}_1^2$ density with respect to Lebesgue measure on $[-r,0]$ (Mohammed and Scheutzow [39], Theorem 4.2). An interesting problem is to classify all finite signed measures ν on $[-r,0]$ for which the hereditary equation (IV) is regular.

Needless to say we do not know a complete characterization of all processes $Z(t) = (Z_0(t), Z_1(t), \ldots, Z_m(t))$, and $\nu(t) = (\nu_0(t), \nu_1(t), \ldots, \nu_m(t))$ for which the hereditary system (I) is regular. On the other hand regularity holds for a large class of linear hereditary systems driven by white noise. Indeed the next result deals with the case: $Q(t) \equiv 0$, $Z_0(t) = t$, $Z_i(t) = W_i(t)$, $i = 1, 2, \ldots, m$ are independent one-dimensional Wiener processes;

$$\nu_0(t,\omega) = \sum_{i=1}^{N} H(0,0,\ldots,0,\overset{\substack{i^{\text{th}}\text{ place}\\ \downarrow}}{\cdot},0,\ldots,0)\delta_{\{-d_i\}} + H(0,0,\ldots,0,\underset{\substack{\uparrow\\ (N+1)^{\text{st}}\text{ place}}}{\cdot},0)\delta_{\{0\}} + h(s)\,ds$$

where $H\colon (\mathbb{R}^n)^{N+1} \times \mathbf{L}^2([-r,0],\mathbb{R}^n) \to \mathbb{R}^n$ is a fixed continuous linear map, $h\colon [-r,0] \to \mathbb{R}^{\kappa\times\kappa}$ is an $n \times n$-matrix-valued $\mathbf{L}^2$ function, ds is Lebesgue measure on $[-r,0]$; $\nu_i(t,\omega)=g_i\delta_{\{0\}}$, g_i, $i=1,2,\ldots,m$ fixed (deterministic) $n\times n$ matrices. This case corresponds to the stochastic linear functional differential system

$$\left.\begin{aligned} dx(t) &= H\big(x(t-d_1),\ldots,x(t-d_N),x(t),x_t\big)\,dt \\ &\quad+\sum_{i=1}^{m} g_i\big(x(t)\big)\,dW_i(t), \qquad t>0 \\ (x(0),x_0) &= (v,\eta)\in\mathbb{R}^n\times\mathbf{L}^2([-r,0]),\mathbb{R}^n) := M_2 \end{aligned}\right\}(V)$$

with several finite delays $0<d_1<d_2<\ldots<d_N\le r$ in the drift term and *no delays in the diffusion coefficient.* Observe that the above equation (V) is defined on the canonical complete filtered Wiener space $(\Omega,\mathcal{F},(\mathcal{F}_t)_{t\ge0},P)$ supporting the m-dimensional Brownian motion $W=(W_1,W_2,\ldots,W_m)$. More specifically, Ω denotes the space of all continuous paths $\omega:\mathbb{R}\to\mathbb{R}^m$ with $\omega(0)=0$ given the compact open topology and the Borel σ-algebra $\mathcal{F}$. For each $t\ge0$, $\mathcal{F}_t$ is the σ-algebra generated by all evaluations $\{\rho_u: u\le t\}$, $\rho_u:\Omega\to\mathbb{R}^m$,

$$\rho_u(\omega):=\omega(u) \qquad u\in\mathbb{R},\ \omega\in\Omega;$$

and P is Wiener measure on Ω.

Theorem 2. (Mohammed [37])
The hereditary system (V) is regular with respect to the state space $\mathcal{E}=M_2:=\mathbb{R}^n\times\mathbf{L}^2([-r,0],\mathbb{R}^n)$. Indeed there is a Borel measurable version $X:\mathbb{R}^+\times\Omega\times M_2\to M_2$ of the trajectory field $\{(x(t),x_t): t\in\mathbb{R}^+, (x(0),x_0)=(v,\eta)\in M_2\}$ with the following properties:

(i) For each $(v,\eta)\in M_2$, $X(t,\cdot,(v,\eta))=(x(t),x_t)$ for all $t\in\mathbb{R}^+$, a.s.

(ii) For each $t\in\mathbb{R}^+$ and $(v,\eta)\in M_2$, $X(t,\cdot,(v,\eta))$ is $\mathcal{F}_t$-measurable and belongs to $\mathbf{L}^2(\Omega,M_2;P)$.

(iii) There is a Borel set $\Omega_0\subset\Omega$ of full Wiener measure such that, for all $\omega\in\Omega_0$, the map $X(\cdot,\omega,\cdot):\mathbb{R}^+\times M_2\to M_2$ is continuous.

(iv) For each $t\in\mathbb{R}^+$ and every $\omega\in\Omega_0$, the map $X(t,\omega,\cdot): M_2\to M_2$ is continuous linear; for each $\omega\in\Omega_0$, the map $\mathbb{R}^+\ni t\mapsto X(t,\omega,\cdot)\in L(M_2)$ is measurable and locally bounded in the uniform operator norm on $L(M_2)$.

(v) For each $t\ge r$ and all $\omega\in\Omega_0$, the map $X(t,\omega,\cdot): M_2\to M_2$ is compact.

The proof of the above theorem hinges on a variational technique which reduces the problem to the solution of a random family of classical hereditary differential systems involving *no stochastic integrals*. Note also the compactness of the flow for $t\ge r$. This fact plays an important role in defining hyperbolicity for (V) and the associated exponential dichotomies in §3 (A), (B). Observe also that in (iv) of the above theorem the map $[r,\infty)\ni t\mapsto X(t,\omega,\cdot)\in L(M_2)$ is continuous for all $\omega\in\Omega_0$.

A non-linear analogue of Theorem 2 also holds under the following conditions: In (V), take $\mathcal{E} := C([-r,0],\mathbb{R}^n)$, H globally Lipschitz, $g_i:\mathbb{R}^n \to \mathbb{R}^n$ C^2 maps satisfying a Frobenius condition

$$Dg_i(v)g_j(v) = Dg_j(v)g_i(v), \qquad 1 \leq i,\, j \leq m,\, v \in \mathbb{R}^n;$$

(Mohammed [33], Theorem (2.1), Chapter (V), §2, p. 121). This latter result is proved in [33] using a non-linear variational method originally due to Sussman [47] and Doss [16] in the non-delay case $r = 0$.

(B) Linear Equations Driven by Semimartingales:

The regularity w.r.t. M_2 of a large class of linear hereditary equations of the form

$$\begin{aligned} dx(t) &= \Big\{ \int_{[-r,0]} \nu(t)(ds)x(t+s) \Big\}\, dt \\ &\quad + dN(t) \int_{-r}^{0} K(t)(s)x(t+s)\, ds + dL(t)\, x(t-) \qquad t > 0 \qquad (VI) \\ x(0) &= v \in \mathbb{R}^n, \quad x(s) = \eta(s), \quad -r < s < 0, \quad r \geq 0 \end{aligned}$$

has recently been established by Mohammed & Scheutzow [39] under the following setting:

In (VI) all processes are defined on a complete filtered probability space $(\Omega, \mathcal{F}, (\mathcal{F}_t)_{t\geq 0}, P)$ satisfying the usual conditions. Denote by $\mathcal{M}([-r,0],\mathbb{R}^{n\times n})$ the space of all $n \times n$-matrix-valued Borel measures on $[-r,0]$ (or $\mathbb{R}^{n\times n}$-valued functions of bounded variation on $[-r,0]$). Give $\mathcal{M}([-r,0],\mathbb{R}^{n\times n})$ the σ-algebra generated by all evaluations. The space $\mathbb{R}^{n\times n}$ of all $n \times n$ matrices $A = (a_{ij})_{i,j=1}^n$ is given the Euclidean norm

$$\|A\| := \Big\{ \sum_{i,j=1}^{n} a_{ij}^2 \Big\}^{1/2} \tag{4}$$

The process $\nu: \mathbb{R} \times \Omega \to \mathcal{M}([-r,0],\mathbb{R}^{n\times n})$ is measurable and $(\mathcal{F}_t)_{t\geq 0}$-adapted. Furthermore, for each $\omega \in \Omega$ and $t \geq 0$, let $\bar{\nu}(t,\omega)$ be the positive measure

$$\bar{\nu}(t,\omega)(A) := |\nu|(t,\omega)\{(A-t) \cap [-r,0]\} \tag{5}$$

for all Borel sets $A \subseteq [-r,\infty)$, with $|\nu|$ the total variation measure of ν w.r.t. the Euclidean norm on $\mathbb{R}^{n\times n}$. For each $\omega \in \Omega$ suppose the positive measure

$$\bar{\bar{\nu}}(\omega)(\cdot) := \int_0^\infty \bar{\nu}(t,\omega)(\cdot)\, dt \tag{6}$$

has a density $d\bar{\nu}(\omega)/ds$ with respect to Lebesgue measure on $[-r,\infty)$ which is locally essentially bounded. If

$$\bar{\bar{\nu}}(t,\omega)(\cdot) := \int_0^t \bar{\nu}(u,\omega)(\cdot)\,du \qquad t \geq 0, \tag{7}$$

suppose further that the map

$$\begin{aligned} [0,\infty) &\to \mathbf{L}^2([-r,0],\mathbb{R}) \\ t &\mapsto \frac{d\bar{\bar{\nu}}(t,\omega)}{ds} \mid [-r,0] \end{aligned}$$

is continuous on $[0,\infty)$ for every $\omega \in \Omega$. It is easy to see that this last condition is satisfied in the deterministic case $\nu(t,\omega) = \nu_0$, $t \geq 0$, $\omega \in \Omega$, for a fixed $\nu_0 \in \mathcal{M}([-r,0],\mathbb{R}^{n\times n})$. The process $N:\mathbb{R}\times\Omega \to \mathbb{R}^{n\times n}$ is an $(\mathcal{F}_t)_{t\geq 0}$-semimartingale and $K:\mathbb{R}\times\Omega \to C^1([-r,0],\mathbb{R}^{n\times n})$ is a measurable $(\mathcal{F}_t)_{t\geq 0}$-adapted process w.r.t. the σ-algebra generated by all evaluations on $C^1([-r,0],\mathbb{R}^{n\times n})$. For a.a. $\omega \in \Omega$, the random field $K(t,\omega)(s)$ is jointly C^1 in $(t,s) \in \mathbb{R}^+ \times [-r,0]$. The $(\mathcal{F}_t)_{t\geq 0}$-semimartingale $L:\mathbb{R}\times\Omega \to \mathbb{R}^{n\times n}$ is assumed to admit a representation $L = M + V$ where M is a *continuous* $(\mathcal{F}_t)_{t\geq 0}$-local martingale and V is an $(\mathcal{F}_t)_{t\geq 0}$-adapted process with a.a. paths right continuous and of bounded variation on compact subsets of $\mathbb{R}^+$.

We then have

Theorem 3. (Mohammed & Scheutzow[39])
Under the above hypotheses, the hereditary system (VI) is regular w.r.t. M_2. In fact its trajectory $\{(x(t),x_t): t \geq 0, (x(0),x_0) \in M_2\}$ has a measurable version $X:\mathbb{R}^+\times\Omega\times M_2 \to M_2$ satisfying assertions (i), (ii), (iv), (v) of Theorem 2 with $\Omega_0 \in \mathcal{F}$ a set of full P-measure. Also for all $\omega \in \Omega_0$ and every $(v,\eta) \in M_2$, the path $X(\cdot,\omega,(v,\eta)):\mathbb{R}^+ \to M_2$ is cadlag.

(C) A Class of Affine Equations:

Consider the affine hereditary system

$$\left.\begin{aligned} dx(t) &= \left\{\int_{[-r,0]} \mu(ds)x(t+s)\right\} dt + dQ(t), \qquad t > 0 \\ x(s) &= \eta(s) \qquad -r \leq s \leq 0. \end{aligned}\right\} (VII)$$

Here Q is an $\mathbb{R}^n$-valued semimartingale on a filtered probability space $(\Omega,\mathcal{F},(\mathcal{F}_t)_{t\geq 0},P)$ with the usual conditions, and $Q(0) = 0$. The memory is driven by a fixed $\mathbb{R}^{n\times n}$-valued Borel measure μ on $[-r,0]$. The initial condition η belongs to the Banach space $D([-r,0],\mathbb{R}^n)$ of all cadlag paths $[-r,0]\to\mathbb{R}^n$ with the supremum norm

$$\|\eta\|_\infty := \sup_{-r\le s\le 0} |\eta(s)|. \tag{8}$$

We shall often take η to be an $\mathcal{F}$-measurable random variable with values in $D([-r,0],\mathbb{R}^n)$ which is *allowed to anticipate the driving noise* Q. (See §3, C).

An essential tool in studying the Lyapunov exponents $\overline{\lim}_{t\to\infty}(1/t)\log\|x_t\|_\infty$ of (VII) is the associated homogeneous deterministic linear hereditary system

$$\left.\begin{aligned} dy(t) &= \left\{\int_{[-r,0]}\mu(ds)y(t+s)\right\}dt, \qquad t>0\\ y_0 &= \eta. \end{aligned}\right\}(VIII)$$

Using the integrated form

$$y(t)=\eta(0)+\int_0^t\int_{[-r,0]}\mu(ds)y(u+s)\,du, \qquad t\ge 0, \qquad (VIII)'$$

we define a strongly continuous semigroup $\left(\tilde{T}(t)\right)_{t\ge 0}$ on the space $D := D([-r,0],\mathbb{R}^n)$ by setting $\tilde{T}(t)\eta := y_t^{(\eta)}$, $t\ge 0$, where $y^{(\eta)}\colon[-r,\infty)\to\mathbb{R}^n$ is the unique solution of (VIII) with initial path $\eta\in D([-r,0],\mathbb{R}^n)$. Denote by $F\colon[-r,\infty)\to\mathbb{R}^{\kappa\times\kappa}$ the fundamental matrix solution of

$$\left.\begin{aligned} \dot{F}(t) &= \int_{[-r,0]}\mu(ds)F(t+s), \qquad t>0,\\ F_0 &= \Delta, \qquad \Delta(s) := \begin{cases} I & s=0\\ 0 & -r\le s<0\end{cases} \end{aligned}\right\}(IX)$$

where $I,0\in\mathbb{R}^{n\times n}$ are the identity and zero $n\times n$ matrices respectively. If we extend Q to all of $\mathbb{R}$ by setting $Q(s)=0$ for all $s\le 0$, then the unique trajectory $\{x_t^{(\eta)}: t\ge 0\}$ of the affine hereditary system (VII) is given by

$$x_t^{(\eta)}(u)=\{\tilde{T}(t)\eta\}(u)+Q_t(u)+\int_0^t \dot{F}(t-s+u)Q(s)\,ds, \quad \text{a.s.}, \qquad (X)$$

for $t\ge 0$, $u\in J:=[-r,0]$, $\eta\in D(J,\mathbb{R}^n)$ (Mohammed and Scheutzow [38], Theorem 1). Alternatively, we have

$$x_t^{(\eta)}(u)=\{\tilde{T}(t)\eta\}(u)+\int_0^t F(t-s+u)\,dQ(s) \qquad t\ge 0,\ u\in J \qquad (XI)$$

These integral representations immediately imply that (VII) is regular with a stochastic flow $X: \mathbb{R}^+ \times \Omega \times D \to D$ given by

$$X(t, \omega, \eta) = x_t^{(\eta)}(\cdot, \omega), \quad \eta \in D([-r, 0], \mathbb{R}^n), \qquad t \geq 0, \omega \in \Omega.$$

This flow has the property that each $X(t, \omega, \cdot): D \to D$ is a continuous affine linear map.

3. Lyapunov Exponents. Hyperbolicity.

(A) Linear Equations. White Noise Case

Let us go back to the setting of §2(A) and reconsider the linear hereditary system (V), viz.

$$\begin{aligned} dx(t) &= H\big(x(t-d_1), \ldots, x(t-d_N), x(t), x_t\big)\, dt \\ &\quad + \sum_{i=1}^{m} g_i\big(x(t)\big)\, dW_i(t), \qquad t > 0 \qquad (V) \\ \big(x(0), x_0\big) &= (v, \eta) \in M_2 := \mathbb{R}^n \times \mathbf{L}^2([-r, 0], \mathbb{R}^n) \end{aligned}$$

The existence of a.s. Lyapunov exponents

$$\lim_{t \to \infty} \frac{1}{t} \log \big\|\big(x(t), x_t\big)\big\|_{M_2}$$

for the above linear hereditary system was studied by Mohammed in [37]. The approach adopted in [37] is to show that the version X of the flow constructed in Theorem 2 is a multiplicative linear cocycle over the canonical Brownian shift $\theta: \mathbb{R} \times \Omega \to \Omega$ on Wiener space:

$$\theta(t, \omega)(u) := \omega(t + u) - \omega(t), \qquad u, t \in \mathbb{R}, \quad \omega \in \Omega.$$

One then uses the compactness of $X(t, \omega, \cdot): M_2 \to M_2$, $t \geq r$, together with an infinite-dimensional version of Oseledec's multiplicative ergodic theorem due to Ruelle ([44], [43]). Indeed we have

Theorem 4. *(Mohammed [37])*
There is an $\mathcal{F}$-measurable set $\hat{\Omega}$ of full P-measure such that $\theta(t, \cdot)(\hat{\Omega}) \subseteq \hat{\Omega}$ for all $t \geq 0$ and

$$X\big(t_2, \theta(t_1, \omega), \cdot\big) \circ X(t_1, \omega, \cdot) = X(t_1 + t_2, \omega, \cdot) \tag{9}$$

for all $\omega \in \hat{\Omega}$ and $t_1, t_2 \geq 0$.

The first step in the proof of the above theorem is to approximate the Brownian motion W in (V) by smooth processes $\{W^k\}_{k=1}^{\infty}$:

$$W^k(t) := k \int_t^{t+(1/k)} W(u)\, du, \qquad t \geq 0,\ k \geq 1,$$

and let $X^k: \mathbb{R}^+ \times \Omega \times M_2 \to M_2$ be the stochastic flow of the following retarded functional differential system with random coefficients:

$$\left.\begin{aligned} dx^k(t) = & \Big\{ H\big(x^k(t-d_1), \ldots, x^k(t-d_N), x^k(t), x^k_t\big) \\ & + \sum_{i=1}^{m} g_i\big(x^k(t)\big)\dot{W}_i^k(t) - \frac{1}{2}\sum_{i=1}^{m} g_i^2\big(x^k(t)\big)\Big\}\, dt, \quad t > 0 \\ & \big(x^k(0), x_0^k\big) = (v, \eta) \in M_2. \end{aligned}\right\} (V^k)$$

It can be shown that if $X: \mathbb{R}^+ \times \Omega \times M_2 \to M_2$ is the flow of (V) constructed in Theorem 2, then

$$\lim_{k\to\infty} \sup_{0 \le t \le T} \left\| X^k(t,\omega,\cdot) - X(t,\omega,\cdot) \right\|_{L(M_2)} = 0 \tag{10}$$

for a.a. $\omega \in \Omega$ and every $0 < T < \infty$ (see Theorem 2 in [37]). The above convergence actually works for all ω in a Borel set $\widehat{\Omega}$ of full Wiener measure which is invariant under $\theta(t,\cdot)$ for all $t \geq 0$. The second step in the proof of Theorem 4 is as follows. We fix $\omega \in \widehat{\Omega}$ and use uniqueness of solutions to (V^k) in order to obtain the cocyle property for (X^k, θ), viz. equation (9) with X replaced by X^k, $k \geq 1$. We then pass to the limit as $k \to \infty$ using the convergence in (10).

The a.s. Lyapunov exponents

$$\lim_{t\to\infty} \frac{1}{t} \log \left\| X\big(t,\omega,(v(\omega),\eta(\omega))\big) \right\|_{M_2}, \text{ a.a. } \omega \in \Omega, \ (v,\eta) \in \mathbf{L}^2(\Omega, M_2)$$

of the system (V) are characterized by the following Oseledec multiplicative ergodic theorem:

Theorem 5. (Mohammed [37])
Let $X: \mathbb{R}^+ \times \Omega \times M_2 \to M_2$ be the flow of (V) given in Theorem 2. Then there exist

(a) a Borel set $\Omega^* \subseteq \Omega$ such that $P(\Omega^*) = 1$ and $\theta(t,\cdot)(\Omega^*) \subseteq \Omega^*$ for all $t \geq 0$,
(b) a fixed (non-random) sequence of real numbers $\{\lambda_i\}_{i=1}^{\infty}$, and
(c) a random family $\{E_i(\omega): i \geq 1, \omega \in \Omega^*\}$ of (closed) finite-codimensional subspaces of M_2, with the following properties:
(i) If the **Lyapunov spectrum** *$\{\lambda_i\}_{i=1}^{\infty}$ is infinite, then $\lambda_{i+1} < \lambda_i$ for all $i \geq 1$ and $\lim_{i\to\infty} \lambda_i = -\infty$; otherwise there is a fixed (non-random) integer $N \geq 1$ such that $\lambda_N = -\infty < \lambda_{N-1} < \cdots < \lambda_2 < \lambda_1$;*
(ii) each map $\omega \mapsto E_i(\omega)$, $i \geq 1$, is $\mathcal{F}$-measurable into the Grassmannian of M_2 (Mañé [28], Thieullen [48]);
(iii) $E_{i+1}(\omega) \subset E_i(\omega) \subset \cdots \subset E_2(\omega) \subset E_1(\omega) = M_2$, $\quad i \geq 1, \omega \in \Omega^*$;
(iv) for each $i \geq 1$, codim $E_i(\omega)$ is fixed independently of $\omega \in \Omega^*$;

(v) *for each* $\omega \in \Omega^*$ *and* $(v,\eta) \in E_i(\omega)\backslash E_{i+1}(\omega)$,

$$\lim_{t\to\infty} \frac{1}{t} \log \|X(t,\omega,(v,\eta))\|_{M_2} = \lambda_i; \tag{11}$$

(vi) *(Top Exponent):*

$$\lambda_1 = \lim_{t\to\infty} \frac{1}{t} \log \|X(t,\omega,\cdot)\|_{L(M_2)} \quad \text{for all } \omega \in \Omega^* \tag{12}$$

(vii) *(Invariance):*

$$X(t,\omega,\cdot)(E_i(\omega)) \subseteq E_i(\theta(t,w)) \quad \text{for all } \omega \in \Omega^*,\ t \geq 0,\ i \geq 1$$

For a proof of the above result see [37] §4, pp. 106–122. The argument in [37] is based on Ruelle's discrete version of Oseledec's multiplicative ergodic theorem in Hilbert space ([44], Theorem (1.1), p. 248 and Corollary (2.2), p. 253). The following strong version of Kingman's subadditive ergodic theorem is also used to construct the shift invariant set Ω^* appearing in Theorem 5 above.

Theorem 6. (Kingman's Subadditive Ergodic Theorem)
Let $f: \mathbb{R}^+ \times \Omega \to \mathbb{R} \cup \{-\infty\}$ *be a measurable process on a complete probability space* $(\Omega, \mathcal{F}, P)$ *such that*
(i) $E \sup_{0\leq u\leq 1} f^+(u,\cdot) < \infty$, $E \sup_{0\leq u\leq 1} f^+(1-u,\theta(u,\cdot)) < \infty$;
(ii) $f(t_1+t_2,\omega) \leq f(t_1,\omega) + f(t_2,\theta(t_1,\omega))$ *for all* $t_1, t_2 \geq 0$ *and* **every** $\omega \in \Omega$.
Then there exist a set $\hat{\hat{\Omega}} \in \mathcal{F}$ *and a measurable* $\tilde{f}: \Omega \to \mathbb{R} \cup \{-\infty\}$ *with the properties:*

(a) $P(\hat{\hat{\Omega}}) = 1, \theta(t,\cdot)(\hat{\hat{\Omega}}) \subseteq \hat{\hat{\Omega}}$ *for all* $t \geq 0$;
(b) $\tilde{f}(\omega) = \tilde{f}(\theta(t,\omega))$ *for all* $\omega \in \hat{\hat{\Omega}}$ *and all* $t \geq 0$;
(c) $\tilde{f}^+ \in \mathbf{L}^1(\Omega, \mathbb{R}; P)$;
(d) $\lim_{t\to\infty} (1/t) f(t,\omega) = \tilde{f}(\omega)$ *for every* $\omega \in \hat{\hat{\Omega}}$.

If θ *is ergodic, then there exist* $f^* \in \mathbb{R} \cup \{-\infty\}$ *and* $\tilde{\tilde{\Omega}} \in \mathcal{F}$ *such that*

(a)′ $P(\tilde{\tilde{\Omega}}) = 1, \theta(t,\cdot)(\tilde{\tilde{\Omega}}) \subseteq \tilde{\tilde{\Omega}}$, $t \geq 0$;
(b)′ $\tilde{f}(\omega) = f^* = \lim_{t\to\infty} (1/t) f(t,\omega)$ *for every* $\omega \in \tilde{\tilde{\Omega}}$.

A proof of Theorem 6 is given in ([37], Lemma 7, pp. 115–117).

The non-random nature of the Lyapunov exponents $\{\lambda_i\}_{i=1}^{\infty}$ of (V) is a consequence of the fact that θ is ergodic. System (V) is said to be *hyperbolic* if $\lambda_i \neq 0$ for all $i \geq 1$. When (V) is hyperbolic the flow statisfies a *stochastic saddle-point property* (or exponential dichotomy) (cf. the deterministic case with $\mathcal{E} = C([-r,0],\mathbb{R}^n)$, $g_i \equiv 0$, $i = 1, \ldots, m$, in Hale [20], Theorem 4.1, p. 181).

Theorem 7. (Mohammed [37])
Suppose the hereditary system (V) is hyperbolic. Then there exist
(a) *a set $\tilde{\Omega}^* \in \mathcal{F}$ such that $\theta(t,\cdot)(\tilde{\Omega}^*) = \tilde{\Omega}^*$ for all $t \in \mathbb{R}$ and $P(\tilde{\Omega}^*) = 1$, and*
(b) *a measurable splitting*

$$M_2 = \mathcal{U}(\omega) \oplus \mathcal{S}(\omega) \qquad \omega \in \tilde{\Omega}^*$$

with the following properties:
(i) $\mathcal{U}(\omega)$, $\mathcal{S}(\omega)$, $\omega \in \tilde{\Omega}^*$, are closed linear subspaces of M_2, $\dim\mathcal{U}(\omega)$ is finite and fixed independently of $\omega \in \tilde{\Omega}^*$.
(ii) The maps $\omega \mapsto \mathcal{U}(\omega)$, $\omega \mapsto \mathcal{S}(\omega)$ are $\mathcal{F}$-measurable into the Grassmannian of M_2.
(iii) For each $\omega \in \tilde{\Omega}^*$ and $(v,\eta) \in \mathcal{U}(\omega)$ there exists $\tau_1 = \tau_1(\omega, v, \eta) > 0$ and a positive δ_1, independent of (ω, v, η) such that

$$\|X(t,\omega,(v,\eta))\|_{M_2} \geq \|(v,\eta)\|_{M_2} e^{\delta_1 t}, \qquad t \geq \tau_1.$$

(iv) For each $\omega \in \tilde{\Omega}^*$ and $(v,\eta) \in \mathcal{S}(\omega)$ there exists $\tau_2 = \tau_2(\omega, v, \eta) > 0$ and a positive δ_2, independent of (ω, v, η) such that

$$\|X(t,\omega,(v,\eta))\|_{M_2} \leq \|(v,\eta)\|_{M_2} e^{-\delta_2 t}, \qquad t \geq \tau_2.$$

(v) For each $t \geq 0$ and $\omega \in \tilde{\Omega}^*$,

$$X(t,\omega,\cdot)\big(\mathcal{U}(\omega)\big) = \mathcal{U}\big(\theta(t,\omega)\big),$$
$$X(t,\omega,\cdot)\big(\mathcal{S}(\omega)\big) \subseteq \mathcal{S}\big(\theta(t,\omega)\big).$$

In particular, the restriction $X(t,\omega,\cdot) \mid \mathcal{U}(\omega){:}\mathcal{U}(\omega) \to \mathcal{U}\big(\theta(t,\omega)\big)$ is a linear homeomorphism onto.

(B) Linear Equations: Semimartingale Noise

We use the general setting and hypotheses in §2(B). The object of this section is to extend Theorems 4, 5, 7 to cover the hereditary system (VI) too:

$$\begin{aligned} dx(t) &= \Big\{ \int_{[-r,0]} \nu(t)(ds)x(t+s) \Big\}\, dt \\ &\quad + dN(t) \int_{-r}^{0} K(t)(s)x(t+s)\, ds + dL(t)\, x(t-) \qquad t > 0 \qquad (VI) \\ x(0) &= v \in \mathbb{R}^n, \quad x(s) = \eta(s), \quad -r < s < 0, \quad r \geq 0 \end{aligned}$$

In order to develop a multiplicative ergodic theory for (VI) we need the following set of hypotheses, which are taken from Mohammed and Scheutzow [39]:

Hypotheses (C):

(i) The processes ν, K are stationary ergodic in the sense that there is a measurable ergodic P-preserving flow $\theta: \mathbb{R} \times \Omega \to \Omega$ such that for each $t \geq 0$, $\theta(t,\cdot)$ is $(\mathcal{F}_t, \mathcal{F}_0)$-measurable and

$$\nu(t,\omega) = \nu\big(0,\theta(t,\omega)\big), \qquad t \in \mathbb{R},\ \omega \in \Omega \tag{13}$$
$$K(t,\omega) = K\big(0,\theta(t,\omega)\big), \qquad t \in \mathbb{R},\ \omega \in \Omega \tag{14}$$

(ii) The processes N, L, M have jointly stationary ergodic increments:

$$\left.\begin{aligned} N(t+h,\omega) - N(t,\omega) &= N\big(h,\theta(t,\omega)\big), & t \in \mathbb{R},\ \omega \in \Omega \\ L(t+h,\omega) - L(t,\omega) &= L\big(h,\theta(t,\omega)\big), & t \in \mathbb{R},\ \omega \in \Omega \\ M(t+h,\omega) - M(t,\omega) &= M\big(h,\theta(t,\omega)\big), & t \in \mathbb{R},\ \omega \in \Omega. \end{aligned}\right\} \tag{15}$$

Semimartingales satisfying Hypothesis (C)(ii) were studied by J. de Sam Lazaro and P.A. Meyer [12], Çinlar, Jacod, Protter and Sharpe [8], Protter [41]. It follows from Hypothesis (C)(ii) that N and L have jointly stationary increments. Conversely, if N and L have jointly stationary increments, one can arrange for (C)(ii) to hold on a suitable probability path space. (See Protter [41], Theorem (2.2), de Sam Lazaro and Meyer [12], Mohammed and Scheutzow [39].)

In view of Theorem 3 we know that equation (VI) is regular w.r.t. M_2 with a measurable flow $X: \mathbb{R}^+ \times \Omega \times M_2 \to M_2$. It will turn out that this flow satisfies Theorems 4, 5 and 7. This is achieved via a construction in [39] based on the following consequence of Hypothesis (C)(ii):

Theorem 8. (Mohammed & Scheutzow [39])
Suppose M satisfies Hypothesis (C)(ii). Then there are an $(\mathcal{F}_t)_{t\geq 0}$-adapted version $\varphi: \mathbb{R}^+ \times \Omega \to \mathbb{R}^{n\times n}$ of the solution to the matrix equation

$$\left.\begin{aligned} d\varphi(t) &= dM(t)\,\varphi(t) \qquad t > 0 \\ \varphi(0) &= I \in \mathbb{R}^{n\times n} \end{aligned}\right\} \tag{XII}$$

and a set $\Omega_1 \in \mathcal{F}$ such that
(i) $P(\Omega_1) = 1$;
(ii) $\theta(t,\cdot)(\Omega_1) \subseteq \Omega_1$ for all $t \geq 0$;
(iii) $\varphi(t_1 + t_2,\omega) = \varphi(t_2,\theta(t_1,\omega))\varphi(t_1,\omega)$ for all $t_1, t_2 \in \mathbb{R}^+$ and every $\omega \in \Omega_1$;

(iv) $\varphi(\cdot,\omega)$ *is continuous for every* $\omega \in \Omega_1$.

A proof of Theorem 8 is given in [39]. The proof is based on a double approximation argument whereby (XII) is replaced by the families of s.d.e.'s

$$\left.\begin{aligned} d\varphi_m^k(t) &= \dot{M}^k(t) f_m(\varphi_m^k(t))\,dt - \tfrac{1}{2} d\langle M\rangle(t) f_m(\varphi_m^k(t)),\ t \in \mathbb{R}^+ \\ \varphi_m^k(0) &= I \in \mathbb{R}^{n\times n}, \end{aligned}\right\} (XII)_m^k$$

$$\left.\begin{aligned} d\varphi_m(t) &= \circ dM(t) f_m(\varphi_m(t)) - \tfrac{1}{2} d\langle M\rangle(t) f_m(\varphi_m(t)),\ t \in \mathbb{R}^+ \\ \varphi_m(0) &= I \in \mathbb{R}^{n\times n}, \end{aligned}\right\} (XII)_m$$

where the $f_m: \mathbb{R}^{n\times n} \to \mathbb{R}^{n\times n}$ are C^∞-bounded approximations of the identity map $\mathrm{id}_{\mathbb{R}^{n\times n}}: \mathbb{R}^{n\times n} \to \mathbb{R}^{n\times n}$ such that $f_m(A) = A$ whenever $\|A\| \le m$; the M^k are smooth $(\mathcal{F}_t)_{t\ge 0}$-adapted mollifiers of M given by

$$M^k(t) := k \int_{t-\frac{1}{k}}^t M(u)\,du, \qquad t \ge 0,$$

and $\circ dM(t)$ denotes Stratonovich differential.

Using results of Mackevičius [27] on S^p-stability of s.d.e.'s, it is shown in ([39], Theorem (3.1)) that the solutions $\varphi^k: \mathbb{R}^+ \times \Omega \to \mathbb{R}^{n\times n}$ of

$$\left.\begin{aligned} d\varphi^k(t) &= \dot{M}^k(t)\varphi^k(t)\,dt - \tfrac{1}{2} d\langle M\rangle(t)\varphi^k(t), \qquad t \in \mathbb{R}^+ \\ \varphi^k(0) &= I \in \mathbb{R}^{n\times n} \end{aligned}\right\} (XII)^k$$

have a subsequence $\{\varphi^{k'}\}_{k'=1}^\infty$ which converges a.s. uniformly on compacta to φ. The multiplicative cocycle property for (φ,θ) follows immediately from the corresponding one for $(\varphi^{k'},\theta)$, $k' \ge 1$.

Under Hypotheses (C) one gets the cocycle property (9) for (X,θ), i.e., Theorem 4 holds true for the linear hereditary system (VI), (Mohammed & Scheutzow [39], Theorem (4.2)(vii)). A key point in proving this fact is to observe that the linear system (VI) is equivalent to the following random family of hereditary linear integral equations

$$\begin{aligned} x(t) = \varphi(t)\Big[v - \int_0^t Z(u)\Big\{&K(u)(0)x(u) - K(u)(-r)x(u-r) \\ &+ \int_{u-r}^u \frac{\partial}{\partial u}(K(u)(s-u))x(s)\,ds\Big\}\,du\Big] \\ &+ \varphi(t)Z(t)\int_{-r}^0 K(t)(s)x(t+s)\,ds \\ + \int_0^t \varphi(t-u,\theta(u,\cdot)) &\int_{[-r,0]} \nu(u)(ds)x(u+s)\,du \qquad (XIII) \\ &+ \int_0^t \varphi(t-u,\theta(u,\cdot))\,dV(u)x(u-) \\ - \int_0^t \varphi(t-u,\theta(u,\cdot))\,d[M,N](u) &\int_{-r}^0 K(u)(s)x(u+s)\,ds, \qquad t \ge 0, \\ x(t) = \eta(t) \quad \text{a.e. } &t \in [-r,0), \end{aligned}$$

where Z is a suitably chosen version of $\int_0^t \varphi^{-1}(u)\,dN(u)$ (Theorem (3.2) in [39]) and $[M,N]$ denotes a version of the $\mathbb{R}^{n\times n}$-valued mutual variation of M and N, viz.

$$[M,N]_{ij} := \sum_{m=1}^{n} [M_{im}, N_{mj}], \quad M = (M_{ij})_{i,j=1}^{n}, \quad N = (N_{ij})_{i,j=1}^{n}.$$

(See Lemma (3.1) in [39].) Observe that the above integral equation *has no stochastic integrals*. This fact contributes to the regularity of the hereditary equation (VI). The cocycle property (9) now follows from the uniqueness of the solution to (XIII) (see the proof of Theorem (4.2) in [39]).

The existence of a discrete non-random Lyapunov spectrum $\{\lambda_i\}_{i=1}^{\infty}$ for the hereditary equation (VI) (cf. Theorem 5) is proved via Ruelle-Oseledec multiplicative ergodic theorem which requires the following integrability property

$$E \sup_{0\le t_1, t_2 \le r} \log^+ \|X(t_1, \theta(t_2,\cdot),\cdot)\|_{L(M_2)} < \infty. \tag{16}$$

In [39] the above integrability property is established under the following set of hypotheses on ν, K, N, L:

Hypotheses (I):

(i) The random variables

$$\sup_{-r\le s\le 2r} \left|\frac{d\overline{\nu}(\cdot)(s)}{ds}\right|^3, \qquad \sup_{0\le t\le 2r, -r\le s\le 0} \|K(t,\cdot)(s)\|^4,$$

$$\sup_{0\le t\le 2r, -r\le s\le 0} \|\frac{\partial}{\partial t}K(t,\cdot)(s)\|^4, \qquad \sup_{0\le t\le 2r, -r\le s\le 0} \|\frac{\partial}{\partial s}K(t,\cdot)(s)\|^4,$$

$$\{|V|(2r,\cdot)\}^4, \quad \{\langle M_{ij}\rangle(2r,\cdot)\}^4, \qquad 1\le i,j\le n,$$

are all integrable. Here $|V|$ is the total variation of V w.r.t. the Euclidean norm $\|\cdot\|$ on $\mathbb{R}^{n\times n}$.

(ii) Let N be of the form $N = N^0 + V^0$ where the local $(\mathcal{F}_t)_{t\ge 0}$-martingale $N^0 = (N_{ij}^0)_{i,j=1}^n$ and the bounded variation process $V^0 = (V_{ij}^0)_{i,j=1}^n$ are such that the random variables

$$\{[N_{ij}^0](2r,\cdot)\}^4, \quad \{[V_{ij}^0](2r,\cdot)\}^4 = \Big(\sum_{0\le s\le 2r} |\Delta V_{ij}^0(s)|^2\Big)^4,$$

$$\{|V_{ij}^0|(2r,\cdot)\}^8, \qquad i,j = 1,2,\ldots,n$$

are integrable. Note that $\Delta V_{ij}^0(s)$ is the jump of V_{ij}^0 at s and $|V_{ij}^0|(2r,\cdot)$ is the total variation of V_{ij}^0 over $[0,2r]$.

(iii) There is a *non-random* time $t_0 > 0$ such that

$$|\langle M_{ij}, M_{kl}\rangle|(t_0, \cdot) \in \mathbf{L}^\infty(\Omega, \mathbb{R}), \qquad i, j, k, l = 1, 2, \ldots, n.$$

The integrability property (16) is a consequence of

$$E \log^+ \sup_{0 \le t_1, t_2 \le r, \|(v,\eta)\| \le 1} |x(t_1, \theta(t_2, \cdot), (v, \eta))| < \infty. \tag{17}$$

The proof of the latter property involves a lengthy argument based on establishing the existence of suitable higher order moments for the coefficients on the right hand side of the random integral equation (XIII). (See Lemmas (5.1), (5.2), (5.3), (5.4), (5.5), (5.6) in [39].)

Since θ is ergodic, the multiplicative ergodic theorem (Ruelle [44]) now gives a fixed discrete set of Lyapunov exponents for the linear system (VI). In fact we have

Theorem 9. (Mohammed and Scheutzow [39])
Under Hypotheses (C) & (I), the statements of Theorems 5 and 7 hold true for the linear hereditary system (VI).

Note that the Lyapunov spectrum of (VI) does not change if one uses the state space $\mathcal{E} = D([-r, 0], \mathbb{R}^n)$ with the supremum norm $\|\cdot\|_\infty$ and drops the hypothesis of the $\mathbf{L}^2$-continuity of $t \mapsto (d\bar{\bar{\nu}}(t, \omega)/ds)|[-r, 0](\omega \in \Omega)$ referred to in §2(B). (See the remark following Theorem (5.3) in [39].)

(C) Affine Systems. Hyperbolicity and Stationary Solutions

1pc Here we consider the affine hereditary system (VII) under the setting and hypotheses of §2(C):

$$\left.\begin{aligned} dx(t) &= \Big\{ \int_{[-r,0]} \mu(ds) x(t+s) \Big\} dt + dQ(t), \qquad t > 0 \\ x(s) &= \eta(s) \qquad -r \le s \le 0. \end{aligned}\right\} (VII)$$

In order to study the Lyapunov spectrum of the affine system (VII) we recall the following classical results of J.K. Hale for the homogeneous $(Q \equiv 0)$ deterministic system (VIII):

$$\left.\begin{aligned} dy(t) &= \int_{[-r,0]} \mu(ds) y(t+s)\, dt, \qquad t > 0 \\ y_0 &= \eta \in D([-r, 0], \mathbb{R}^n). \end{aligned}\right\} (VIII)$$

Recall that $\tilde{T}(t): D([-r, 0], \mathbb{R}^n) \to D([-r, 0], \mathbb{R}^n)$, $t \ge 0$, is the strongly continuous semigroup given by the trajectories of (VIII). Consider the complexification

$$\tilde{T}_{\mathbb{C}}(t)(\eta) := \tilde{T}(t)(\operatorname{Re}\eta) + i\tilde{T}(t)(\operatorname{Im}\eta), \qquad \eta \in D([-r, 0], \mathbb{C}^n),$$

of $\tilde{T}(t)$ and its restriction $T_{\mathbb{C}}(t)$ to the space of continuous maps $C([-r,0], \mathbb{C}^n)$. Note that $T_{\mathbb{C}}(t)$ is simply the complexification of

$$T(t) := \tilde{T}(t)|C([-r,0],\mathbb{R}^n).$$

Denote by A the infinitesimal generator of the strongly continuous semigroup $(T_{\mathbb{C}}(t))_{t\geq 0}$. Then the spectrum, $\sigma(A)$, of A is discrete and consists entirely of eigenvalues with real parts bounded above (Hale [20], pp. 168–170). Indeed $\sigma(A)$ coincides with the complex roots λ of the characteristic equation

$$\det\left[\lambda I - \int_{[-r,0]} e^{\lambda s}\mu(ds)\right] = 0 \tag{18}$$

(Hale [20], pp. 168–170). It follows from the above equation that $\lambda \in \sigma(A)$ iff $\bar{\lambda} \in \sigma(A)$; and for every $\beta \in \mathbb{R}$ the sum of the generalized eigenspaces corresponding to all λ's such that $\operatorname{Re}\lambda \geq \beta$ is finite-dimensional ([20], p. 168). If $\mathbb{R}^{n^*}$ stands for the space of all n-row vectors, we shall let A^* be the formal adjoint of A in $C^* := C([0,r],\mathbb{R}^{n^*})$ with respect to the continuous bilinear form $(\cdot,\cdot): C^* \times D \to \mathbb{R}$,

$$(\psi,\varphi) := \psi(0)\varphi(0) + \int_{[-r,0]}\int_s^0 \psi(\xi - s)\mu(ds)\varphi(\xi)\, d\xi, \tag{19}$$

$\psi \in C^*$, $\varphi \in D$. Then $\sigma(A^*) = \sigma(A)$ ([20], p. 169). For a given finite set Λ of conjugate pairs of eigenvalues of A, denote by $E_\Lambda(E_\Lambda^*)$ the sum of the corresponding real generalized eigenspaces of $A(A^*$, resp.) corresponding to the eigenvalues in Λ ([38]). Then $E_\Lambda(E_\Lambda^*)$ is a finite-dimensional real subspace of $C(C^*$, resp.) Pick bases $\{\varphi_i\}_{i=1}^d$, $\{\psi_i\}_{i=1}^d$ of E_Λ, E_Λ^* such that $(\psi_i,\varphi_j) = \delta_{ij}$, $1 \leq i,j \leq d = \dim E_\Lambda = \dim E_\Lambda^*$. Let $B = (B_{ij})_{i,j=1}^d$ be the $d \times d$ matrix representation of $A \mid E_\Lambda$ with respect to $\{\varphi_i\}_{i=1}^d$. The space D admits a $(\tilde{T}(t))_{t\geq 0}$-invariant topological splitting

$$D = E_\Lambda \oplus E'_\Lambda$$

where $E'_\Lambda := \{\varphi : \varphi \in D, (\psi,\varphi) = 0 \text{ for all } \psi \in E_\Lambda^*\}$ ([38]). If $\eta \in D$, we let η^{E_Λ}, $\eta^{E'_\Lambda}$ denote its projections on E_Λ, E'_Λ, respectively. Applying these projections to both sides of (X) and writing $E = E_\Lambda$, $E' = E'_\Lambda$ we get

$$\left.\begin{aligned} x_t^E &= \tilde{T}(t)(\eta^E) + Q_t^E + \int_0^t \dot{F}(t-s+\cdot)^E Q(s)\,ds \\ x_t^{E'} &= \tilde{T}(t)(\eta^{E'}) + Q_t^{E'} + \int_0^t \dot{F}(t-s+\cdot)^{E'} Q(s)\,ds, \end{aligned}\right\}(XIV)$$

$t \geq 0$, $\eta \in D$ ([38], Theorem 3).

Define the d-dimensional stochastic process

$$y(t) = (\Psi, x_t) := \begin{bmatrix} (\psi_1, x_t) \\ (\psi_2, x_t) \\ \vdots \\ (\psi_d, x_t) \end{bmatrix} \in \mathbb{R}^d,$$

with $\Psi := \begin{bmatrix} \psi_1 \\ \psi_2 \\ \vdots \\ \psi_d \end{bmatrix}$. Then one gets

$$x_t^E = \sum_{j=1}^{d} \varphi_j Y_j(t) := \Phi Y(t), \qquad \Phi := (\varphi_1, \ldots, \varphi_d),\, t \geq 0 \tag{20}$$

$$dY(t) = BY(t)\,dt + \Psi(0)\,dQ(t), \qquad t > 0 \tag{XV}$$

$$Y(t) = e^{tB}(\Psi, \eta) + \int_0^t Be^{(t-s)B}\Psi(0)Q(s)\,ds + \Psi(0)Q(t),\, t \geq 0 \tag{21}$$

(Mohammed and Scheutzow [38], Theorem 6).

By extending the estimate on the complementary subspace in ([20], Theorem 4.1, p. 181) to cover all cadlag initial paths (Mohammed and Scheutzow [38], Theorem 4), one gets:

Theorem 10.
For each $\eta \in D$, $\lambda(\eta) := \lim_{t\to\infty} (1/t) \log \|\tilde{T}(t)\eta\|_\infty$ exists and the set of Lyapunov exponents $\{\lambda(\eta): \eta \in D\}$ of the homogeneous system (VIII) coincides with the set $\{Re\, \lambda: \lambda \in \sigma(A)\}$ together with possibly $-\infty$.

We now give an Oseledec theorem which characterizes the a.s. Lyapunov exponents of the affine hereditary system (VII). The proof of the following theorem may be found in (Mohammed and Scheutzow [38] §4, Scheutzow [46], C13, Theorem 1, p. 160–161).

Theorem 11.
Let $\beta_1 > \beta_2 > \beta_3 > \ldots$ be an ordering of the real parts of all eigenvalues in $\sigma(A)$. Fix $m \geq 1$ and let $E = E_\Lambda$ where $\Lambda = \{\beta_i\}_{i=1}^m$. Define Φ, B, Ψ, E' as before. Let $\beta < \beta_m$ and assume that $|Q(t)| = o(e^{(\beta+\epsilon)t})$ for all $\epsilon > 0$ as $t \to \infty$ a.s. Let Y^* stand for the d-dimensional process ($d = \dim E$)

$$Y^*(t) = -\int_t^\infty \mathrm{B}e^{(t-s)B}\Psi(0)Q(s)\,ds + \Psi(0)Q(t). \tag{22}$$

For each $1 \le j \le m$ suppose E_j is the sum of generalized subspaces corresponding to the eigenvalues with real parts $\{\beta_i\}_{i=1}^j$. Assume that E_j' is the complementary subspace to E_j for $1 \le j \le m$. Take $E_0 = \{0\}$. Then, for a.a. $\omega \in \Omega$, one has

$$\lim_{t\to\infty} \frac{1}{t} \log \|x_t(\omega)\|_\infty = \beta_j \quad \text{if } x_0(\omega) \in \Phi Y^*(0,\omega) + E_{j-1}' \backslash E_j',\ 1 \le j \le m, \tag{23}$$

and

$$\overline{\lim_{t\to\infty}} \frac{1}{t} \log \|x_t(\omega)\|_\infty \le \beta \quad \text{if } x_0(\omega) \in \Phi Y^*(0,\omega) + E_m'. \tag{24}$$

The key to the proof of the above theorem is to identify the Lyapunov exponents of the projection $\{x_t^E : t \ge 0\}$ with those of $\{Y(t): t \ge 0\}$ and then observe that

$$Y(t) = e^{tB}(Y(0) - Y^*(0)) + Y^*(t), \qquad t \ge 0.$$

Sufficient conditions for equality in (24) are given in the following theorem. Note here that one does not require $Q(t)$ to be zero for $t < 0$.

Theorem 12. (Mohammed and Scheutzow [38])
Assume all the conditions and notations of Theorem 11. Suppose also that $|Q(t)| = o(e^{-(\beta+\varepsilon)|t|})$ for some $\varepsilon > 0$ as $t \to -\infty$ a.s. Let E^β be the sum of generalized eigenspaces of A corresponding to all eigenvalues with real parts greater than or equal to β. Define the process $Z^*(t) \in (E^\beta)'$ by

$$Z^*(t) = \int_{-\infty}^t \dot{F}(t - s + \cdot)^{(E^\beta)'} Q(s)\, ds + Q_t^{(E^\beta)'}, \qquad t \ge 0. \tag{25}$$

Let $\omega \in \Omega$ be such that $\overline{\lim_{t\to\infty}} (1/t) \log \inf_{v \in E^\beta} \|Z^*(t,\omega,v)\| \ge \beta$ and suppose that $x_0(\omega) \in \Phi Y^*(0,\omega) + E_m'$. Then

$$\overline{\lim_{t\to\infty}} \frac{1}{t} \log \|x_t(\omega)\|_\infty = \beta.$$

Remarks

(i) Under the conditions of Theorem 12 the statements (23) and (24) can be modified by replacing $\Phi Y^*(0,\omega)$ with $\eta(\omega) := Z^*(0,\omega) + \Phi Y^*(0,\omega)$, which is independent of the choice of m.

(ii) Y^* and Z^* may also be represented as

$$Y^*(t) = -\int_t^\infty \mathrm{B} e^{(t-s)B} \Psi(0)(Q(s) - Q(t))\, ds, \qquad t \ge 0, \tag{26}$$

$$Z^*(t) = \int_{-\infty}^t \tilde{T}(t-s) \Delta^{(E^\beta)'} dQ(s), \qquad t \ge 0. \tag{27}$$

When Q has stationary increments, these representations imply that Y^* and Z^* are stationary processes. In fact Y^* is the only stationary solution of the s.o.d.e. (XV) (Mohammed and Scheutzow [38]).

We now consider the *hyperbolic case* when $\operatorname{Re}\lambda \neq 0$ for all $\lambda \in \sigma(A)$. In this case, the following result (Mohammed and Scheutzow [38], Theorem 20) establishes the existence of a unique stationary solution for the affine hereditary system (VII).

Theorem 13.
Suppose that Q is cadlag and has stationary increments. Assume that the characteristic equation

$$\det\left(\lambda I - \int_{[-r,0]} e^{\lambda s}\mu(ds)\right) = 0 \tag{18}$$

has no roots on the imaginary axis; i.e. the homogeneous equation (VIII) has no zero Lyapunov exponents. Suppose also that

$$\overline{\lim_{t\to\pm\infty}} \frac{1}{|t|}\log|Q(t)| < |\operatorname{Re}\lambda| \qquad \text{a.s.}$$

for all characteristic roots λ of (18). Then there is a unique D-valued random variable η such that the trajectory $\{x_t^{(\eta)}: t \geq 0\}$ of (VII) is a D-valued stationary process. The random variable η is measurable with respect to the σ-algebra generated by $\{Q(t): t \in \mathbb{R}\}$.

If E is the sum of all generalized eigenspaces of A corresponding to all $\lambda \in \sigma(A)$ with $\operatorname{Re}\lambda > 0$, then in Theorem 13, the projection $\eta^E(\eta^{E'})$ is measurable with respect to the σ-algebra generated by $Q(t)$, $t \geq 0$ ($Q(t)$, $t \leq 0$, respectively). (See Mohammed and Scheutzow [38], Theorem 20). Furthermore if Q has *independent* increments (e.g., Q is Brownian motion or a Poisson process), then the projections ${x_t^{(\eta)}}^E$, ${x_t^{(\eta)}}^{E'}$, $t \geq 0$, are stationary and indpendent processes.

We conclude this section by discussing *p-th moment Lyapunov* exponents

$$\overline{\lim_{t\to\infty}} \frac{1}{t}\log E\|x_t^{(\eta)}\|_\infty^p, \qquad p \geq 1, \tag{28}$$

of (VII). The following result is proved in ([38], Remark (iii) following Theorem 21) by looking at the moment exponents of the projections ${x_t^{(\eta)}}^E$, ${x_t^{(\eta)}}^{E'}$, where E is the sum of the generalized eigenspaces corresponding to all eigenvalues $\lambda \in \sigma(A)$ with the largest real part β_1.

Theorem 14.
Let β_1 be the top a.s. Lyapunov exponent of (VII) and fix $p \geq 1$. Assume that $Q(t) \in \mathbf{L}^p(\Omega, \mathbb{R}^n)$ for all $t \geq 0$, $|Q(t)| = o(e^{(\beta_1-\epsilon)t})$ a.s. as $t \to \infty$ for some $\epsilon > 0$ and $\overline{\lim_{t\to\infty}}(1/t)\log E|Q(t)|^p < p\beta_1$. If $Y^*(0)$ is not a.s. constant, then

$$\lim_{t\to\infty}\frac{1}{t}\log E\|x_t^{(\eta)}\|_\infty^p = \lim_{t\to\infty}\frac{1}{t}\log E\|x_t^{(\eta)} - Ex_t^{(\eta)}\|_\infty^p = p\beta_1,$$

for all $\eta \in D$.

Under the mild non-degeneracy condition that $Y^*(0)$ is not a.s. constant, the above theorem asserts the existence of *only* one p-th moment exponent which is independent of all *random* (possibly *anticipating*) initial conditions in D. This result is in agreement with the affine linear finite-dimensional non-delay case ($r = 0$) (Arnold, Oeljeklaus and Pardoux [3], Baxendale [5], Arnold, Kliemann and Oeljeklaus [2]).

Note also the following interesting fact in connection with Theorem 14. The affine hereditary system (VII) may be viewed as a *finite-dimensional* stochastic perturbation of the *infinitely degenerate* deterministic homogeneous system (VIII) with *countably many* Lyapunov exponents. However, these finite-dimensional perturbations provide noise that is generically rich enough to account for a *single* moment Lyapunov exponent in the affine system (VII).

Remark:
More work needs to be done in order to characterize p-th moment exponents for general linear hereditary systems (I) with $Q \equiv 0$. In the white noise case with an asymptotically stable linear drift and a small diffusion, estimates on the mean square moment exponent

$$\overline{\lim_{t\to\infty}}\frac{1}{t}\log E\|x_t\|_\infty^2$$

may be found in (Mohammed, Scheutzow and Weizsäcker [40], Mohammed [34], [33], Theorems (4.2) & (4.3), pp. 208–222). Similar estimates in a rather special case with a small discrete delay appear in (Mao [29]).

4. Examples. Upper Bounds on the Top Exponent.

The examples in this section are all one-dimensional and linear. Regularity of the equations is established and estimates on the top a.s. Lyapunov exponent λ_1 are given. Details of the computations are incorporated in ongoing joint work of the author with M. Scheutzow and will appear elsewhere.

Example 1: (A Linear Delay Equation with Poisson Noise)

Consider the one-dimensional linear delay equation

$$\left.\begin{aligned} dx(t) &= x((t-1)-)\,dN(t), \qquad t>0 \\ x_0 &\in D = D([-r,0],\mathbb{R}) \end{aligned}\right\} (XVI)$$

The process $N(t) \in \mathbb{R}$ is a Poisson process with i.i.d. inter-arrival times $\{T_i\}_{i=1}^{\infty}$ which are exponentially distributed with the same parameter μ. The jumps $\{Y_i\}_{i=1}^{\infty}$ of N are i.i.d. and independent of all the T_i's. Writing

$$j(t) := \sup\Big\{j \geq 0 : \sum_{i=1}^{j} T_i \leq t\Big\},$$

and

$$N(t) = \sum_{i=1}^{j(t)} Y_i$$

it is easy to see that (XVI) can be solved a.s. in steps giving

$$x^{(\eta)}(t) = \eta(0) + \sum_{i=1}^{j(t)} Y_i x\Big(\big(\sum_{j=1}^{i} T_j - 1\big) - \Big) \quad \text{a.s.} \tag{29}$$

Observe that $\{x_t : t \geq 0\}$ is a Markov process in the state space D (with the supremum norm $\|\cdot\|_\infty$). Furthermore the above relation implies that (XVI) is regular in D; i.e. it admits a measurable flow $X: \mathbb{R}^+ \times \Omega \times D \to D$ with $X(t,\omega,\cdot)$ continuous linear for all $t \geq 0$ and a.a. $\omega \in \Omega$ (cf. the singular equation (II) in §2(A)).

The a.s. Lyapunov spectrum of (XVI) may be characterized directly (without appealing to the Oseledec theorem) by interpolating between the sequence of random times:

$$\tau_0(\omega) := 0,$$

$$\tau_1(\omega) := \inf\Big\{n \geq 1 : \sum_{j=1}^{k} T_j \notin [n-1, n] \quad \text{for all} \quad k \geq 1\Big\},$$

$$\tau_{i+1}(\omega) := \inf\Big\{n > \tau_i(\omega) : \sum_{j=1}^{k} T_j \notin [n-1, n] \quad \text{for all} \quad k \geq 1\Big\}, \qquad i \geq 1.$$

(For details see Scheutzow [46], pp. 162–166.)

Theorem 15. (Scheutzow [46])
Let $\xi \in D$ stand for the constant path $\xi(s) = 1$ for all $s \in [-1,0]$. Suppose $E \log \|X(\tau_1(\cdot),\cdot,\xi)\| \leq \infty$ (possibly $= -\infty$). Then the a.s. Lyapunov spectrum

$$\lambda(\eta) := \lim_{t\to\infty} \frac{1}{t} \log \|X(t,\omega,\eta)\|_\infty, \qquad \eta \in D,\, \omega \in \Omega$$

of (XVI) is $\{-\infty, \lambda_1\}$ where

$$\lambda_1 = \frac{1}{E\tau_1} \cdot E \log \|X(\tau_1(\cdot),\cdot,\xi)\|_\infty.$$

In fact,

$$\lim_{t\to\infty} \frac{1}{t} \log \|X(t,\omega,\eta)\|_\infty = \begin{cases} \lambda_1 & \eta \notin \operatorname{Ker} X(\tau_1(\omega),\omega,\cdot) \\ -\infty & \eta \in \operatorname{Ker} X(\tau_1(\omega),\omega,\cdot) \end{cases}$$

If N has J jumps in $[0,\tau_1]$ and $P(\alpha \leq |Y_1 + 1| \leq \beta) = 1$ for some positive α, β, then

$$\frac{EJ \log \alpha}{E\tau_1} \leq \lambda_1 \leq \frac{EJ \log \beta}{E\tau_1}.$$

If $P(Y_1 = -1) > 0$, then $\lambda_1 = -\infty$.

The computations underlying the proof of the above theorem also work for the one-dimensional hereditary equation

$$dx(t) = \left\{ \int_{[-r,0]} x(t+s)\, d\mu(s) \right\} dN(t) \qquad (XVII)$$

where N is as before and μ is a deterministic finite signed measure on $[-r,0]$ with support bounded away from zero (Scheutzow [46], pp. 166–167).

Example 2:

The one-dimensional hereditary equation

$$dx(t) = \{\nu x(t) + \mu x(t-r)\}\, dt + \left\{ \int_{-r}^{0} x(t+s)\sigma(s)\, ds \right\} dW(t), \qquad t > 0 \qquad (XVIII)$$

with real constants ν, μ is a special case of (VI) in §2(B). If $\sigma\colon [-r,0] \to \mathbb{R}$ is a C^1 deterministic function, then it follows from Theorem 3 (§2(B)) that (XVIII) is regular w.r.t. M_2. Observe that the process $\int_{-r}^{0} x(t+s)\sigma(s)\, ds$ has C^1 paths in t and so the stochastic differential dW w.r.t. the one-dimensional Brownian motion W in (XVIII) may be interpreted in the Itô or Stratonovich sense *without changing the solution* x. Taking (Stratonovich) differentials of the process $\log \rho(t)$,

$$\rho(t)^2 := x(t)^2 + \int_{t-r}^{t} x(u)^2\, du, \qquad t > 0, \qquad (30)$$

and analyzing the resulting expression one gets the following theorem:

Theorem 16. (Mohammed)
In (XVIII) let δ_0 be the unique solution of the equation

$$2(\nu + \delta) + \mu^2 e^{2\delta r} + 1 = 0. \tag{31}$$

If λ_1 is the top a.s. Lyapunov exponent of (XVIII) (as given by Theorems 9 & 5(vi)), then $\lambda_1 \leq -\delta_0$.

Details of the proof of the above theorem will appear elsewhere.

Example 3:

Let M be a one-dimensional, sample-continuous square integrable martingale with stationary ergodic increments. From the ergodic theorem we have the fixed (non-random) a.s. limit

$$\beta := \lim_{t\to\infty} \frac{\langle M\rangle(t)}{t}.$$

E.g., if M is standard Brownian motion, then $\beta = 1$. Consider the one-dimensional hereditary equation

$$dx(t) = \{\nu x(t) + \mu x(t-r)\}\, dt + x(t)\, dM(t), \qquad t > 0. \tag{XIX}$$

This equation satisfies Hypotheses (C), (I) in §3(B). So (XIX) is regular w.r.t. M_2 (Theorem 3). Furthermore an analysis of the process in (30) gives the following estimate for λ_1:

Theorem 17. *(Mohammed)*
In (XIX) define δ_0 as in Theorem 16. Then the top a.s. Lyapunov exponent λ_1 of (XIX) satisfies

$$\lambda_1 \leq -\delta_0 + \frac{\beta}{16}.$$

The estimate for λ_1 in the above theorem is clearly not sharp even when $M = W$, one-dimensional standard Brownian motion (cf. the non-delay case $\mu = 0$).

In the special case $M = \sigma W$ for a fixed real σ, the above bound may be sharpened to

$$\lambda_1 \leq \inf\{\kappa(\alpha, \delta) \colon \alpha > 0,\ \delta \in \mathbb{R}\}$$

where

$$\kappa(\alpha, \delta) := -\delta + \frac{1}{16\sigma^2}\left(\mu^2 e^{2\delta r}\alpha + \frac{1}{\alpha} + 2\nu + 2\delta + \sigma^2\right)^2.$$

The proof of this fact was the result of joint discussion involving S.T. Ariaratnam, L. Arnold, P. Baxendale, H. Crauel, W. Kliemann, N. Sri Namachchivaya, M. Pinsky and V. Wihstutz. Observe that the above estimate agrees with $\lambda_1 = \nu - \frac{1}{2}\sigma^2$ in the non-delay case $\mu = 0$.

It is not clear under what conditions on the parameters ν, μ, β the hereditary equation (XIX) becomes hyperbolic.

Acknowledgment

The author is very grateful to Victor Mizel for his helpful comments on a preliminary version of the manuscript.

REFERENCES

[1] Arnold, L., *Stochastic Differential Equations: Theory and Applications*, John Wiley and Sons, Inc. New York (1974)

[2] Arnold, L., Kliemann, W. and Oeljeklaus, E. *Lyapunov exponents of linear stochastic systems*, in *Lyapunov Exponents*, Springer Lecture Notes in Mathematics **1186** (1989), 85–125

[3] Arnold, L., Oeljeklaus, E. and Pardoux, E., *Almost sure and moment stability for linear Itô equations*, in *Lyapunov Exponents*, Springer Lecture Notes in Mathematics **1186** (ed. L. Arnold and V. Wihstutz) (1986), 129–159

[4] Arnold, L. and Wihstutz, V. (eds.)*Lyapunov Exponents: Proceedings of a Workshop*, Bremen, November 1984, Springer Lecture Notes in Mathematics **1186** (1986)

[5] Baxendale, P.H., *Moment stability and large deviations for linear stochastic differential equations*, in Ikeda, N. (ed.) *Proceedings of the Taniguchi Symposium on Probabilistic Methods in Mathematical Physics*, Katata and Kyoto (1985), 31–54, Tokyo: Kinokuniya (1987)

[6] Bellman, R., and Cooke, K. *Differential Difference Equations*, Academic Press (1963)

[7] Berger, M.A. and Mizel, V.J. *Volterra equations with Itô integrals I and II*, J. of Integral Equations **2** (1980), 187–245, **4** (1980), 319–337

[8] Çinlar, E., Jacod, J., Protter, P. and Sharpe, M. *Semimartingales and Markov processes*, Z. Wahrsch. Verw. Gebiete **54** (1980), 161–219

[9] Coleman, B. and Mizel, V.J. *Norms and semigroups in the theory of fading memory*, Arch. Rat. Mech. Ana. **2** (1966), 87–123

[10] Coleman, B. and Mizel, V.J. *On the stability of solutions of functional-differential equations*, Arch. Rat. Mech. Ana. **30**, 3 (1968), 173–196

[11] Corduneanu, C. and Lakshmikantham, V. *Equations with unbounded delay: A survey*, Nonlinear Analysis **4**, 5 (1980), 831–877

[12] de Sam Lazaro, J. and Meyer, P.A. *Questions de théorie des flots*, Seminaire de Probab. IX, Springer Lecture Notes in Mathematics **465**, (1975), 1–96

[13] Delfour, M.C. and Mitter, S.K. *Hereditary differential systems with constant delays I. General case.*, J. Differential Eqns. **12** (1972), 213–255, *II. A class of affine systems and the adjoint problem*, J. Differential Eqns. **18** (1975), 18–28

[14] Dellacherie, C. and Meyer, P.A. *Probabilités et Potential*, 2e'me ed., **1**:Chapters 1–4, **2**:Chapters 5–8, Hermann, Paris (1980)

[15] Doleans-Dade, C. *On the existence and unicity of solutions of stochastic integral equations*, Z. Wahrsch. Verw. Gebiete **36** (1976), 93–101

[16] Doss, H. *Liens entre équations différentielles stochastiques et ordinairés*, Ann. Inst. Henri Poincaré **XIII**, 2 (1977), 99-125

[17] El'sgol'tz, L.E. *Introduction to the Theory of Differential Equations with Deviating Arguments*, Holden-Day, Inc. (1966)

[18] Flandoli, F. and Schaumlöffel, K.-U. *Stochastic parabolic equations in bounded domains: Random evolution operator and Lyapunov exponents*, Stochastics and Stochastic Reports **29**, 4 (1990), 461–485

[19] Hale, J.K. *Sufficient conditions for stability and instability of autonomous functional differential equations*, J. Differential Equations **1** (1965), 452–482

[20] Hale, J.K. *Theory of Functional Differential Equations*, Springer-Verlag (1977)

[21] Itô, K. and Nisio, M. *On stationary solutions of a stochastic differential equation*, J. Math. Kyoto University **4–1** (1964), 1–75

[22] Jacod, J. *Equations differentielles stochastiques linearies: La methode de variation des constantes*, Seminaire de Probabilités XVI, Lecture Notes in Mathematics **920**, Springer-Verlag (1982), 442–446

[23] Kolmanovskii, V.B. and Nosov, V.R. *Stability of Functional Differential Equations*, translated from Russian, Academic Press, London (1986)

[24] Krasovskii, N. *Stability of Motion*, Moscow (1959), translated by J.L. Brenner, Stanford University Press (1963)

[25] Kushner, H. *On the stability of processes defined by stochastic differential-difference equations*, J. Differential Equations **4** (1968), 424–443

[26] Leandre, R. *Flot d'une equation differentielle stochastique avec semimartingale directrice discontinue*, Seminaire de Probabilités XIX, Lecture Notes in Mathematics **1123**, Springer-Verlag (1984), 271-275

[27] Mackevičius, V. *$\mathcal{S}^p$-stability of solutions of symmetric stochastic differential equations*, Lietuvos Matematikos Rinkinys **T. 25**, 4 (1985), 72–84 (in Russian); English translation: Lithuanian Math. J. (1989), 343–352

[28] Mañé, R. *Lyapunov exponents and stable manifolds for compact transformations*, Springer Lecture Notes in Mathematics **1007**, Springer-Verlag (1983), 522–577

[29] Mao, X. *Exponential stability for delay Itô equations*, Proceedings of IEEE International Conference on Control and Applications (April 1989)

[30] Métivier, M. *Semimartingales, a Course on Stochastic Processes*, Walter de Gruyter, Berlin, New York (1982)

[31] Métivier, M. and Pellaumail, J. *Stochastic Integration*, Academic Press (1980)

[32] Mizel, V.J. and Trutzer, V. *Stochastic hereditary equations: existence and asymptotic stability*, Journal of Integral Equations **7** (1984), 1–72

[33] Mohammed, S.-E.A. *Stochastic Functional Differential Equations*, Research Notes in Mathematics **99**, Pitman Advanced Publishing Program, Boston-London-Melbourne (1984)

[34] Mohammed, S.-E.A. *Stability of linear delay equations under small noise*, Proceedings of the Edinburgh Mathematical Society **29** (1986), 233–254

[35] Mohammed, S.-E.A. *Non-linear flows for linear stochastic delay equations*, Stochastics **17**, 3 (1986), 207–212

[36] Mohammed, S.-E.A. *Unstable invariant distributions for a class of stochastic delay equations*, Proceedings of the Edinburgh Mathematical Society **31** (1988), 1–23

[37] Mohammed, S.-E.A. *The Lyapunov spectrum and stable manifolds for stochastic linear delay equations*, Stochastics and Stochastic Reports **29** (1990), 89–131

[38] Mohammed, S.-E.A. and Scheutzow, M.K.R. *Lyapunov exponents and stationary solutions for affine stochastic delay equations*, Stochastics and Stochastic Reports **29**, 2 (1990), 259–283

[39] Mohammed, S.-E.A. and Scheutzow, M.K.R. *Lyapunov exponents of linear stochastic functional differential equations driven by semimartingales. Part I: The multiplicative ergodic theory*, preprint (1990)

[40] Mohammed, S.-E.A., Scheutzow, M.K.R. and Weizsäcker, H.v. *Hyperbolic state space decomposition for a linear stochastic delay equation*, SIAM Journal on Control and Optimization **24**, 3 (1986), 543–551

[41] Protter, Ph.E. *Semimartingales and measure-preserving flows*, Preprint, Statistics Department, Purdue University (1985)

[42] Protter, Ph.E. *Right-continuous solutions of stochastic integral equations*, J. Multivariate Anal. **7** (1977), 204–214

[43] Ruelle, D. *Ergodic theory of differentiable dynamical systems*, I.H.E.S. Publications **50** (1979), 275–305

[44] Ruelle, D. *Characteristic exponents and invariant manifolds in Hilbert space*, Annals of Mathematics **115** (1982), 243–290

[45] Scheutzow, M.K.R. *Qualitative behaviour of stochastic delay equations with a bounded memory*, Stochastics **12** (1984), 41–80

[46] Scheutzow, M.K.R. *Stationary and Periodic Stochastic Differential Systems: A study of Qualitative Changes with Respect to the Noise Level and Asymptotics*, Habilitationsschrift, University of Kaiserslautern, W. Germany (1988)

[47] Sussman, H.J. *On the gap between deterministic and stochastic ordinary differential equations*, Ann. Prob. **6**, 1 (1978), 19–41

[48] Thieullen, P. *Fibres dynamiques asymptotiquement compacts exposants de Lyapunov. Entropie. Dimension.*, Ann. Inst. Henri Poincaré, Anal. Non Linéaire **4**, 1 (1987), 49–97

[49] Weizsäcker, H.v. and Winkler, G. *Stochastic Integrals, An Introduction*, Vieweg Advanced Lectures in Mathematics, Friedr. Vieweg and Sohn, Braunschweig/Wiesbaden (1990)

Salah-Eldin A. Mohammed

Department of Mathematics
Southern Illinois University at
Carbondale, IL 62901

Department of Mathematics
Carnegie Mellon University
Pittsburgh, PA 15213-3890

Convergence in Distribution of a Markov Process Generated by I.I.D. Random Matrices

Arunava Mukherjea

1. The aim of this paper is to study a discrete time Markov process (η_n) on the state space $S = \{0,1\}^V$, V a countable set (of sites), where the transition rule is governed by a $|V|$-dimensional random matrix X with nonnegative integers as entries. Thus, we have:

$$\eta_{n+1} = X_{n+1}\eta_n, \quad n \geq 0 \tag{1.1}$$

where (X_j) is an i.i.d sequence of copies of X (independent of η_0). We are interested in the asymptotic behavior of (η_n). We will also discuss in the last section a finite dimensional analog of this problem when $S = [0,\infty)^V$, $|V| = d < \infty$, and X is a $d \times d$ nonnegative matrix. But, in what follows (in the first four sections), the discussion is restricted to the infinite dimensional context.

We will assume in Sections 2 through 4 that the random matrix X is a random element in the set $\mathcal{S}$ of all $|V|$-dimensional matrices with nonnegative integers as entries such that the sum of non-zero entries in any row of a matrix in $\mathcal{S}$ is bounded. Notice that $\mathcal{S}$ is a topological semigroup with respect to usual matrix multiplication and metric topology induced by the metric d given by

$$d(A,B) = \sum_{i\in V} \frac{1}{2^i} \sum_{j\in V} |A_{ij} - B_{ij}|.$$

Let us also consider product topology in S with $\{0,1\}$ a discrete space. The natural map $\Phi : \mathcal{S} \times S \to S$ defined by $\Phi(A,\eta) = A\eta$, where

$$\begin{aligned} A\eta(i) &= 1 \text{ iff } \sum_j A_{ij}\eta(j) > 0, \\ &= 0, \quad \text{otherwise.} \end{aligned}$$

Notice that for A, B in $\mathcal{S}$ and $\eta \in S$,

$$(AB)\eta = A(B\eta);$$

also, $A_k \to A$ in $\mathcal{S}$ and $\eta_k \to \eta$ in S imply that $A_k\eta_k \to A\eta$ in S, so that Φ is continuous.

In equation (1.1), if η_0 has a distribution $\mu \in P(S)$, the Borel probability measures on the compact space S, and if $Q \in P(\mathcal{S})$, the Borel probability measures on the metric topological semigroup S, is the distribution of X, then the distribution of η_n is $Q^n * \mu$, where Q^n is the usual nth convolution product of Q (with itself). Here, for $Q_1 \in P(S)$ and $\nu \in P(S)$, $Q_1 * \nu$ is defined, as usual, by

$$Q_1 * \nu_1(A) = \int \nu(\{\eta \in S : f \cdot \eta \in A\}) \cdot Q_1(df)$$
$$= \int Q_1\{f \in S : f\eta \in A\}\nu(d\eta) \tag{1.2}$$

Notice that for Q_1, Q_2 in $P(S)$ and $\nu \in P(S)$, $(Q_1 * Q_2) * \nu = Q_1 * (Q_2 * \nu)$; also, if $Q_n \to Q_0$ weakly in $P(S)$ and $\nu_n \to \nu$ weakly in $P(S)$, then $Q_n * \nu_n \to Q_0 * \nu$ weakly in $P(S)$.

In Section 2, we give a reasonably complete solution of the problem of determining most of the (at least the family of absolutely continuous) Q-invariant probability measures for <u>any</u> Q in $P(S)$. A similar result is discussed in the third chapter of [3] in a different setting. We are not sure, however, if the proofs in [3] are useful in our setting. Our proofs are necessarily different.

In Section 3, to allow inclusion of well-known models such as a discrete time contact process, a discrete time voter model or an oriented bond percolation model (see [3]), we assume that $V = Z$ (only for convenience, we take Z rather than general Z^d) and that the random matrix X satisfies the following assumption

(*) Given $i \in Z$ and a p-tuple $n_1 < n_2 < \ldots < n_p$ and a q-tuple $s_1 < s_2 < \ldots < s_q$ in Z, the probability given by

$$P_{mr} \equiv \sum_{u \in Z} \text{Prob} \; \{\text{all the non-zero elements on each of the } i\text{th}, (i+n_1)^{\text{th}}, \ldots, (i+n_p)^{\text{th}} \text{ rows of } X \text{ appear on exactly } (q+1) \text{ different columns, the } u\text{th}, (u+s_1)^{\text{th}} \ldots \text{and the } (u+s_q)^{\text{th}}, \text{ each of these columns containing at least one non-zero entry}\} \tag{1.3}$$

is the same for every i in Z. Note that the probability $P_{mr}(i, u)$, that is, the probability inside the summation in (1.3) is exactly the probability of transition from a configuration ξ_n at time n in S where the only 1's are at sites marked as ith, $(i + n_1)^{th}, \ldots,$ and $(i + n_p)^{th}$ to a configuration $\xi_{n+1} = \xi_n X_{n+1}$ at time $n + 1$, where the only 1's are at sites marked as uth, $(u + s_1)^{th}, \ldots,$ and $(u + s_q)^{th}$, for u in Z. The assumption (1.3) is a translation invariance assumption and this includes the case when $p+1 = 1$, that is, when the configuration ξ_n has a "1" only at the ith site, and the case when $q+1=0$ (or 1), that is, when the configuration ξ_{n+1} has no "1" at any of its sites (or "1" at a single site only).

In Section 3, we will study the problem of convergence in distribution of the chain (η_n) given in (1.1) using the transition matrix P given in (1.3), and in Section 4, we study the ergodic (and nonergodic behavior) of (η_n) in two simple special cases. Though we have not discussed any serious "non-ergodicity" result in the infinite dimensional context, reasonably complete such results in the finite dimensional context are given in Section 5.

2. In this section, we describe all Q-invariant probability measures μ in $P(S)$ (i.e. $Q*\mu=\mu$), where $Q \in P(S)$. Note that all weak limits of the sequence $(Q^n*\mu)$ are Q-invariant. In what follows, δ_ξ will denote the unit mass at the singleton $\{\xi\} \subset S$. 1 and 0 will denote the two elements in S, given by

$1(i) = 1$ for each $i \in V$, and

$0(i) = 0$ for each $i \in V$.

The we have the following theorem, which is the best possible in the present context.

Theorem 2.1. *The sequence $Q^n * \delta_1$ converges weakly to some probability measure λ in $P(S)$ which has the following properties:*

(i) $\lambda \neq \delta_0 \Rightarrow \lambda\{0\} = 0$;

(ii) λ is an extreme point in the compact (in the weak topology) convex set $\mathcal{J}(Q)$ of Q-invariant probability measures in $P(S)$;

(iii) if $\nu \in P(S)$, $\nu \neq \lambda$, $\nu \neq \delta_0, \nu = Q*\nu$, then ν is not an extreme point when one of the following conditions hold:

(a) $\nu\{0\} > 0$

(b) ν has a non-zero absolutely continuous component with respect to λ.

Thus, $\{\mu \in \mathcal{J}(Q) | \mu << \lambda\} = \{\lambda\}$; *also,* $\mathcal{J}(Q) = \{\alpha\lambda + (1-\alpha)\nu : 0 \leq \alpha \leq 1,\ \nu \perp \lambda,\ Q*\nu = \nu\}$.

(iv) $\lambda = \delta_0$ *iff* $\mathcal{J}(Q) = \{\delta_0\}$. *(This is the ergodic case.)*

[Note that when Q is the unit mass at the identity matrix, then $\mathcal{J}(Q) = P(S)$, $\lambda = \delta_1$ and the extreme points in $\mathcal{J}(Q)$ are all those elements in $P(S)$ which have one-point supports. A simple example of a Q where λ is a product measure on S and $\lambda \neq \delta_0$ is given at the end of Section 4.]

Proof. The simplest way to prove that $Q^n*\delta_1$ converges weakly is perhaps the following. First note that for finite $A \subset V$, write:

$$I(A) = \{\eta \in S : \eta(i) = 0 \ \forall i \in A\} \quad \text{and}$$
$$\mathcal{J}(A) = \{f \in \mathcal{S} : \sum_j f(i,j) = 0 \ \forall i \in A\}.$$

Then we have

$$Q^n * \delta_1(I(A)) = Q^n(\mathcal{J}(A)).$$

Note that $\mathcal{J}_A \mathcal{S} \subset \mathcal{J}_A$ so that

$$Q^{n+1}(\mathcal{J}_A) = \int Q^n\{f \in \mathcal{S} : fg \in \mathcal{J}_A\} Q(dg) \geq Q^n(\mathcal{J}_A),$$

which means that $\forall A$ (finite) $\subset V$,

$$\lim_{n \to \infty} \quad Q^n * \delta_1(I(A)) \quad \text{exists.} \tag{2.1}$$

Since $P(S)$ is compact in the weak topology, every subsequence of $(Q^n * \delta_1)$ has a weak limit point. If λ_1 and λ_2 are two such limit points, then it follows from (2.1) (after noting that the indicator function of $I(A)$ is a continuous function on S) that λ_1 and λ_2 coincide on all subsets $I(A)$ for all finite $A \subset V$. It follows that $\lambda_1 = \lambda_2$, proving that the weak limit λ of $Q^n * \delta_1$ exists.

Now suppose that $\lambda \neq \delta_0$. If $\lambda\{0\} > 0$, it is clear that for some positive integer N,

$$Q^n \quad (\{\text{the zero matrix}\}) > 0.$$

Since { the zero matrix } is an ideal of the semigroup $\mathcal{S}$, a simple Borel–Cantelli type of argument shows that

$$\lim_{n \to \infty} \quad Q^n(\{\text{the zero matrix}\}) = 1, \quad \text{implying that} \quad \lambda = \delta_0.$$

Now we can prove that λ is an extreme point in the compact (weakly) convex subset $\mathcal{J}(Q)$ of Q-invariant (i.e. $Q * \mu = \mu$) probability measures in $P(S)$. Write, if possible,

$$\begin{aligned} &\lambda = \alpha\lambda_1 + (1-\alpha)\lambda_2, \ 0 < \alpha < 1, \\ &\lambda_1 \in P(S), \ \lambda_2 \in P(S), \ Q * \lambda_1 = \lambda_1 \quad \text{and} \quad Q * \lambda_2 = \lambda_2. \end{aligned} \tag{2.2}$$

It is then immediate that for $i = 1, 2$, and any finite $A \subset V$,

$$\lambda_i(I(A)) = Q^n * \lambda_i(I(A)) \geq Q^n(\mathcal{J}(A)) = Q^n * \delta_1(I(A)) \tag{2.3}$$

for all positive integers n. Since $I(A)$ is both open and closed, it follows that $\lambda_i(I(A)) \geq \lambda(I(A))$, $i = 1, 2$. Then (2.2) implies that

$$\lambda_1(I(A)) = \lambda_2(I(A)) = \lambda(I(A))$$

for all finite $A \subset V$ so that $\lambda_1 = \lambda_2 = \lambda$. This proves that λ is an extreme point in $\mathcal{J}(Q)$.

Not let $\nu \in \mathcal{J}(Q)$, $\nu \neq \delta_0$, $\nu \neq \lambda$. Suppose that

$$\nu(\{0\}) = \alpha > 0 \tag{2.4}$$

Define the probability measure ν_0 in $P(S)$ by

$$\nu_0\{0\} = \beta, \quad 0 < \beta < \alpha;$$
$$\nu_0(\mathcal{B}\backslash\{0\}) = \frac{1-\beta}{1-\alpha}\nu(\mathcal{B}\backslash\{0\}), \ \mathcal{B} \text{ Borel} \subset S.$$

Then we have:

$$\nu = \frac{\alpha-\beta}{1-\beta}\delta_0 + \frac{1-\alpha}{1-\beta}\nu_0 \tag{2.5}$$

It follows from (2.5) that

$$\begin{aligned} \nu &= Q * \nu \\ &= \frac{\alpha-\beta}{1-\beta}Q * \delta_0 + \frac{1-\alpha}{1-\beta}Q * \nu_0 \\ &= \frac{\alpha-\beta}{1-\beta}\delta_0 + \frac{1-\alpha}{1-\beta}Q * \nu_0 \end{aligned} \tag{2.6}$$

We have from (2.5) and (2.6) that

$$Q * \nu_0 = \nu_0, \quad \nu \neq \nu_0 \tag{2.7}$$

(2.5), (2.6) and (2.7) show that ν is *not* an extreme point.

Now let $\nu \in \mathcal{J}(Q)$ such that the Lebesgue decomposition

$$\nu = \nu_1 + \nu_2, \ \nu_1 << \lambda, \quad \nu_2 \perp \lambda \tag{2.8}$$

is such that $\nu_1 > 0$, $\nu_2 > 0$. Let $\nu_1(S) = r$, $\nu_2(S) = 1 - r \equiv s$ where $0 < r < 1$. Then we have from (2.8) that for any positive integer m,

$$\begin{aligned} \nu &= Q^m * \nu \\ &= rQ^m * \left(\frac{1}{r}\nu_1\right) + (1-r)Q^m * \left(\frac{1}{s}\nu_2\right) \\ &= r\frac{1}{m}\sum_{n=1}^{m} Q^n * \left(\frac{1}{r}\nu_1\right) + (1-r)\frac{1}{m}\sum_{n=1}^{m} Q^n * \left(\frac{1}{s}\nu_2\right). \end{aligned} \tag{2.9}$$

Notice that for $\mu \in P(S)$,

$$\lim_{m\to\infty} \left\| \left[\frac{1}{m}\sum_{n=1}^{m} Q^n * \mu\right] - Q * \left[\frac{1}{m}\sum_{n=1}^{m} Q^n * \mu\right] \right\| = 0$$

proving that every weak limit point of the sequence $\left[\frac{1}{m}\sum_{n=1}^{m} Q^n * \mu\right]$ is in $\mathcal{J}(Q)$. Thus, if (m_k) is a subsequence of positive integers such that

$$(w)\frac{1}{m_k}\sum_{n=1}^{m_k} Q^n * \left[\frac{1}{r}\nu_1\right] \to \nu_1^* \in P(S)$$

and

$$(w)\frac{1}{m_k}\sum_{n=1}^{m_k} Q^n * \left[\frac{1}{s}\nu_2\right] \to \nu_2^* \in P(S),$$

then we have:

$$\nu = r\nu_1^* + (1-r)\nu_2^*, \tag{2.10}$$

where $Q * \nu_1^* = \nu_1^*$ and $Q * \nu_2^* = \nu_2^*$.

We now show that

$$\nu \neq \nu_2^* \quad \text{and} \quad \nu_1^* << \lambda \tag{2.11}$$

To establish (2.11), note that $\nu_1 << \lambda$. Thus, for a Borel set $\mathcal{A} \subset S$ and any positive integer n,

$$\begin{aligned} \lambda(\mathcal{A}) = 0 &\Longrightarrow Q^n * \lambda(\mathcal{A}) = 0 \\ &\Longrightarrow \lambda\{\eta \in S : f\eta \in \mathcal{A}\} = 0 \quad \text{for almost all} \quad f\,(w.r.t\,Q^n) \\ &\Longrightarrow \nu_1\{\eta \in S : f\eta \in \mathcal{A}\} = 0 \quad \text{for almost all} \quad f\,(w.r.t\,Q^n) \\ &\Longrightarrow Q^n * \nu_1(\mathcal{A}) = 0. \end{aligned} \tag{2.12}$$

(Note that it does not yet follow that $\nu_1^*(\mathcal{A}) = 0$, whenever $\lambda_1(\mathcal{A}) = 0$.). Since $\nu_2 \perp \lambda$, there exists $\mathcal{B}$ such that

$$\lambda(\mathcal{B}) = 0, \quad \nu_2(\mathcal{B}^c) = 0.$$

Then we have from (2.8) that

$$\nu(\mathcal{B}) = \nu_2(\mathcal{B}) = \nu_2(S) = s. \tag{2.13}$$

Now for any closed subset $\mathcal{B}_c \subset \mathcal{B}$, $\lambda(\mathcal{B}_c) = 0$, and it follows from (2.12) that $\nu(\mathcal{B}_c) = Q^n * \nu(\mathcal{B}_c) + Q_n * \nu_2(\mathcal{B}_c) = Q^n * \nu_2(\mathcal{B}_c)(\forall_n \geq 1)$, which implies that

$$\nu_2^*(\mathcal{B}_c) \geq \lim_{n\to\infty} \sup \frac{1}{n}\sum_{m=1}^{n}\left[Q^m * \left[\frac{1}{s}\nu_2\right]\right](\mathcal{B}_c) = \frac{1}{s}\nu(\mathcal{B}_c).$$

By the regularity of ν_2^* and ν, we have from (2.13) that $\nu_2^*(\mathcal{B}) = 1$, so that $\nu_2^* \perp \lambda$ and $\nu_1^*(\mathcal{B}) = 0$. If we replace $\mathcal{B}$ above by $\mathcal{B} \cup \mathcal{A}$, where $\mathcal{A}$ is as in (2.12), then it follows as before that $\nu_1^*(\mathcal{B} \cup \mathcal{A}) = 0$ implying that $\nu_1^*(\mathcal{A}) = 0$. Thus, (2.11) is established. It follows that the measure ν in (2.8) is not an extreme point in $\mathcal{J}(Q)$.

Finally, we suppose that $\nu \in \mathcal{J}(Q)$, $\nu \neq \lambda$ and $\nu << \lambda$. We prove that ν is then not an extreme point in $\mathcal{J}(Q)$. By the Radon–Nikodym Theorem, there exists a nonnegative λ-integrable function on S such that

$$\nu(E) = \int_E f(x)\lambda(dx), \quad E \subset S. \tag{2.14}$$

Notice that

$$\lambda\{x \in S | f(x) = 0\} = 1$$

implies that $\nu \equiv 0$, so that

$$\lambda\{x \in S \,| f(x) > 0\} > 0.$$

Thus, there exists $\alpha > 0$ such that

$$\lambda(\mathcal{B}_\alpha) > 0, \tag{2.15}$$

where $\mathcal{B}_\alpha = \{x \in S \,| f(x) > \alpha\}$. Notice that for $\mathcal{E} \subset \mathcal{B}_\alpha$,

$$(\nu - \alpha\lambda)(\mathcal{E}) \geq 0 \quad \text{and} \quad (\nu - \alpha\lambda)(\mathcal{B}_\alpha) > 0 \tag{2.16}$$

Define the positive measures ν_1 and ν_2 by:

$$\begin{cases} \nu_1(\mathcal{E}) & = (\nu - \alpha\lambda)(\mathcal{E} \cap \mathcal{B}_\alpha) + \nu(\mathcal{E} \cap \mathcal{B}_\alpha^c), \\ \nu_2(\mathcal{E}) & = \alpha\lambda(\mathcal{E} \cap \mathcal{B}_\alpha) \end{cases} \tag{2.17}$$

Then we have:

$$\nu = \nu_1 + \nu_2.$$

Let $r = \nu_1(S)$, $s = 1 - r = \nu_2(S)$. Then $0 < r < 1$ and $0 < s < 1$. Proceeding as before,

$$\nu = r \cdot \frac{1}{n} \sum_{m=1}^{n} Q^m * \left[\frac{1}{r}\nu_1\right] + s \cdot \frac{1}{n} \sum_{m=1}^{n} Q^m * \left[\frac{1}{s}\nu_2\right]$$

and then

$$\nu = r\nu_1^* + (1-r)\nu_2^*, \tag{2.18}$$

where ν_1^* and ν_2^* are in $\mathcal{J}(Q)$, and for some subsequence (n_k) of positive integers,

$$\nu_1^* = (w) \lim_{k \to \infty} \frac{1}{n_k} \sum_{m=1}^{n_k} Q^m * \left[\frac{1}{r}\nu_1\right]$$

and

$$\nu_2^* = (w) \lim_{k \to \infty} \frac{1}{n_k} \sum_{m=1}^{n_k} Q^m * \left[\frac{1}{s}\nu_2\right].$$

Now we show that $\nu_2^* = \lambda$. From (2.15), $0 < \lambda(\mathcal{B}_\alpha) \le 1$. If $\lambda(\mathcal{B}_\alpha) = 1$, then $\nu_2 = \alpha\lambda$ so that $s = \alpha$, and $\frac{1}{s}\nu_2 = \lambda$, and, therefore, $\nu_2^* = \lambda$. Let us suppose that

$$0 < \lambda(\mathcal{B}_\alpha) < 1.$$

Define:

$$\lambda_1(\mathcal{E}) = \lambda(\mathcal{E} \cap \mathcal{B}_\alpha) \quad \text{and}$$
$$\lambda_2(\mathcal{E}) = \lambda(\mathcal{E} \cap \mathcal{B}_\alpha^c) \quad \text{for all Borel} \quad \mathcal{E}.$$

Then,

$$\lambda = \lambda_1 + \lambda_2 \tag{2.19}$$

Notice that $\alpha\lambda_1 = \nu_2$. Let $t_\alpha = \lambda(\mathcal{B}_\alpha)$. Then $0 < t_\alpha < 1$. Proceeding as before, we obtain from (2.18) that

$$\lambda = t_\alpha \lambda_1^* + (1 - t_\alpha)\lambda_2^*,$$

where

$$\lambda_1^* = (w) \lim_{k \to \infty} \frac{1}{s_k} \sum_{m=1}^{s_k} Q^m * \left[\frac{1}{t_\alpha}\lambda_1\right]$$

and $\lambda_2^* = (w) \lim_{k\to\infty} \frac{1}{s_k} \sum_{m=1}^{s_k} \sum Q^m * \left[\frac{1}{1 - t_\alpha}\lambda_2\right]$, where (s_k) is some subsequence of the sequence (n_k) chosen earlier. Notice that $\alpha\lambda_1 = \nu_2$ so that $\alpha t_\alpha = s$ or $\frac{1}{s}\nu_2 = \frac{1}{t_\alpha} \cdot \lambda_1$. This means that $\lambda_1^* = \nu_2^*$. Since λ is an extreme point in $\mathcal{J}(Q)$ and λ_1^*, λ_2^* are both in $\mathcal{J}(Q)$, it follows that $\lambda = \nu_2^*$. Since $\nu \neq \lambda$, (2.18) implies that ν is *not* an extreme point in $\mathcal{J}(Q)$.

The final part of the theorem is immediate, since for every finite $A \subset V$ and any $\nu \in \mathcal{J}(Q)$,

$$\nu\{\eta \in S \,|\eta(i) = 0 \quad \text{for} \quad i \in A\}$$
$$\ge \lambda\{\eta \in S \,|\eta(i) = 0 \quad \text{for} \quad i \in A\},$$

so that $\lambda = \delta_0 \Rightarrow \nu = \delta_0$. ∎

3. In this section, we study convergence in the distribution of the Markov process (η_n) in (1.1). We will assume that $V = Z$, and the random matrix X satisfies the condition (1.3). Let $\mu \in P(S)$ be the distribution of η_0 so that the distribution of η_n is $Q^n * \mu$.

First, we introduce a dual Markov chain with a countable state space and transition matrix P (as described in (1.3)). Consider the state space E given by $E = \{0, 1, 01, 02, 011, 03, 021, 012, 0111, 04, 031, 022, 0211, 013,$

0121, 0112, 01111, 05, ...}. Here "0n" represents the $(2^{n-1}+2)^{\text{th}}$ state, and the states described in order between "0n" and "0$(n+1)$" are the states $0(n-1)1, 0(n-1)2, 0(n-1)11, 0(n-3)3, 0(n-3)21, \ldots, \underbrace{0111\ldots1}_{n}$.

The state "$0n_1n_2\ldots n_k$" precedes state "$0s_1s_2\ldots s_{k+m}$" ($k>1$, $m\geq 0$) if either

(i) $n_1+n_2+\ldots+n_k < s_1+s_2+\ldots+s_{k+m}$ or

(ii) $n_1+n_2+\ldots+n_k = s_1+s_2+\ldots+s_{k+m}$ and for the first i for which $n_i \neq s_1$, $n_i > s_i$. Let us now describe the transition matrix P. The first row is (1 0 0 0...). Suppose now that m corresponds to the state labelled as "1" of "$0m_1m_2\ldots m_p$" (with $p\geq 1$) and r corresponds to the state labelled as "0," "1" or "$0t_1t_2\ldots t_q$" (with $q \geq 1$). Consider again the equation (given in Section 1)

$$\xi_{n+1} = \xi_n X_{n+1} \tag{3.1}$$

We say that an element ξ in S is in configuration "0," "1" or "$0m_1m_2\ldots m_p$" according as

(i) $\xi_n(j)=0 \quad \forall j \in Z$,

(ii) $\xi_n(j)=0$ for all but a single j in Z, or

(iii) $\xi_n(j)=1$ only when $j\in\{i, i+m_1, i+m_1+m_2, \ldots, i+m_1+\ldots+m_p\}$, occurs, for some i.

Then P_{mr} is the probability of transition from a configuration ξ_n (at time n) that corresponds to the labelling of the mth state for some i to a configuration ξ_{n+1} at time $(n+1)$ that corresponds to the labelling of the rth state for any u. Notice that we have defined P_{mr} in (1.3), where $m_1+m_2+\ldots+m_i = n_i, t_1+t_2+\ldots+t_i = s_1$ (for $i\geq 1$).

The transition $P_{mr}(i,u)$

ξ_n : 1 1 1 1

i $\quad i+m_1$ $\quad i+m_1+m_2$ $\quad i+m_1+\cdots+m_p$

$m \equiv$ "$0m_1m_2\ldots m_p$"

(Zeros at every other site)

ξ_{n+1} : 1 1 1 1

u $\quad u+s_1$ $\quad u+s_2$ $\quad u+s_q$

$= \xi_n X_{n+1}$

$r \equiv$ "$0t_1\,t_2\ldots t_q$"

(Zeros at every other site)

is the probability of transition from the first configuration to the second configuration above. Then $P_{mr} = \sum_{u \in Z} P_{mr}(i,u)$, which is independent of i, by our assumption (1.3).

Now Q is the distribution of X and $\mu \in P(S)$ is the distribution of η_0. Then we have

$$\begin{aligned} Q^{n+1} * \mu\{\xi \in S : \xi(i) = \xi(i+n_1) = \ldots = \xi(i+n_p) = 0\} \\ = \int Q^n * \mu\{\xi \in S : (g\cdot\xi)(i) = (g\cdot\xi)(i+n_1) \\ = (g\cdot\xi)(i+n_p) = 0\}\ Q(dg) \end{aligned} \tag{3.2}$$

Let us now write:

$$a_m(n) = \inf_i Q^n * \mu\{\xi \in S : \xi(i) = \xi(i+n_1) = \ldots = \xi(i+n_p) = 0\} \tag{3.3}$$

$$b_m(n) \equiv \sup_i Q^n * \mu\{\xi \in S : \xi(i) = \xi(i+n_1) = \ldots = \xi(i+n_p) = 0\},$$

where the mth state is labelled as

$$\text{“}0m_1m_2\ldots m_p\text{”}, \quad m_1 + \ldots + m_i = n_i, \quad i \geq 1.$$

Then the routine is to verify (using (3.2)) that

$$a_m(n+1) \geq \sum_{r=0}^{\infty} P_{mr} a_r(n)$$

and

$$b_m(n+1) \leq \sum_{r=0}^{\infty} P_{mr} b_r(n),$$

where $a_0(n) = b_0(n) = 1$ for $n \geq 0$. Thus, if $a(n)$ is the column vector $(a_m(n))$ and $b(n)$ is the column vector $b_m(n)$, then we have:

$$a(n+1) \geq P\,a(n) \quad \text{and} \quad b(n+1) \leq P\,b(n) \tag{3.4}$$

Notice that we have from (3.3) that

$$a_m(0) = \inf_i \mu\{\xi \in S : \xi(i) = \xi(i+n_1) = \ldots = \xi(i+n_p) = 0\} \tag{3.5}$$

and

$$b_m(0) = \sup_i \mu\{\xi \in S : \xi(i) = \xi(i+n_1) = \ldots = \xi(i+n_p) = 0\}$$

It follows from (3.4) that

$$P^n a(0) \le a(n) \le b(n) \le P^n b(0) \tag{3.6}$$

In what follows, we assume that

$$P_{m0} > 0 \quad \forall m > 0, \tag{3.7}$$

that is, given any k tuple $(r_1, r_2, \ldots, r_k)$ in Z, Pr(the r_1th, the r_2th,..., and the r_kth rows of X are zero rows)$\equiv \delta_k > 0$.

Under assumption (3.7), all states in E except the first state are inessential (with respect to P) so that

$$\lim_{n \to \infty} (P^n)_{mr} = 0 \quad \forall r > 0 \tag{3.8}$$

Moreover,

$$\left(P^{n+1}\right)_{m0} \ge P^n_{m0} + \sum_{r>0} P^n_{mr} P_{r0}$$

so that

$$\lim_{n \to \infty} P^n_{mr} \quad \text{exists} \quad \forall m \ge 0, \forall r \ge 0.$$

Suppose now that

$$\lim_{n \to \infty} \sup_i \mu\{\xi \in S : \xi(i) = 0 = \xi(i+n)\} = 0 \tag{3.9}$$

It follows from (3.5), (3.6) and (3.9) that

$$\lim_{n \to \infty} \sum_{r=0}^{\infty} P^n_{mr} a_r(0) = \lim_{n \to \infty} \sum_{r=0}^{\infty} P^n_{mr} b_r(0) = \lim_{n \to \infty} P^n_{mo},$$

which exists for each $m \ge 0$. This means that $\lim_{n\to\infty} a_m(n) = \lim_{n\to\infty} b_m(n)$ exists $\forall m \ge 0$, and thus, for any $p+1$ tuple

$$0 \le n_1 \le n_2 \le \ldots \le n_p$$

in Z, under (3.9),

$$\lim_{n \to \infty} Q^n * \mu\{\xi \in S : \xi(i) = \xi(i+n_1) = \ldots = \xi(i+n_p) = 0\}$$

exists and this limit is independent of μ. It follows easily that $Q^n * \mu$ converges weakly for all μ satisfying (3.9).

Let us now assume (instead of (3.9)) that

$$1 > \lim_{n\to\infty} \sup \left[\sup_i \mu\{\xi \in S \mid \xi(i) = 0 = \xi(i+n)\}\right] \tag{3.10}$$

holds. [For example, $\mu = \alpha\delta_1 + (1-\alpha)\beta$, $0 < \alpha < 1$ and $\beta \in P(S)$, satisfies (3.10).] Suppose also that for some μ satisfying (3.10), $Q^n * \mu$ converges weakly to δ_0. Then, we have:

$$\lim_{n\to\infty} b_m(n) = 1 \quad \text{for} \quad m \geq 1 \tag{3.11}$$

By condition (3.10), there exists $\delta > 0$ and some $m_0 \geq 1$ such that $m \geq m_0 \Rightarrow b_m(0) < 1 - \delta$.

From (3.6), we have: for $m \geq 0$,

$$b_m(n) \leq \sum_{r=0}^{\infty} P^n_{mr} b_r(0), \tag{3.12}$$

so that given $\epsilon > 0$, there exists $n_0 \geq 1$ such that $n \geq n_0 \Rightarrow$

$$(1-\epsilon)\sum_{r=0}^{\infty} P^n_{mr} \leq (1-\delta) \sum_{r\geq m_0+1} P^n_{mr} + \sum_{r=0}^{m_0} P^n_{mr}$$

so that we have:

$$\sum_{r\geq m_0+1} P^n_{mr} \leq \frac{\epsilon}{\delta - \epsilon}, \tag{3.13}$$

uniformly in $n \geq n_0$. It follows from (3.8) and (3.13) that $\lim_{n\to\infty} P^n_{m0} = 1$ for all $m \geq 0$. Using (3.6), it follows then easily that for any $\beta \in P(S)$, $Q^n * \beta \to \delta_0$ weakly.

Now notice that $\lambda = (w) \lim_{n\to\infty} Q^n * \delta_1$ is δ_0 iff P is ergodic. This follows immediately from (3.6). Thus we have proven the following theorem.

Theorem 3.1. *Consider the Markov chain (given in (1.1)) (η_n) in S such that*

$$\eta_{n+1} = X_{n+1}\eta_n, \qquad n \geq 0$$

where (X_i) is an i.i.d. sequence of copies of X (independent of η_0), where X satisfies conditions (1.3) and (3.7). Let μ be the distribution of η_0. Then the following results hold:

(i) $\lambda = (w) \lim_{n\to\infty} Q^n * \delta_1$ is δ_0 iff P is ergodic.

(ii) If μ satisfies (3.9), then $Q^n * \mu$ converges weakly.

(iii) If μ satisfies (3.10) and $Q^n * \mu$ converges weakly to δ_0 (for some μ in $P(S)$), then $\lambda = \delta_0$ (see (i)) and consequently, the chain (η_n) is ergodic for all initial distributions.

(iv) Suppose μ and Q are such that $\sum_r P_{mr}b_r(0)$, $m \geq 0$ converges uniformly in m. Then $Q^n * \mu$ converges weakly as $n \to \infty$.

(v) $\lambda \neq \delta_0$ iff there is a vector $\epsilon = (\epsilon_0, \epsilon_1, \epsilon_2, \ldots)$, $\epsilon_0 = 1, 0 \leq \epsilon_i \leq 1$, $\epsilon \neq 1$ such that $P\epsilon^T \leq \epsilon^T$.

(vi) Suppose that μ is such that $\lim_{r\to\infty} a_r(0) = \lim_{r\to\infty} b_r(0)$ exists. Then $Q^n * \mu$ converges weakly as $n \to \infty$.

(vii) Suppose that μ is translation invariant, that is, $a_r(0) = b(r)(0)$ for $r \geq 0$. Then $Q^n * \mu$ converges weakly iff $\lim_{n\to\infty} \sum_{r\geq 0} P^n_{mr} a_r(0)$ exists for every $m \geq 0$. ∎

Remark. The proofs of (iv) , (v) and (vi) are easily omitted. In fact, (ii) and (vi) follow from (iv); however, the conditions in (ii) and (vi) are more readily understandable and involve only μ. The above theorem can be generalized to Z^d, for any positive integer d, with virtually no difficulty.

4. This section is meant for the reader who is a newcomer to interacting particle systems. In this section, we consider a general discrete time contact process by choosing the random matrix X suitably and then present simple situations when $\lambda = \delta_0$ and $\lambda \neq \delta_0$, determining λ completely in the latter case. [Recall that $\lambda = (w) \lim_{n\to\infty} Q^n * \delta_1$.] We also present a lower bound in the general context for

$$\lambda\{\eta \mid \eta(i) = 0 \quad \forall i \in A\}$$

for all finite A. Even though in this section we do not tackle the really difficult problems, we believe our approach here is new.

Consider, as before, the set of all configurations given by $S = \{0,1\}^V$, V a countable set. Assume that for each i in V, there is a well-defined finite subset of V, called the neighborhood of i and denoted by $N(i)$, such that its cardinality is independent of i. The $|V|$-dimensional random matrix X is then given by $X \in \{0,1\}^{V\times V}$ such that

(*i*) $X_{ij} = 0$ if $j \notin N(i), i \in V$,
$= 0$ or 1 if $j \in N(i)$;

(*ii*) the rows of X are independent;

(*iii*) for $i \in V$, $Pr\left[\sum_j X_{ij} = k\right] = \delta_k$ independent of i, where $0 \leq k \leq r$, $r = |N(i)|$, $0 < \delta_0 < 1$ and $\delta_0 + \delta_1 + \ldots + \delta_r = 1$. (4.0)

Let Q be the distribution of X in S. The contact process (η_m) is then given by

$$\eta_{m+1} = X_{n+1}\eta_n, \quad n \geq 0, \tag{4.1}$$

where (X_i) is a sequence of i.i.d copies of X, independent of η_0, a random element in S.

A natural physical interpretation for (4.1) is that V represents a countable set of sites, each site being occupied by an individual, and η_n represents, at time n, the health status of all these individuals. Thus, $\eta_n(i) = 1$ or 0 accordingly as the individual at the ith site is infected or not at time n. Condition (iii) in (4.0) implies that the probability for our invidual at site i to get infected increases with the number of its infected neighbors. Notice that a healthy individual can get infected only if he has at least one infected neighbor.

Let F_k be the class of all finite subsets of V with cardinality $k \geq 0$. Write for $n \geq 1$, $k \geq 1$,

$$\begin{aligned} a_k^{(n)} &= \inf_{A \in F_k} Q^n\{f : \sum_j f_{ij} = 0 \quad \text{for} \quad i \in A\}, \\ a_0^{(n)} &= 1. \end{aligned} \tag{4.2}$$

The we have for $A \in F_k$,

$$\begin{aligned} Q^{n+1} * \delta_1\{\eta \in S : \eta(i) = 0 \quad &\text{for} \quad i \in A\} \\ = Q^{n+1}\{f : \sum_j f_{ij} = 0 \quad &\text{for} \quad i \in A\} \\ = \int Q^n\{f : \sum_s g_{is} \sum_j f_{sj} = 0 \quad &\text{for} \quad i \in A\} Q(dg) \\ \geq \sum_j E_{kj} a_j^{(n)}, \end{aligned}$$

where the stochastic matrix E has state space the set of all nonnegative integers and it is defined by

$$E_{kj} = Pr\ \{\text{there are exactly } j \text{ non-zero entries on } k \text{ given rows of } X\}.$$

Then we have: for $n \geq 1$

$$a^{(n+1)} \geq Ea^{(n)} \geq E^n a^{(1)}. \tag{4.3}$$

Notice that

$$E_{00} = 1 \quad \text{and} \quad E_{k0} > 0, \quad k \geq 0.$$

It follows that

$$\lim_{n \to \infty} E_{ij}^n \quad \text{exists for all} \quad i \geq 0, j \geq 0, \quad \text{and for} \quad j > 0, \text{ this limit is } 0. \tag{4.4}$$

Using multinomial distribution theory, it is easily shown that for $k \geq 0$,

$$\sum_{j=0}^{kr} E_{kj} \cdot j = k[\delta_1 + 2\delta_2 + \ldots + r\delta_r]. \tag{4.5}$$

Now consider the condition

$$\delta_1 + 2\delta_2 + \ldots + r\delta_r \leq 1. \tag{4.6}$$

Then we have for all $n \geq 1$, from (4.5) and (4.6), that

$$\sum_{j=0}^{\infty} E_{kj}^{n} \cdot j \leq k, \quad k \geq 0. \tag{4.7}$$

It follows from (4.7) that for $n \geq 1$,

$$\sum_{j=k+1}^{\infty} E_{kj}^{n}(j-k) \leq \sum_{j=1}^{k} E_{kj}^{n}(k-j),$$

where the right side tends to zero as $n \to \infty$, by (4.4). Consequently, for $k \geq 0$,

$$\lim_{n \to \infty} E_{k0}^{n} = 1.$$

Then we have from (4.3) that

$$\lim_{n \to \infty} a_k^{(n)} = 1, \quad \text{for} \quad k \geq 0.$$

Thus, $\lambda = \delta_0$ so that the contact process (4.2) dies out under condition (4.6).

A lower bound for

$$\lambda\{\eta \in S : \eta(j) = 0 \quad \text{for} \quad j \in A\} \tag{4.8}$$

when $A \in F_k$, is now given below. We assume condition (4.6) does not hold below. Then it is elementary to show that there exists $1 > \alpha > 0$ such that

$$\sum_{i=0}^{r} \alpha^i \delta_i = \alpha \tag{4.9}$$

(For example, when $r = 2$, $\alpha = \delta_0/\delta_2$ satisfies (4.9) when $\delta_0 < \delta_2$.) It follows after routine calculations that for $m \geq 0$,

$$\sum_{j=0}^{\infty} E_{mj} \cdot \alpha^j = \alpha^m. \tag{4.10}$$

Writing v for the vector for which $v_i = \alpha^i$, we then have

$$Ev = v \tag{4.11}$$

It follows from (4.11) that for $i \geq 0$,

$$\lim_{n \to \infty} E^n_{i0} = \alpha^i. \tag{4.12}$$

It is now clear that for $i \geq 1$,

$$\alpha^i \leq \inf_{|A| = i} \lambda\{\eta \in S : \eta(j) = 0 \quad \text{for} \quad j \in A\}. \tag{4.13}$$

Let us now consider the extreme (no interaction) situation, the branching process-like case, when for $i \neq j$, $N(i) \cap N(j) = \phi$. Then, as in (4.2), we define $b_k^{(n)}$ by replacing the "inf" in (4.2) by "sup." As in (4.3), it follows easily that

$$b^{(n+1)} \leq Eb^{(n)} \leq E^n b^{(1)}, \tag{4.14}$$

for the same matrix E there. By condition (iii) in (4.0), $a_k^{(1)} = b_k^{(1)} = \delta_0^k$. Then we have from (4.3) and (4.14) that

$$a^{(n+1)} = b^{(n+1)} = E^n a^{(1)}, \quad n \geq 1. \tag{4.15}$$

It follows from (4.12) and (4.14) that for $A \subset V$ with $|A| = k$, $k \geq 1$,

$$\lambda\{\eta \in S \mid \eta(j) = 0 \quad \text{for} \quad j \in A\} = \alpha^k, \tag{4.16}$$

where $0 < \alpha < 1$ and α is the unique solution of (4.9), which exists when $E\left[\sum_j X_{ij}\right] > 1$.

5. In this section, we look into the corresponding finite dimensional problem in the more general context of $S \equiv [0, \infty)^V$, rather than $\{0,1\}^V$, where $|V| = d < \infty$, and $\mathcal{S}$ is the set of $d \times d$ non-negative matrices (a topological semigroup under multiplication and usual topology) such that

$$\mathcal{S} = cl\left[U_{n=1}^{\infty} \quad S_Q^n\right], \tag{5.0}$$

where $Q \in P(\mathcal{S})$. Note that none of the distributions Q^n has mass outside $\mathcal{S}$.

Let us introduce the condition (KS), one of the two main conditions used in [4] (see also [5]), below:

$$(KS) : \mathcal{S} \quad \text{contains a strictly positive matrix.} \tag{5.1}$$

The natural map $\Phi : \mathcal{S} \times S \to S$ is now defined by $\Phi(A, \eta) = A\eta$, where $(A\eta)(i) = \sum_{j=1}^{d} A_{ij}\eta(j)$. As before, Φ is continuous and we define $Q * \mu$ for $Q \in P(\mathcal{S})$ and $\mu \in P(S)$, and this convolution is continuous as a map from $P(\mathcal{S}) \times P(S)$ into $P(S)$ with respect to weak toplogy. Then we have the following simple, but surprising result.

Proposition 5.1. *Suppose that Q satisfies (5.1) and that $\mu \neq \delta_0$, where $0 \equiv (0, 0, \ldots, 0)$ and $\mu \in P(S)$. Then, (Q^n) is tight iff $(Q^n * \mu)$ is tight.* ∎

Proof. The "only if" part is trivial. For the "if" part, let $\mathcal{S} \cap \mathcal{J} \neq \phi$, where $\mathcal{J}$ is the set of all $d \times d$ strictly positive matrices. Let I_k be the set

$$\{x \equiv (x_1, x_2, \ldots, x_d) | \quad \text{for} \quad 1 \leq i \leq d \quad x_i > \frac{1}{k}\}.$$

Let m be the positive integer such that $Q^m(\mathcal{J}) > 0$. Since $\mu \neq \delta_0$, there exists $\delta > 0$ such that

$$Q^m * \mu\{x \in S : \quad \text{for} \quad 1 \leq i \leq d,\ x_i > 0\} = 2\delta, \tag{5.2}$$

since for $A \in \mathcal{J}$ and $x \in S$, $x \neq 0, (Ax)_i > 0$ for each $i, 1 \leq i \leq d$. It follows from (5.2) that there is a positive integer k such that

$$Q^m * \mu(I_k) > \delta. \tag{5.3}$$

Now notice that for any compact subset $K \subset S$, the set KI_k^{-1}, defined by

$$KI_k^{-1} = U\{Kx^{-1} | x \in I_k\},$$

where

$$Kx^{-1} = \{A \in \mathcal{S} : Ax \in K\},$$

is a compact subset of $\mathcal{S}$. If (Q^n) is not tight, then we have:

$$\sup_{\substack{K \subset S \\ K \text{ compact}}} \quad \lim_{n \to \infty} \inf Q^n(KI_k^{-1}) = 1 - 2u < 1. \tag{5.4}$$

It follows from (5.3) and (5.4) that for any $K \subset S$, with K compact,

$$\begin{aligned} Q^{n+m} * \mu(K) &= \int Q^n(Kx^{-1}) Q^m * \mu(dx) \\ &= \int_{I_k} + \int_{I_k^c} \\ &\leq (1 - u) Q^m * \mu(I_k) + Q^m * \mu(I_k^c) \\ &= 1 - uQ^m * \mu(I_k) \\ &< 1 - u\delta < 1, \end{aligned}$$

for infinitely many n. This means that $(Q^n * \mu)$ is not tight. ∎

We remark that when Q does not satisfy (5.1), it is very easy to construct examples of Q and $\mu, \mu \neq \delta_0$ such that $(Q^n * \mu)$ is tight, whereas (Q^n) is not. For example, take Q to be the probability measure on

$$\left\{ \begin{pmatrix} a & 0 \\ 0 & 0 \end{pmatrix} : a > 0 \right\}$$

induced by $\sum_{n=0}^{\infty} (1/2^{n+1}) m|_{(n,n+1]}$, where m is the Lebesgue measure. Take $\mu \in P(R^2)$ such that

$$\operatorname{Supp} \mu \subset \{(0,b) \mid b \geq 0\}.$$

Then, (Q^n) converges vaguely to the zero measure, whereas $Q^n * \mu = \delta_0$.
Let us now state, without proof, the following theorem.

Theorem 5.2. *Suppose that (Q^n) is tight. Then there is a unique weak limit point Q_0 of the sequence (Q^n) such that $Q_0 = Q_0 * Q_0$. The sequence $(Q^n * \mu)$ converges weakly to $Q_0 * \mu$ iff $(Q * Q_0) * \mu = Q_0 * \mu$.*

Let us mention that a necessary and sufficient condition for the sequence (Q^n) to be tight is given in [6], where $Q \in P(S)$. Let us now give an "extinction" and a "survival" result.

Theorem 5.3. *Suppose that the sequence (Q^n) is tight. If S contains the zero matrix, then $Q^n * \mu$ converges weakly to δ_0. Suppose that S does not contain the zero matrix and Q satisfies (5.1). Then S must be compact and for $\mu \neq \delta_0$, the sequence $Q^n * \mu$ converges weakly to some $\lambda \in P(S)$, where*

$$\lambda\{x \in S \mid x_i > 0 \quad \text{for} \quad 1 \leq i \leq d\} > 0.$$

This theorem is essentially given in [6]. Let us not give a theorem which characterizes *all* Q-invariant probability measures in the presence of (5.1). (A similar result for the invertible matrices and under certain conditions on μ is given in [1].)

Theorem 5.4. *Suppose (5.1) holds and that $Q * \mu = \mu, \mu \in P(S)$, $Q \in P(S)$. Then either $\mu = \delta_0$ or μ is $Q_0 * \beta$, where $\beta \in P(S)$, and Q_0 is the weak limit of $\frac{1}{n}\sum_{k=1}^{n} Q^k$ so that $Q * Q_0 = Q_0 * Q = Q_0 = Q_0 * Q_0$. (The weak limit Q_0 always exists whenever such a $\mu, \mu \neq \delta_0$, exists.) The Q-invariant measure μ has the following invariance property:*

If $\mathcal{B}$ is a Borel subset of S, $A \in \mathcal{S}$ with minimal rank in $\mathcal{S}$, then for any $B \in \mathcal{S}$,

$$\mu\{x \in S : Ax \in \mathcal{B}\} = \mu\{x \in S : (AB)x \in \mathcal{B}\}.$$

This theorem can be easily proved using our Proposition 5.1 and the fact that Q_0 exists whenever (Q^n) is tight. The last invariance property of μ can be established by exploiting the Choquet–Deny type equation

$$Q_0 * \mu = \mu, \tag{5.5}$$

the idempotence of Q_0, and the fact that the support of Q_0 is the set of all matrices in $\mathcal{S}$ with the minimal rank. This theorem can be well understood by looking at the solution of a similar Choquet–Deny equation in $\mathcal{S}$. We present it next.

Theorem 5.5. *Let Q_1 and Q_2 be in $P(\mathcal{S})$ such that*
(i) $Q_1 = Q_1 * Q_2$; (5.6)
(ii) Q_1 { *The matrices in $\mathcal{S}$ with zero columns*} $= 0$;
(iii) $S(Q_2)S(Q_1) \subset cl\,[\bigcup_{n=1}^{\infty} S(Q_2)^n] \equiv \mathcal{S}_1$, say,). (5.7)
[Here, $S(Q_i) \equiv$ the support of Q_i.] When (ii) holds, Q_1 satisfies (5.6) iff Q_1 is of the form $Q' * Q_0$, where $Q' \in P(\mathcal{S})$, and $Q_0 = Q_0 * Q_0 = Q_0 * Q_2 = Q_2 * Q_0$ is the weak limit of $\frac{1}{n}\sum_{m=1}^{n} Q_2^m$. When (ii) and (iii) both hold, then every solution Q_1 of (5.6) satisfies: $Q_1 = Q_1 * Q_1$, and moreover, $\delta_A Q_1 = \delta_A * Q_0$, where A is any matrix with minimal rank in $\mathcal{S}_1$ (see (5.7)) and δ_A is the unit mass at the singleton $\{A\}$. If (iii) is replaced by the stronger condition:
(iv) $S(Q_1) \subset \mathcal{S}_1$,
then under (i), (ii) and (iv), the product representations of Q_1 and Q_0 are as follows: Let e be an idempotent matrix of minimal rank in $\mathcal{S}_1$. Then, there exist α, α_1, β, γ in $P(\mathcal{S})$ such that

$$Q_0 = \alpha * \beta * \gamma;$$
$$Q_1 = \alpha_1 * \beta * \gamma,$$

where $S(\beta)$ is a finite group such that $S(\beta) = e\mathcal{S}_1 e$ and β is the uniform distribution of this group, $S(\gamma) \subset$ {the idempotent elements in $e\mathcal{S}_1$}, $S(\alpha_1) \subset S(\alpha) \subset$ {the idempotent elements in $\mathcal{S}_1 e$}.

Proof. Suppose that (i) and (ii) hold. Given $\epsilon > 0$, let $\mathcal{B}$ be a compact subset of $\mathcal{S}$ such that $Q_1(\mathcal{B}) > 1-\epsilon$ and $\mathcal{B} \subset$ {the matrices on $\mathcal{S}$ with no zero columns}. It is easy to verify that the set $\mathcal{B}^{-1}\mathcal{B} \equiv \bigcup_{C \in \mathcal{B}}\{A \in \mathcal{S} : CA \in \mathcal{B}\}$

is compact; also,

$$1-\epsilon < Q_1(\mathcal{B}) = \int Q_2^n\{A \in S : CA \in \mathcal{B}\}Q_1(dC)$$
$$\leq Q_2^n(\mathcal{B}^{-1}\mathcal{B}) + \epsilon.$$

This means that (Q_2^n) is tight and, therefore, there exists

$$Q_0 \in P(S) \ni \frac{1}{n}\sum_{m=1}^{n} Q_2^m \to Q_0,\ Q_0 = Q_0 * Q_0.$$

Also,

$$Q_0 = Q_2 * Q_0 = Q_0 * Q_2 \tag{5.8}$$

so that

$$Q_1 = Q_1 * Q_0. \tag{5.9}$$

By (5.7), (5.8) and (5.9), we have:

$$S(Q_0)S(Q_1) \subset S(Q_0). \tag{5.10}$$

Write: $Q_0 * Q_1 \equiv Q_3$. Then, we have:

$$Q_3 * Q_0 = Q_3 = Q_0 * Q_3 \tag{5.11}$$

and $S(Q_3) \subset S(Q_0)$, by (5.10). Since $S(Q_3)$ is an ideal of $S(Q_0)$ by (5.11) and $S(Q_0)$, Q_0 being idempotent, is simple, we have:

$$S(Q_3) = S(Q_0) \tag{5.12}$$

By Proposition 4.7 [7, p. 35], it follows that

$$Q_3 = Q_3 * Q_3. \tag{5.13}$$

For $A \in S(Q_0)$ and any Borel subset $\mathcal{B} \subset S$,

$$Q_3\{B \in S : BA \in \mathcal{B}\}$$
$$= \int Q_0\{B \in S : (BC)A \in \mathcal{B}\}Q_3(dC)$$
$$= Q_0\{B \in S : BA \in \mathcal{B}\},$$

since $Q_0 = Q_0 * Q_0$; therefore,

$$Q_0(\mathcal{B}) = \int Q_0\{B : BA \in \mathcal{B}\}Q_3(dA)$$
$$= \int Q_3\{B : BA \in \mathcal{B}\}Q_3(dA)$$
$$= Q_3(\mathcal{B}).$$

Thus, we have:

$$Q_0 * Q_1 = Q_0 \quad \text{and} \quad Q_1 * Q_0 = Q_1.$$

This means that

$$\begin{aligned} Q_1 = Q_1 * Q_0 &= (Q_1 * Q_0) * (Q_0 * Q_1) \\ &= (Q_1 * Q_0 * Q_0) * Q_1 \\ &= (Q_1 * Q_0) * Q_1 = Q_1 * Q_1. \end{aligned}$$

Also, for $A \in S(Q_0)$ and any Borel set $\mathcal{B} \subset \mathcal{S}$,

$$\begin{aligned} &Q_1\{B \in \mathcal{S} : AB \in \mathcal{B}\} \\ &= \int Q_0\{B \in \mathcal{S} : (AC)B \in \mathcal{B}\} Q_1(dC) \\ &= \int Q_0\{B \in \mathcal{S} : AB \in \mathcal{B}\} Q_1(dC) \\ &= Q_0\{B \in \mathcal{S} : AB \in \mathcal{B}\} \end{aligned}$$

so that

$$A.S(Q_1) = A.S(Q_0). \tag{5.14}$$

The rest of the theorem follows easily from (5.14) since Q_1 and Q_0 are both idempotent and under (iv), $S(Q_1) \subset S(Q_0)$. Note that the product representations of idempotent probabilities are discussed in Theorem 3.16, p. 25 [7]. ∎

Before we end this paper, we present a theorem, without proof, that provides a relevance of what we discussed above to attrractors (and fractals).

Theorem 5.6. *Consider a family $\mathcal{F}$ of affine maps on R^d such that one of the two following conditions holds:*

(a) Each T in $\mathcal{F}$ is of the form $T(x) = Ax + b$, where A is a $d \times d$ nonnegative matrix and b is a nonnegative d-vector; also, $\mathcal{F}_0 \equiv \{T \in \mathcal{F} :$ T is a constant map, that is, the A-part of T is a zero matrix$\}$ is nonempty.

(b) Each T in $\mathcal{F}$ is of the form $T(x) = Ax + b$, where A is a $d \times d$ real (not necessarily nonnegative) matrix and b is a real d-vector such that there are fixed positive numbers α and $r (r < 1)$, independent of A and b for T in $\mathcal{F}$, with the property that $\|A\| < r$ and $\|b\| < \alpha$.

Let Q be a distribution on $\mathcal{F}$ (such that in the case of (a), $Q(\mathcal{F}_0) > 0$). (Actually this restrictive condition can be avoided. See Theorem 5.7 (iv).)

Then the attractor $\mathcal{A}(x)$ corresponding to the function system $(\mathcal{F}, R^d)$, defined by

$$\mathcal{A}(x) \equiv \{y \in R^d : Pr(W_n(x) \in N(y)\ i.o.) > 0 \quad \text{for every open} \quad N(y) \quad \text{containing} \quad y\},$$

where (W_n) is the random walk induced by Q and taking values in the closed multiplicative (with respect to composition of functions) semigroup generated by $\mathcal{F}$, is independent of x in R^d (R^d to be replaced by $(R^+)^d$, in case of (a)), and is the support of the unique Q-invariant probability on R^d ($(R^+)^d$, in case of (a)).

In Theorem 5.6(b), we have used contraction maps and the attractor is found to be the support of the unique Q-invariant probability. In two papers, one by J. H. Elton (Ergod. Th. and Dynam. Sys. 7, 1987) and the other by M. F. Barnsley, S. G. Demko, J. H. Elton and J. S. Geronimo (Ann. Inst. Henri Poincaré **24**(3), 1988), a Markov process with transition probability

$$P(x, B) = \sum_{i=1}^{N} p_i(x) I_B(w_i x),$$

where $0 \le p_i(x) \le 1$, $\sum_{i=1}^{N} p_i(x) = 1$, and each w_i is a Lipschitz map from a metric space (in which sets of finite diameter are relatively compact), is studied. The w_i's in these papers are assumed to be contractive on the average (and some conditions are imposed on the p_i's) to show the existence of a unique invariant probability for the Markov process and to prove an ergodic theorem for this process. As will be clear from our Theorem 5.7(ii), "average contractivity" is really a form of tightness. When the family of affine maps is infinite, a form of "uniform average contractivity" has been used by Berger and Soner (J. Th. Prob. 1, No. 3, 1988) to obtain an ergodic theorem for random affine maps. We show here that in the context of the case (a) in Theorem 5.6, the right conditions for the existence of a unique invariant measure ν are that the sequence (Q^n) be tight and that the rank of the matrices in mS, where mS is the set of all matrices in S with the minimal rank, be one; under these conditions, we present also an ergodic theorem like that of Elton and identify almost all of the attractor set $\mathcal{A}(x)$.

Theorem 5.7. *Consider a family $\mathcal{F}$ of affine maps on $(R^d)^+$ such that each T in $\mathcal{F}$ is of the form*

$$T(x) = A(T)x + b(T),$$

where $A(T)$ is a $d \times d$ nonnegative matrix and $b(T)$ is a $d \times 1$ vector. We identify each such T with the $(d+1) \times (d+1)$ nonnegative matrix

$$\theta(T) \equiv \begin{bmatrix} A(T) & b(T) \\ 0 & 0 \end{bmatrix}.$$

This gives an isomorphism θ between $\mathcal{F}$ (with composition of maps) and a set of $(d+1)\times(d+1)$ nonnegative matrices, the last row in each of which is $(0,0,\ldots,0,1)$. Thus the vector $T(x)$ is the vector y, where $\begin{bmatrix} y \\ 1 \end{bmatrix} = \begin{bmatrix} A(T) & b(T) \\ 0 & 1 \end{bmatrix} \begin{bmatrix} x \\ 1 \end{bmatrix}$. *Let Q be a distribution on $\mathcal{F}$. Let S be the closed (w.r.t the usual topology for matrices) multiplicative (usual matrix multiplication) semigroup generated by the matrices in $\mathcal{F}$. Then the following results hold:*

(i) If (Q^n) is tight, then for any $\mu \in P((R^d)^+)$,

$$Q*(Q_0*\mu) = Q_0*\mu,$$

where $Q_0 = (w)\lim \frac{1}{n}\sum_{k=1}^{n} Q^k$.

Furthermore, if the minimal rank of the matrices in S is one (which holds iff there are $T_1, T_2, \ldots, T_m$ in $\mathcal{F}$ such that $\|A(T_1)A(T_2)\ldots A(T_n)\| < 1$ for some positive integer n), then the invariant probability $\nu, \nu = Q*\nu$, is unique, and given by

$$\nu(B) = Q_0 \quad \left\{ \begin{bmatrix} 0 & z \\ 0 & 1 \end{bmatrix} \in S : z \in B \right\}, \quad \textit{where } Q_0 \textit{ is as before.}$$

(ii) Suppose that the following three conditions hold:
(a) $Q\{T \in \mathcal{F} : \theta(T)$ has a zero row or a zero column$\} = 0$;
(b) then there exists a positive integer N such that $Q^N\{T \in S :$ for $1 \le i \le d$, the ith row of $A(T)$ is strictly positive whenever $b_i(T)$ is zero$\} > 0$;

(Note that condition (b) holds whenever S contains some T for which $A(T)$ is strictly positive.)

(c) there exists $\mu \in P((R^d)^+)$ such that $Q*\mu = \mu$ and μ is not the unit mass at $0 = (0,0,\ldots,0)$. Then the sequence (Q^n) is tight. Furthermore, if μ in (c) is unique, then there are maps $T_1, T_2, \ldots, T_m$ in $\mathcal{F}$ such that

$$\|A(T_1)A(T_2)\ldots A(T_m)\| < 1 \tag{5.15}$$

(iii) Suppose that (Q^n) is tight and (5.15) holds. Let (X_i) be i.i.d. on $\mathcal{F}$ with distribution Q. Then for any x in $(R^d)^+$ and for any bounded continuous function f on $(R^d)^+$, we have:

$$\lim_{n\to\infty} \frac{1}{n}\sum_{k=1}^{n} f(X_k X_{k-1}\ldots X_1 x) = \int f\, d\mu$$

almost surely. (For this part, $(R^d)^+$ can be replaced by R^d with no change in the statement of the result.) Here, μ is the unique Q-invariant probability on $(R^d)^+$, which exists by (i).

(iv) Suppose that (Q^n) is tight, (5.15) holds, and also Q is such that Q^n does not converge weakly to the unit mass at the zero map (that is, the map T where $T(x) = 0$ for all x) and condition (b) in (ii) holds. Then for $x \in (R^d)^+$ with $x > 0$, the attractor $\mathcal{A}(x)$, as defined in Theorem 5.6, contains the set

$$\mathcal{A}_1(x) = \{A \cdot x : A \in \mathcal{S} \quad \text{and for each} \quad B \in \mathcal{S},\ \text{rank}\, A \leq \text{rank}\, B\}$$

and in $y \in \mathcal{A}(x) - \mathcal{A}_1(x)$, then $y_i = 0$ for some i, $1 \leq i \leq d$. Also the invariant probability μ for Q is unique and $S(\mu) = \mathcal{A}_1(x)$ for $x > 0$.

Proof. Assertion (i) is clear from earlier discussions. The uniqueness of the invariant measure follows from the fact that $S(Q_0) = m(\mathcal{S})$, the set of matrices in $\mathcal{S}$ with minimal rank, and, when this rank is one, then for $T \in m(\mathcal{S})$ and $x \in R^d$, the vector $T(x)$ is independent of x since the matrix $A(T)$ is then the zero matrix. Notice also that $Q_0 * \mu = \mu$ implies that $Q_0 * \mu = \mu$.

Proof of (ii). Let us write

$$J = \{T \in \mathcal{F} : \theta(T) \quad \text{has a zero row or a zero column}\}.$$

By assumption (a), $Q(J) = 0$. It is easily verified that $Q^n(J) = 0$ for each positive integer n and J^c, identified with $\theta(J^c)$, is a multiplicative semigroup. Also, if we consider the set

$$\begin{aligned} I \equiv \{T \in J^c : \ &\text{in}\, \theta(T), \text{ for } 1 \leq i \leq d, \text{ the } i\text{th row of} \\ &A(T) \quad \text{is strictly positive whenever} \quad b_i(T)\, \text{is zero}\}, \end{aligned}$$

then I is an ideal of J^c. [We are always working inside $\mathcal{S}$.] Then by assumption (b), we have:

$$\lim_{n \to \infty} \quad Q^n(I) = 1 \tag{5.16}$$

(Notice that $Q^n(J^c) = 1$ for all n, and $Q^n(I)$ is an increasing sequence in n; also, $Q^n(I^c) \leq [Q(I^c)]^n$.)

Now we assume condition (c) so that we have:

$$Q * \mu = \mu,\ \mu \neq \delta_{\{0\}},\ \mu \in P((R^d)^+). \tag{5.17}$$

Suppose that $\mu\{0\} = p > 0$. Then for $\epsilon > 0$ and n sufficiently large,

$$\begin{aligned} 0 < p &= Q^n * \mu\{0\} \\ &= \int Q^n \left\{ \begin{bmatrix} A & b \\ 0 & 1 \end{bmatrix} \in \mathcal{S} Ax + b = 0 \right\} \mu(dx) \\ &= \int_{\{0\}} + \int_{\{0\}^c} \leq \int_{\{0\}} + p\epsilon \\ &\quad + \int_{\{0\}^c} Q^n \left\{ \begin{bmatrix} A & b \\ 0 & 1 \end{bmatrix} \in I : Ax + b = 0 \right\} \mu(dx). \end{aligned}$$

Notice that

$$\begin{bmatrix} A & b \\ 0 & 1 \end{bmatrix} \in I \quad \text{and} \quad Ax + b = 0 \Rightarrow x = 0.$$

Thus, it follows that

$$\lim_{n \to \infty} Q^n \left\{ \begin{bmatrix} A & 0 \\ 0 & 1 \end{bmatrix} : A \quad \text{is a} \quad d \times d \quad \text{nonnegative matrix} \right\} = 1 \quad (5.18)$$

Define λ by

$$\lambda = [\mu - p \cdot \delta_{\{0\}}]/(1 - p).$$

Then $\lambda \in P((R^d)^+)$ and $\lambda\{0\} = 0$. Also, we have: for $n \geq 1$,

$$Q^n * \lambda = [\mu - p \cdot Q^n * \delta_{\{0\}}]/(1 - p). \tag{5.19}$$

Let us now define the set $\mathcal{B}$ by

$$\mathcal{B} = \{x \in (R^d)^+ : x > 0\}.$$

Then using (5.16) and the fact that $\lambda\{0\} = 0$, it follows that

$$\lim_{n \to \infty} Q^n * \lambda(\mathcal{B}) = 1. \tag{5.20}$$

It follows from (5.19) and (5.20) that

$$\mu(\mathcal{B}) - pQ^n * \delta_{\{0\}}(\mathcal{B}) \to 1 - p$$

as $n \to \infty$, and therefore, by (5.18),

$$\mu(\mathcal{B}) = 1 - p, \quad \lambda(\mathcal{B}) = 1. \tag{5.21}$$

Let $\mathcal{H}$ be a compact subset of $\mathcal{B}$ such that $\lambda(\mathcal{H}) > 1 - \epsilon$. Then for n sufficiently large,

$$Q^n * \lambda(\mathcal{H}) \geq 1 - \epsilon,$$

and it follows that for n large,

$$Q^n \left\{ \begin{bmatrix} A & b \\ 0 & 1 \end{bmatrix} \in \mathcal{S} : Ax + b \in \mathcal{H} \quad \text{for some} \quad x \in \mathcal{H} \right\} > 1 - 2\epsilon.$$

It is easily verified that the set above inside the parenthesis is compact, and therefore, the sequence (Q^n) is tight.

We prove the last part of (ii). Since (Q^n) is tight,

$$\frac{1}{n} \sum_{k=1}^{n} Q^k \quad \text{converges weakly to} \quad Q_0,$$

and $S(Q_0) = m(S)$. Suppose that the minimal rank of the matrices in S is greater than 1. Then there exists $p > 0$ such that

$$Q_0\left\{\begin{bmatrix} A & b \\ 0 & 1 \end{bmatrix} \in S : \max_{i,j} A_{ij} > p\right\} > p. \tag{5.22}$$

We now assume that the invariant measure μ in condition (c) is unique. This means that

$$\mu = Q_0 * \delta_{\{x\}} \tag{5.23}$$

for any $x \in (R^d)^+$. Let $\mathcal{D}$ be a compact subset of $(R^d)^+$ such that

$$\mu(\mathcal{D}) > 1 - \frac{p}{2}. \tag{5.24}$$

Consider the point x given by

$$x = (2m/p, 2m/p, \ldots, 2m/p), \tag{5.25}$$

where m is such that

$$\max\{y_i : 1 \leq i \leq d,\ y \in \mathcal{D}\} < m.$$

Using (5.23), (5.24) and (5.25), we have:

$$1 - \frac{p}{2} < Q_0 * \delta_{\{x\}}(\mathcal{D}) \leq Q_0\left\{\begin{bmatrix} A & b \\ 0 & 1 \end{bmatrix} \in S : \max_{i,j} A_{ij} < p\right\},$$

which contradicts (5.22). This means that there must exist in S at least one matrix with rank one. This implies (5.15).

Proof of (iii). It is clear from earlier discussions that in this case there is a *unique* invariant probability $\mu \in P((R^d)^+)$ such that $Q * \mu = \mu$. Let f be a bounded continuous function on $(R^d)^+$. We can then define, for any $x \in (R^d)^+$, a bounded continuous function on S such that

$$g(A) = f(Ax), \quad A \in S.$$

One can then follow the paper of Furstenberg and Kifer (Israel J. Math. **46**, 1983) to establish the conclusion in (iii).

Proof of (iv). The proof of this case is not straightforward. We sketch the proof only briefly. First, it is not difficult to verify that for $x > 0$, the attractor set $\mathcal{A}(x)$ coincides with the set

$$\{Ax : Pr(W_n \in N(A) i.o.) > 0 \quad \text{for every open} \quad N(A) \ni A\}.$$

[Recall that $W_n = X_n X_{n-1} \dots X_1$, where the sequence (X_i) is i.i.d. with values in $\mathcal{F}$ and distribution Q.]

Next, in can be shown that when (Q^n) is tight, then the closure of the union of the supports of all weak*-limit points of (Q^n) is exactly the set $m(\mathcal{S})$ of all matrices in $\mathcal{S}$ with minimal rank. This means that the set of all positive recurrent states (that is, those matrices A in $\mathcal{S}$ for which $\lim_{n\to\infty} \sup Q^n(N(A)) > 0$ for every open $N(A)$ containing A) of the random walk (W_n) is exactly $m(\mathcal{S})$.

Now notice that if $m(\mathcal{S})$ has rank one and has more than one element, then condition (b) in (ii) implies that $m(\mathcal{S})$ contains at least one element with no zero row, since then the element $B \cdot A$, where $A \in M(\mathcal{S})$, $A \neq \begin{bmatrix} 0 & 0 \\ 0 & 1 \end{bmatrix}$ and $B \in I$ (I has been defined earlier), is in $m(\mathcal{S})$ and has no zero row. Also, we can prove that the set

$$\{A \in \mathcal{S} : A \quad \text{has no zero row and} \quad \sum_{n=1}^{\infty} Q^n(N(A)) = \infty$$
$$\text{for every open set} \quad N(A) \quad \text{containing} \quad A\}$$

is the smallest ideal of the multiplicative semigroup

$$\mathcal{S}_0 = \{A \in \mathcal{S} : A \quad \text{has no zero row}\}$$

and as such, must consist only of matrices in $\mathcal{S}_0$ with the minimal rank. It follows that every recurrent state for the random walk (W_n) that has no zero row must be contained in $m(\mathcal{S})$. Also, the support of the unique invariant probability μ, $Q * \mu = \mu$, $\mu = Q_0 * \delta_{\{x\}}$, $x \in (R^d)^+$ is exactly the set

$$\{Ax : A \in m(\mathcal{S}) \ = \{y \in (R^d)^+ : \begin{bmatrix} 0 & y \\ 0 & 1 \end{bmatrix} \in m(\mathcal{S})\}.$$

The rest is clear. ∎

Here is an example illustrating Theorem 5.6.

Example 5.7. Take $d = 2$ and

$$\mathcal{F} = \{f_1, f_2, f_3, f_4, f_5\},$$

where

$$f_1(x,y) = \left[\frac{x}{2}, \frac{y}{2}\right], \quad f_2(x,y) = \left[\frac{x+1}{2}, \frac{y}{2}\right],$$

$$f_3(x,y) = \left[\frac{x}{2}, \frac{y+1}{2}\right], \quad f_4(x,y) = (2x, y), f_5(x,y) = \left[\frac{1}{2}, \frac{1}{2}\right].$$

Note that f_i (for each i) can be identified with a unique 3×3 nonnegative matrix A_i such that

$$(f_i(x,y),1)^T = A_i \begin{bmatrix} x \\ y \\ 1 \end{bmatrix}.$$

For example, $f_2 \leftrightarrow \begin{bmatrix} \frac{1}{2} & 0 & \frac{1}{2} \\ 0 & \frac{1}{2} & 0 \\ 0 & 0 & 1 \end{bmatrix}$. Let Q be a probability distribution such that $S(Q) = \{f_1, f_2, f_3, f_4, f_5\}$. Then the attractor of $\mathcal{F}$ is verified to be

$$\{(x,y) : (x,y,c) = A \cdot \begin{bmatrix} a \\ b \\ c \end{bmatrix}, \quad \text{for} \quad A \in m(\mathcal{S})\}, \quad \text{where}$$

(a, b, c) is any point in R^3 and $m(\mathcal{S})^1$ is the set of all matrices with minimal rank in $\mathcal{S}$, the closed multiplicative semigroup generated by $S(Q)$. A computer graph of this attractor is shown below. (See Diagram 5.2.) If f_4 and f_5 were dropped from the set $\mathcal{F}$, then the attractor would be what is popularly known as Sierspinsky gasket. (See Diagram 5.1.) If $\mathcal{F}$ above were chosen to be the set

$$\{T_1, T_2, T_3, T_4\},$$

where the T_i's are given by:

$$T_1(x,y) = (.856x + .0414y + .07, -.025x + .858y + .147),$$
$$T_2(x,y) = (.244x - .385y + .393, .176x + .224y + .102),$$
$$T_3(x,y) = (-.144x + .39y + .527, .181x + .259y - .014),$$
$$T_4(x,y) = (.486, .031x + .216y + .05),$$

Diagram 5.1.

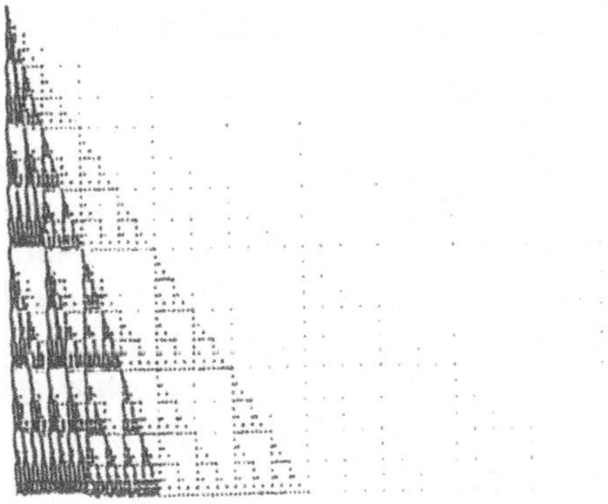

Diagram 5.2.

then the attractor of the system of these four maps (as maps from R^2 into R^2) is shown in Diagram 5.3. This last example and its diagram are due to M. Barnsley and his co-workers and they will illustrate Theorem 5.6.

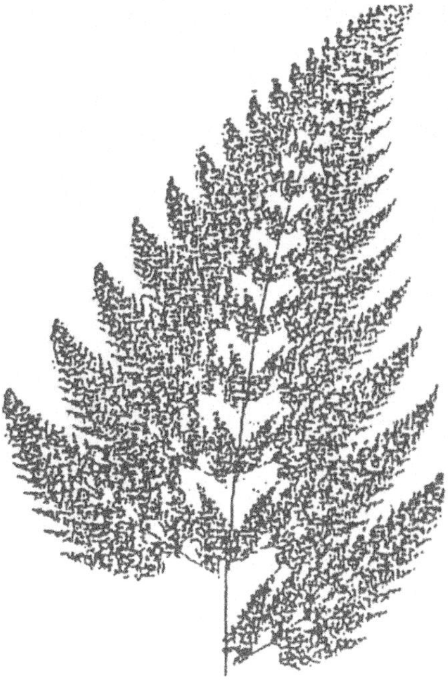

Diagram 5.3.

Acknowledgment. The author thanks R. W. R. Darling for many discussions on this subject. His recent paper [2] is relevant in the present context.

REFERENCES

1. Bougerol, P., *Tightness of products of random matrices and stability of linear stochastic systems,* Ann. Prob., **15**(1987), 40–74.
2. Darling, R. W. R., *The Lyapunov exponent for products of infinite dimensional random matrices,* (1990), (preprint).
3. Liggett, T. M., *Interacting Particle Systems,* (1985), Springer–Verlag, New York.
4. Kesten, H. and Spitzer, F., *Convergence of distribution of products of random matrices,* Z. Wahrscheinlichkeitstheorie verw. Gebeite, **67** (1984), 363–386.
5. Mukherjea, A., *Convergence in distribution of products of random matrices: a semigroup approach.* Trans. Amer. Math. Soc., **303**(1), (1987), 395–411.
6. Mukherjea, A., *Tightness of products of i.i.d. random matrices.* Prob. Theo. and Rel. Fields, **87**(1991), 389–401.
7. Mukherjea, A. and Tserpes, N. A., *Measures on topological semigroups: Convolution products and random walks,* Springer–Verlag LNM, **547** (1976).

A. Mukherjea
University of South Florida,
Tampa, Florida 33620-5700.

PART IV

Invariant Measures in Real and White Noise-Driven Systems

REMARKS ON ERGODIC THEORY OF STOCHASTIC FLOWS AND CONTROL FLOWS

FRITZ COLONIUS AND WOLFGANG KLIEMANN

ABSTRACT. Stochastic systems and control systems with values in a state space M can be considered as dynamical systems on the space $\mathcal{U} \times M$, where $\mathcal{U}$ denotes the space of admissible control functions for control systems, and the trajectory space of an underlying noise process for stochastic systems. Invariant probability measures for these flows are the main topic of this paper: We show that their support is contained in sets $\mathcal{D} \subset \mathcal{U} \times M$, which are the lifts of so-called control sets $D \subset M$ to invariant sets in $\mathcal{U} \times M$. Several results on the characterization of control sets D are given, together with criteria for the existence of invariant measures μ on $\mathcal{U} \times M$ with supp $\mu \subset \mathcal{D}$. The case of Markovian stochastic systems is treated in some detail. Because of the importance in applications, we prove rather complete results for two classes of systems: linearized systems, which play a crucial role in the theory of Lyapunov exponents for stochastic and control flows, and general nonlinear systems with one dimensional state space, which are important in stochastic bifurcation theory.

1. INTRODUCTION: COMMON TECHNIQUES FOR STOCHASTIC, CONTROL AND DYNAMICAL SYSTEMS

Common ideas and approaches in the theories of continuous time dynamical systems and of (Markovian) stochastic systems go back at least to the 1930's, when Kolmogorov introduced rigorously the generator of certain diffusion processes as a second order (elliptic) operator. With Itō's formulation of stochastic calculus, and Dynkin's characterization of those operators that are generators of diffusion processes, many techniques that had been developed for ordinary differential equations and their flows could be carried over to stochastic differential equations, yielding in particular results on invariant sets, recurrence and transience, invariant measures, ergodicity, stability, and stochastic perturbations of

1980 *Mathematics Subject Classification* (1985 *Revision*). 93C10, 60H10, 58F11, 54H20.

Key words and phrases. dynamical systems, control systems, stochastic systems, Markov processes, control sets, chain control sets, invariant measures.

Research supported in part by NSF grant no. DMS 8813976 and DFG grants no. Co124/6-1 and Co124/8-1 .

(deterministic) differential equations. (It should be noted that corresponding results for discrete time systems have been discovered at the same time or even earlier (see e.g. Kolmogorov (1931)), and that this correspondence also works for the newer developments in our area, but the discrete time situation is not the topic under consideration in this paper.)

The first important connection between (optimal) control theory and stochastic systems was probably the celebrated duality between the linear quadratic control problem and the Kalman-Bucy filter (1961), which led to an almost parallel development of linear optimal control for deterministic and stochastic control problems (compare e.g. the books of Kwakernaak and Sivan (1972) or Hijab (1987)). New ideas in nonlinear control theory, in particular the geometric approach of Brockett, Lobry, and Sussmann, and the so called support theorem of Stroock and Varadhan led since the 1970's to control theoretic descriptions of the supports of transition probabilities and invariant measures, and to characterizations of ergodicity, recurrence and transience. Some general uniqueness results on invariant measures for stochastic systems were then obtained using Hörmander's theorem (1967) on the characterization of hypoelliptic operators. (Again, similar approaches with corresponding results also hold for the discrete time case.)

Recently, the common viewpoint of flows of dynamical systems, stochastic systems and control systems is leading to a further unification and cross fertilization of these areas. While the theory of smooth flows for (time independent) vector fields is classical (see e.g. the textbook of Nemytskii and Stepanov (1949)), the theory for stochastic flows has been developed in the 1980's by Kunita, Elworthy, Baxendale and others. The corresponding concept of control flows was introduced by Colonius and Kliemann (1990[a]). Besides questions about invariant measures and (smooth or measurable) ergodicity, which will be treated later on in this paper, it is in particular linearization techniques that have led to common developments, based on Oseledeč's multiplicative ergodic theory (1968), which can be interpreted as a result for dynamical systems, stochastic systems, or even control systems. In particular the areas of Lyapunov exponents, bifurcation theory, chaos, hyperbolicity and strange attractors are active research fields at the moment for systems with noise and for systems with control inputs, as these and other conference proceedings (e.g. Arnold et al. (1991)) demonstrate.

In this paper we will utilize ideas from ergodic theory of dynamical systems (see e.g. Mañé (1987)) to discuss existence and possible supports of invariant measures for control and stochastic flows, where it is assumed that the stochastic flows are defined over a probability space Ω, which is the trajectory space of an underlying (stationary) driving process. Over the same space, now viewed as the space of admissible control functions, control flows can be defined, and their interplay yields the results in Section 4, mainly with the Krylov-Bogolyubov construction of invariant probability measures. In Section 2, the precise concepts of stochastic and control flows are defined, and some examples are given. Since it turns out that the so-called 'control sets' and 'chain control sets' play a crucial role in the analysis of invariant measures, Section 3 is devoted to the study of dynamical properties of control flows, in particular to lifts of control and chain control sets. Two classes of examples, which are central in linearization techniques and in the study of bifurcation behavior, are treated in detail in Sections 3 and 4. An application of the theory developed here to linearized systems and their Lyapunov spectrum will appear elsewhere (Colonius and Kliemann (1991[a])).

But first of all it seems useful to describe a little bit more precisely the current state of common techniques in ergodic theory of dynamical, stochastic and control systems, and so the remainder of this introductory section is devoted to this topic.

Dynamical systems and Markovian stochastic systems

Ordinary differential equations with generator X, a smooth vector field, on a smooth manifold M, give rise to (local) flows and (local) one parameter groups of diffeomorphisms, describing the solutions of the differential equation. (From now on we will assume that all systems are complete, i.e. the explosion time of all trajectories is $\pm\infty$.) Generators of (Markov) diffusion processes given by stochastic differential equations $dx = X_0(x)dt + \sum_{i=1}^{m} X_i(x) \circ dW_i$ on M (where $(W_1 \ldots W_m)$ is a vector of independent standard Wiener processes, and "$\circ$" denotes the symmetric or Stratonovič stochastic differential) are second order (elliptic) operators $\mathcal{A} = X_0 + \frac{1}{2}\sum_{i=1}^{m} X_i^2$, acting on the space of bounded measurable (or C^∞-) functions. An associated one parameter semigroup (for $t \geq 0$, i.e. forward in time) is given by the transition probabilities $P(t, x, B)$, see e.g. Ethier and Kurtz (1986) for these facts in a much more general context. Qualitative theory, in particular ergodic theory, for dynamical and stochastic systems analyzes the long term behavior (i.e. $t \to +\infty$) of these (semi-)groups, without solving the equations explicitly. For problems like invariant sets, recurrence and transience, stability, etc. Lyapunov functions are one appropriate tool. The stochastic version of this theory is described e.g. in some detail in the books by Hasminskii (1969) or Friedman (1975), where it becomes clear that stochastic Lyapunov functions are a convenient tool for the generator $\mathcal{A}$ as are ordinary Lyapunov functions for the vector field X. For the existence of invariant probability measures of the above semigroups, a Krylov-Bogolyubov construction works in both cases: If the time averages of the semigroups yield a tight family of probability measures, then any accumulation point is an invariant measure (see e.g. Nemytskii and Stepanov (1949) for the deterministic, and Hasminskii (1969) for the stochastic case). Compactness of the state space M is in any case sufficient for the existence of at least one invariant measure (compare Kunita (1971) for the stochastic case). Ergodic theorems can now be developed in complete analogy, because for the stochastic case there is a one-to-one correspondence of invariant sets in the trajectory space and the state space for Markov processes. Lyapunov functions are again a convenient tool to obtain criteria for the existence of invariant measures etc. in terms of the vector field(s), see e.g. Bhatia and Szegö (1970) and Hasminskii (1969). For the problem of analyzing stochastic perturbations of (deterministic) vector fields via Markov theory, large deviation approaches have been developed (see e.g. Freidlin and Wentzell (1984) or also Zeeman (1988) for related ideas), combining qualitative theory for ordinary differential equations with that for Markov diffusion processes. Finally it should be mentioned that for non-flat manifolds the behavior of solutions of ordinary differential equations and stochastic differential equations depend on the (global) geometry of M; for many we mention only Emery (1990).

Control systems and Markovian stochastic systems

Besides the connections via optimal control theory (see the remarks above) it is in particular the support theorems that allow the use of control theoretic results for the analysis of stochastic systems: If we replace the Wiener processes in a stochastic differential equation by admissible control functions with values in $\mathbb{R}^m$, we arrive at the control system $\dot{x} = X_0(x) + \sum_{i=1}^{m} u_i(t)X_i(x)$. The support theorem says that the closure of the trajectory space of this control system is the support of the diffusion measure induced by the stochastic differential equation on this space (see e.g. Wong and Zakai (1969), Stroock and Varadhan (1972), Kunita (1974), Ikeda and Watanabe (1981)). In particular, if the distribution $\Delta_{\mathcal{L}}$, generated by the Lie algebra of vector fields $\mathcal{L} = \mathcal{LA}\{X_0, \ldots, X_m\}$ in the tangent bundle TM is integrable, then the control system and the stochastic system live on the maximal integral manifolds of $\Delta_{\mathcal{L}}$ and the support of the transition probabilities $P(t, x, B)$ and of invariant measures are described by accessible and control sets of the control system (see e.g. Brockett (1973), Sussmann and Jurdjevic (1972), Clark (1973), Elliott (1973), Kunita (1978), Kliemann (1987)). Together with Hörmander's characterization of hypoelliptic operators (Hörmander (1967)), one obtains uniqueness results for invariant measures on control sets and corresponding ergodic theorems (Kliemann (1987), Arnold and Kliemann (1987[a])). This is also the starting point for various results on the controllability of stochastic systems with input (see e.g. Zabczyk (1981), Ehrhardt and Kliemann (1982), Varsan (1982)). It should be mentioned that, among others, also the theory of large deviations for stochastic differential equations has an interpretation in terms of control theoretic concepts, which becomes particularly clear from Azencott's (1980) formulation, see also Arnold and Kliemann (1987[b]).

Dynamical, stochastic, and control flows

So far we have considered questions in the qualitative theory of stochastic systems that could be analyzed using the Markov semigroup (i.e. the generator) of a stochastic differential equation. Problems that are concerned with the long term behavior of different trajectories relative to each other (e.g. convergence or divergence of trajectories), i.e. properties of the multipoint motion, do not only depend on the generator $\mathcal{A}$, but on the stochastic flow induced by a stochastic differential equation (see e.g. Baxendale (1986[b]) for example). The theory of flows for stochastic differential equations was developed by Kunita (1984), Elworthy (1978), Baxendale (1980), and others. In particular, Baxendale (1986[b]) has shown that the stochastic differential equation for the 2-point motion (and its generator) are sufficient to construct the corresponding stochastic flow. Once a stochastic flow is given, it is of particular importance to characterize those flows that are associated to Markov processes, and to determine those invariant measures of the flow that are also invariant under the corresponding Markov semigroup (in general, a Markov stochastic flow can have invariant measures that are not Markovian). These are questions of appropriate measurability, and they are treated e.g. in Crauel (1987, 1990). Problems concerning invariant probability measures of stochastic flows (see e.g. LeJan (1986)) and associated control flows are treated in Section 4 of this paper. The unified formulation of systems as flows does not only allow a common approach to basic problems of ergodic theory, but also, via linearization techniques, to the analysis of local behavior of nonlinear systems: Given the linearization with respect to a stationary situation (e.g. a rest point or a stationary solution), Oseledeč's multiplicative ergodic theory (1968) describes the Lyapunov exponents and the corresponding invariant subspaces of the linearized system, and these subspaces can be projected down to the state space M as stable, unstable, or center manifolds (see e.g. Boxler (1989) or Dahlke (1989)). For linear stochastic systems (i.e. linearizations around rest points) the theory of Lyapunov exponents is fairly complete (see e.g. Arnold and Wihstutz (1986), Arnold and

Kliemann (1987^b)). (Note that for linear systems all trajectories are 'compared' with the unique steady state 0, and hence the theory of stochastic flows is not really needed.) For nonlinear stochastic systems basic properties have been proved e.g. by Carverhill (1985^a, 1985^b), Kifer (1986), Carverhill et al. (1986) (see also Arnold et al. (1991)), connections with stability, the existence of Lyapunov functions and invariant measures are discussed in Baxendale and Stroock (1988), and Baxendale (1990), and the relation to fiber entropy is analyzed e.g. in Ledrappier and Young (1985, 1988) and Crauel (1991). Based on the concept of Lyapunov exponents, Arnold and Boxler (1989, 1990) have defined a concept of stochastic bifurcations. Similar developments for control flows can be found e.g. for linear systems in Colonius and Kliemann (1990^c, 1991^b), and some basic properties for the nonlinear situation in Colonius and Kliemann (1990^a). Connections between the stochastic and the controlled situation are given in Arnold and Kliemann (1987^b), Colonius and Kliemann (1990^d), see also Baxendale and Stroock (1988). These connections lead e.g. to a common (deterministic and stochastic) concept for the analysis and stabilization of uncertain linear systems (see e.g. Willems and Willems (1983), Colonius and Kliemann (1990^d, 1990^e)).

This is, of course, only a very brief review, neglecting in particular the recent developments in the theory of dynamical systems, of discrete time systems (like products of random matrices, and iterated function systems), and of infinite dimensional systems. But it shows, how ideas from these three fields penetrate into the other areas, creating a common toolbox, and new applications and examples.

2. Systems and Associated Flows

The unifying point of view, which enables us to use common concepts and techniques for dynamical systems, stochastic systems and control systems, is the concept of topological flows:

Let S be a complete metric space, T a time set (e.g. $T = \mathbb{R}, \mathbb{R}^+, \mathbb{Z}, \mathbb{N}$), and $\Psi: T \times S \to S$ a continuous map, then (S, Ψ) is a flow (or continuous dynamical system) if $\Psi_{t+s} = \Psi_t \circ \Psi_s$, where $\Psi_t: S \to S$ denotes the map $\Psi_t(x) = \Psi(t,x)$ for all $t \in T$.

The most prominent examples of flows are those generated by (ordinary) differential equations: Let M be a smooth manifold and X a smooth complete vectorfield on M. Denote by $\psi(t,x)$ the solution of $\dot{x} = X(x)$ at time t with $\psi(0,x) = x$, then $\Psi: T \times M \to M$, $\Psi(t,x) = \psi(t,x)$ is a flow on M (with $T = \mathbb{R}$). In this case, the Ψ_t, $t \in \mathbb{R}$ are even diffeomorphisms. If X is not complete, then one obtains a local flow on M, and local flows are in $1-1$ correspondence with the vectorfields on M, see e.g. Boothby (1975), Chapter IV.4. If M is compact, then all vector fields are complete. If the vectorfield is time dependent, then the solutions of $\dot{x} = X(t,x)$ will, in general, not generate a flow on M, because $\psi(t+s,x,X(\tau,\cdot)) = \psi(t,\psi(s,x,X(\tau,\cdot)),X(\tau+s,\cdot))$, where $\psi(t,x,X(\tau,\cdot))$ denotes the solution of $\dot{x} = X(\tau,x)$ at time t with $\psi(0,x,X(\tau,\cdot)) = x$. For time dependent differential equations one therefore has to take the time shift into account, when formulating an associated flow. This will become important, when we discuss control systems and stochastic systems.

Next consider the nonlinear control system on a smooth manifold M

$$\dot{x} = X_0(x) + \sum_{i=1}^{m} u_i(t)X_i(x) \tag{1}$$

where $X_0,\ldots,X_m$ are given smooth vectorfields on M, and $(u_i)_{i=1\ldots m} =: u \in \mathcal{U} := \{u\colon \mathbf{R} \to U \subset \mathbf{R}^m$, locally integrable$\}$. Typical questions asked in control theory are e.g.: Given $x,y \in M$, does there exist a time $t \geq 0$ and a control $u \in \mathcal{U}$ such that $\varphi(t,x,u) = y$ (controllability); here $\varphi(t,x,u)$ is the solution of (1) at time t using the control function u such that $\varphi(0,x,u) = x$. Or given a rest point $x^0 \in M$ of $\dot{x} = X_0(x)$, does there exist a control function u such that x^0 becomes stable for (1), etc. (see e.g. Wonham (1979), Isidori (1989) or Nijmeijer and van der Schaft (1990) for a discussion of control theoretic problems).

If we want to formulate control systems as flows, we have to take the dependence of the solutions of (1) on the functions $u \in \mathcal{U}$ into account:

Denote by $\theta\colon \mathbf{R} \times \mathcal{U} \to \mathcal{U}$, $\theta_t u(\cdot) = u(t+\cdot)$ the usual time shift, and define

$$\phi\colon \mathbf{R} \times \mathcal{U} \times M \to \mathcal{U} \times M, \quad \phi(t,u,x) = (\theta_t u, \varphi(t,x,u)). \tag{2}$$

Then $\theta_{t+s} = \theta_t \circ \theta_s$, $\varphi(t+s,x,u) = \varphi(t,\varphi(s,x,u),\theta_s u)$, and therefore $\phi_{t+s} = \phi_t \circ \phi_s$. In particular we obtain (compare Colonius and Kliemann (1990^a, Lemma 3.3)): If $U \subset \mathbf{R}^m$ is compact and convex, equip $\mathcal{U}$ with the weak *-topology of $L^\infty(\mathbf{R},\mathbf{R}^m) = \left(L^1(\mathbf{R},\mathbf{R}^m)\right)^*$. Then (2) is a continuous dynamical system on the separable, complete metric space $\mathcal{U} \times M$, called the control flow associated with (1).

Now the typical objects of control theory for (1), like reachable sets etc., are projections of similar objects for (2), e.g. denote by $\mathcal{O}^+(x) = \{y \in M;$ there exist $t \geq 0$ and $u \in \mathcal{U}$ with $\varphi(t,x,u) = y\}$ the reachable set (or forward orbit) of $x \in M$ for (1), then $\mathcal{O}^+(x) = \bigcup_{u\in\mathcal{U}} \pi_M \phi(t \geq 0, x, u)$, where $\pi_M\colon \mathcal{U} \times M \to M$ is the projection onto the second component, and similarly for $\mathcal{O}^-(x)$, the set of points, from which x can be reached. In Section 3. we will discuss in more detail several connections between control theoretic concepts for (1) and notions from the theory of dynamical systems for (2).

Stochastic flows are abstractly defined in the following way: Let $(\Omega, \mathcal{F}, P)$ be a probability space and $\theta_t : \Omega \to \Omega$, $t \in T$, a family of measurable maps such that $\theta_{t+s} = \theta_t \circ \theta_s$ and $\theta_t P = P$ for all $t \in T$, i.e. P is a θ-invariant probability measure. Let M be a Polish space (separable, complete, metric) and $\mathcal{B}$ its Borel σ-algebra. A stochastic flow is then a measurable map

$$(3) \qquad \phi : T \times \Omega \times M \to \Omega \times M, \quad \phi(t, w, x) = (\theta_t w, \varphi(t, w, x))$$

such that $\varphi(t, \cdot, w) : M \to M$ is a homeomorphism for all (t, w),
and $\varphi(t+s, x, w) = \varphi(t, \varphi(s, x, w), \theta_t w)$ for all $s, t \in T$.
Again, this implies $\phi_{t+s} = \phi_t \circ \phi_s$.

In the present context, two types of stochastic flows are of particular interest:
Random differential equations: Let $\{\xi_t,\ t \in \mathbf{R}\}$ be a stationary (ergodic) stochastic process taking values in some Polish space $(N, \mathcal{N})$, where $\mathcal{N}$ denotes the Borel σ-algebra of N. Let Ω be the trajectory space of $\{\xi_t,\ t \in \mathbf{R}\}$, and construct the measure P from the finite dimensional distributions of $\{\xi_t,\ t \in \mathbf{R}\}$ via the Kolmogorov construction. Denote by $\zeta : \mathbf{R} \times \Omega \to N$ the evaluation map $\zeta(t, w) = w(t)$, then $\{\zeta_t,\ t \in \mathbf{R}\}$ is a stationary (ergodic) stochastic process with the same finite dimensional distributions as $\{\xi_t,\ t \in \mathbf{R}\}$, and the probability measure P is θ-invariant, where θ is the shift on Ω (compare e.g. Rozanov (1967, Chapter 4.)). Now let X be a measurable map from N into the smooth vectorfields on a smooth manifold M, and consider the random differential equation $\dot{x} = X(x, \zeta_t(w))$ on M, and assume that all its solutions $\varphi(t, x, w)$ do not explode in finite time (this holds e.g. if M is compact). Then $\phi = (\theta, \varphi)$ defines a stochastic flow.

Stochastic differential equations: Let M be a smooth manifold and $X_0 \dots X_m$ smooth vector fields on M. Consider the stochastic differential equation $dx = X_0(x)dt + \sum_{i=1}^{m} X_i(x) \circ dW_i$, where the W_i are independent standard Wiener processes and "$\circ$" denotes the symmetric or Stratonovič stochastic integral. Denote by $(\Omega, \mathcal{F}, P)$ the Wiener space of continuous functions into $\mathbf{R}^m$, vanishing at zero, with the Wiener measure P. P is invariant (and ergodic) with respect to the Wiener shift on Ω, defined by $\theta_t w(\cdot) = w(t + \cdot) - w(t)$. Under certain regularity conditions on the vector fields $X_0 \dots X_m$, $\phi = (\theta, \varphi)$ defines a stochastic flow (of diffeomorphisms on M) for $t \geq 0$, where $\varphi(t, x, w)$ denote the (pathwise) solution of the stochastic differential equation (see e.g. Kunita (1984, Chapter II)). If we extend the Wiener process backwards in time for $t \leq 0$ with an independent copy, we obtain with the same construction a stochastic flow for $t \in \mathbf{R}$.

We have seen that stochastic flows and control flows are flows of homeomorphisms (or even diffeomorphisms) over a shift space of trajectories (the trajectories of an underlying stochastic process, or the admissible control functions, respectively) i.e. skew product flows. If they are defined over the same function space (and with the same dynamics on M), then the difference is basically that for stochastic flows the trajectory space carries an additional shift invariant probability measure. In this sense one can talk of the control system associated with a random or stochastic differential equation (in the latter case one has to consider admissible controls, which vanish at zero), and vice versa each control system has associated with it a class of stochastic flows, determined by all θ-invariant probability measures on $(\mathcal{U}, \theta)$. The goal of this paper is to discuss ergodic properties of such stochastic flows, using control theoretic concepts and aspects of the theory of dynamical systems, applied to control flows. This point of view will be discussed in the next section.

3. Dynamical Properties of Control Flows

In this section we will characterize several properties of control systems using concepts from the theory of dynamical systems. This will enable us to analyze invariant measures and their supports in Section 4. While most of the theory developed here also works for the discrete time case (i.e. $T = \mathbf{Z}$ or $\mathbf{N}$), we will restrict ourselves to $T = \mathbf{R}$.

We will need the following concepts (see e.g. Mañé (1987) and Conley (1978)):

Definition 3.1. Let (S, Ψ) be a continuous dynamical system. For $x \in S$ the *limit set* $\omega(x)$ is defined as $\omega(x) = \{y \in S;$ there exists $t_k \to \infty$ with $\Psi(t_k, x) \to y\}$. (S, Ψ) is *topologically transitive*, if there exists $x \in S$ with $\omega(x) = S$, and *topologically mixing*, if for any two open sets $V_1, V_2 \subset S$ there exist $T_0 \in \mathbf{R}$, $T_1 > 0$ such that for all $n \in \mathbf{N}$ $\Psi(-nT_1 + T_0, V_1) \cap V_2 \neq \phi$.

A closed Ψ-invariant subset $W \subset S$ is called a *maximal topologically mixing set* if $(W, \Psi|_W)$ is topologically mixing and every closed Ψ-invariant set $W' \supset W$, for which $(W', \Psi|_{W'})$ is topologically mixing, satisfies $W' = W$. Analogously maximal topologically transitive sets are defined.

Consider the nonlinear control system on a paracompact, C^∞ Riemannian manifold M

$$\dot{x} = X_0(x) + \sum_{i=1}^{m} u_i(t) X_i(x) \tag{1}$$

where $X_0, \ldots, X_m$ are C^∞ vector fields, $(u_i) = u \in \{u : \mathbf{R} \to U,$ locally integrable$\}$, $U \subset \mathbf{R}^m$ is compact and convex. We define for such a control system:

Definition 3.2. The *positive orbit* of (1) from $x \in M$ is given by

$$\mathcal{O}^+(x) = \{y \in M;\ \text{there is } t \geq 0 \text{ and } u \in \mathcal{U} \text{ such that } y = \varphi(t, x, u)\}.$$

$D \subset M$ is called a *control set* of (1) if (i) $D \subseteq \overline{\mathcal{O}^+(x)}$ for all $x \in D$, (ii) for all $x \in D$ there exists $u \in \mathcal{U}$ such that $\varphi(t, x, u) \in D$ for all $t \geq 0$, and (iii) D is maximal with respect to these properties.

The system (1) is *completely controllable* if $\mathcal{O}^+(x) = M$ for all $x \in M$.

In order to avoid degenerate situations we will assume that $\mathcal{O}^+(x)$ (and also the negative orbit $\mathcal{O}^-(x) = \{y \in M;$ there is $t \geq 0$ and $u \in \mathcal{U}$ such that $x = \varphi(t, y, u)\}$) have nonvoid interior in M. To ensure this property, it is convenient to assume

$$\text{(H)} \qquad \dim \mathcal{LA}\{X_0 + \sum u_i X_i,\ (u_i) \in U\}(x) = \dim M \text{ for all } x \in M,$$

where for a set $\mathcal{X}$ of vector fields $\mathcal{LA}\{\mathcal{X}\}$ denotes the Lie algebra generated by $\mathcal{X}$, and $\mathcal{LA}\{\mathcal{X}\}(x)$ is the linear subspace of $T_x M$, the tangent space at x, which is spanned by $\mathcal{LA}\{\mathcal{X}\}$, compare Isidori (1989) and Nijmeijer and van der Schaft (1990) for a detailed discussion of (H). Here it suffices to note the following consequences:

- If $\mathcal{V} \subset \mathcal{U}$ is a dense subset, then the control sets defined via $\mathcal{V}$ are the same as those defined via $\mathcal{U}$. This applies in particular to the continuous, piecewise constant, or periodic control functions in $\mathcal{U}$.
- If $D \subset M$ is a control set with int $D \neq \phi$, then for all $x \in D$, all $y \in \text{int } D$ there exist $t \geq 0$ and $u \in \mathcal{U}$ with $\varphi(t, x, u) = y$, i.e. we have precise controllability in int D. In particular $\overline{\mathcal{O}^+(x)} = M$ for all $x \in M$ implies $\mathcal{O}^+(x) = M$ for all $x \in M$, i.e. complete controllability.

Consider now the control flow induced by (1) on $\mathcal{U} \times M$

$$\text{(2)} \qquad \phi: \mathbf{R} \times \mathcal{U} \times M \to \mathcal{U} \times M, \quad \phi = (\theta, \varphi).$$

We lift the control sets $D \subset M$ with nonvoid interior to ϕ-invariant sets on $\mathcal{U} \times M$ via

$$\text{(3)} \qquad \mathcal{D} = cl\{(u, x) \in \mathcal{U} \times M;\ \varphi(t, x, u) \in \text{int } D \text{ for all } t \in \mathbf{R}\},$$

where the closure is taken with respect to the weak*-topology on $\mathcal{U}$ and the manifold topology on M. If we are looking for properties of ϕ that are related to the control structure of (1), then these must be properties of $\mathcal{D}$. And, since the shift θ is not affected by the dynamics of (1), θ has to enjoy these properties as well. It turns out that topological mixing (and transitivity) are the appropriate concepts:

Theorem 3.3. (i) *The shift* $(\mathcal{U}, \theta)$ *is topologically transitive and mixing.*

(ii) *Let* $D \subset M$ *be a control set with int* $D \neq \phi$. *Then* $\mathcal{D}$, *defined by* (3), *is a maximal topologically mixing (and a maximal topologically transitive) set with*

$$\text{(4)} \qquad \text{int } D = \text{int } \pi_M \mathcal{D}, \text{ and } \overline{D} = \pi_M \mathcal{D}.$$

(iii) *If* $\mathcal{D} \subset \mathcal{U} \times M$ *is a maximal topologically mixing (or transitive) set of* $(\mathcal{U} \times M, \phi)$ *with int* $\pi_M \mathcal{D} \neq \phi$, *then there exists a control set* $D \subset M$, *which satisfies* (4).

The proof is given in Colonius and Kliemann (1990[a], Proposition 2.6, Theorem 3.8, and Corollary 3.9). Note, in particular, that the lifted control sets $\mathcal{D} \subset \mathcal{U} \times M$ are topologically mixing and transitive. This leads to the following characterization of complete controllability:

Corollary 3.4. *Under the assumptions above, the following statements are equivalent:*

(i) *The control system* (1) *is completely controllable on* M.
(ii) *The dynamical system* $(\mathcal{U} \times M, \phi)$ *is topologically mixing.*
(iii) *The dynamical system* $(\mathcal{U} \times M, \phi)$ *is topologically transitive.*

The ergodic theory of stochastic flows and control flows deals with invariant measures of $(\mathcal{U} \times M, \phi)$ and their properties. In particular, the supports of invariant measures and therefore the ω-limit sets have to be characterized. It was shown in Colonius and Kliemann (1989, Proposition 3.6) that $\pi_M \omega(u,x) \subset M$ always has nonvoid intersection with some control set D. But it does not follow that $\pi_M \omega(u,x) \subset D$, nor that D is unique. This leads to the concept of chain control sets $E \subset M$ and their lifts to $\mathcal{E} \subset \mathcal{U} \times M$, which always contain entire ω-limit sets. (For the connection of chain control sets with subbundle decompositions of linear flows on vector bundles see Colonius and Kliemann (1990[b]).)

Definition 3.5. Let (S, Ψ) be a continuous dynamical system on a metric space (S, d). For $\varepsilon > 0$ and $T > 0$ an (ε, T)-*chain* from $x \in S$ to $y \in S$ consists of a sequence $x_0, \ldots, x_n \in S$ and a sequence $t_0, \ldots, t_{n-1}$ in $\mathbb{R}$ such that $x_0 = x$, $x_n = y$, $t_j \geq T$ and $d(\Psi(t_j, x_j), x_{j+1}) \leq \varepsilon$ for $j = 0, \ldots, k-1$.

For $A \subset S$ define the *chain limit set* by

$$\Omega(A) = \{y \in S;\ \text{for all } \varepsilon > 0,\ T > 0 \text{ there exists } x \in A \text{ such that there is an } (\varepsilon, T) - \text{chain from } x \text{ to } y\},$$

and the *chain recurrent set* as $\mathcal{CR} = \{x \in S;\ x \in \Omega(x)\}$.

The system (S, Ψ) is called *chain recurrent*, if $S = \mathcal{CR}$, and *chain transitive*, if $y \in \Omega(x)$ for all $x, y \in S$.

Recall that (S, Ψ) is chain transitive iff it is chain recurrent and S is connected, and for A closed, $\Omega(A)$ is closed, invariant and contains $\omega(x)$ for all $x \in A$, compare Conley (1978).

For the control system (1) we define the corresponding concept using chain control sets:

Definition 3.6. A set $E \subset M$ is called a *chain control set* of (1), if

(i) for all $x, y \in E$ and all $\varepsilon > 0$, $T > 0$ there are $n \in \mathbf{N}$, $x_0, \ldots, x_n \in M$, $u_0, \ldots, u_{n-1} \in \mathcal{U}$, and $t_0, \ldots, t_{n-1} \geq T$ with $x_0 = x$, $x_n = y$ and $d(\varphi(t_j, x_j, u_j), x_{j+1}) < \varepsilon$ for $j = 0, \ldots, n-1$,

(ii) for all $x \in E$ there exists $u \in \mathcal{U}$ such that $\varphi(t, x, u) \in E$ for all $t \in \mathbf{R}$, and

(iii) E is maximal with respect to these properties.

The system (1) is *completely chain controllable*, if M is the chain control set.

For the control flow $(\mathcal{U} \times M, \phi)$ we again lift the chain control sets $E \subset M$ to $\mathcal{U} \times M$ via

$$\mathcal{E} = \{(u, x) \in \mathcal{U} \times M;\ \varphi(t, x, u) \in E \text{ for all } t \in \mathbf{R}\}. \tag{5}$$

The analogue of Theorem 3.3 for chain control sets reads:

Theorem 3.7. (i) *The shift $(\mathcal{U}, \theta)$ is chain transitive.*

(ii) *$E \subset M$ is a chain control set of (1) iff $\mathcal{E}$ is a maximal invariant chain transitive set of $(\mathcal{U} \times M, \phi)$.*

(iii) *(1) is completely chain controllable iff $(\mathcal{U} \times M, \phi)$ is chain transitive iff $(\mathcal{U} \times M, \phi)$ is chain recurrent.*

For a proof of (i) and (ii) see Colonius and Kliemann (1990[a], Lemma 4.6 and Theorem 4.9), (iii) is an easy consequence of (ii).

We now have two control structures on M that are related to ergodic properties of $(\mathcal{U} \times M, \phi)$. For the remainder of this section we discuss these control structures and their relations in some more detail.

Lemma 3.8. (i) *Chain control sets are pairwise disjoint, closed, and connected.*

(ii) *Control sets are pairwise disjoint and connected. They are closed if they are invariant, i.e. $\overline{D} = \overline{\mathcal{O}^+(x)}$ for all $x \in D$.*

(iii) *Invariant control sets have nonvoid interior.*

(iv) *If* $\operatorname{int} D \neq \phi$ *for some control set* D, *then* $\mathcal{O}^+(x) \supset \operatorname{int} D$ *for all* $x \in D$, *i.e. we have precise controllability in* $\operatorname{int} D$.

Remark 3.9. (i) For each control set D there exists a (unique) chain control set $E \subset M$ with $D \subset E$.

(ii) Several control sets may be contained in one chain control set.

(iii) There may be points in a chain control set that are in no control set.

While (i) is obvious from the definitions, (ii) and (iii) can be seen from the following example.

Example 3.10. Consider the control system on the circle $\mathbf{S}^1$

$$\dot{x} = -\sin^2 x + a\cos^2 x - u\cos^2 x, \quad x \in \mathbf{R} \bmod 2\pi, \quad a > 0,$$

with $U = [A, a] \subset \mathbf{R}$. There are four control sets $D_1 = [0, \arctan(a - A)^{1/2}]$, $D_2 = (\pi - \arctan(a - A)^{1/2}, \pi)$, $D_3 = D_1 + \pi$, $D_4 = D_2 + \pi$, compare Arnold and Kliemann (1983, Theorem 4.8) for a general technique to compute control sets for systems with one-dimensional state space. There exists, however, only one chain control set $E = \mathbf{S}^1$. Furthermore, the points in $\mathbf{S}^1 \setminus \bigcup_{i=1}^{4} D_i$ are in no control set, but in E. It is true in general that two control sets D_1 and D_2 with $\overline{D_1} \cap \overline{D_2} \neq \phi$ are in one chain control set. Control sets are chain control sets, if they have a certain isolation property, compare Colonius and Kliemann (1990[a], Section 4).

Control sets and chain control sets are ordered in the following way: Let D_1 and D_2 be control sets, then we define

(6) $\qquad D_1 \prec D_2$ if there exist $x \in D_1$ and $y \in D_2$ with $y \in \overline{\mathcal{O}^+(x)}$.

Lemma 3.11. (i) $\prec$ *defined by* (6) *is an order on the control sets of* (1).

(ii) *The closed (i.e. invariant) control sets are maximal elements of* $\prec$, *the open control sets are minimal elements.*

(iii) *If* M *is compact, then* (1) *has at least one closed and one open control set. In this case the maximal (minimal) elements are exactly the closed (open) control sets.*

Proof. (i) For a control set D, $D \prec D$ is obvious from the definition. $D_1 \prec D_2$ and $D_2 \prec D_1$ means that there exist $x_1 \in D_1$, $x_2 \in D_2$ with $x_2 \in \overline{\mathcal{O}^+(x)}$, and also $y_2 \in D_2$, $y_1 \in D_1$ with $y_1 \in \overline{\mathcal{O}^+(y_2)}$. Since $x_i, y_i \in D_i$ for $i = 1, 2$, we have $x_i \in \overline{\mathcal{O}^+(y_i)}$ and $y_i \in \overline{\mathcal{O}^+(x_i)}$. Therefore $y_2 \in \overline{\mathcal{O}^+(x_1)}$ and $x_1 \in \overline{\mathcal{O}^+(y_2)}$, and hence x_1 and y_2 are in the same control set by maximality. Finally, if $D_1 \prec D_2$

and $D_2 \prec D_3$, then there are $x \in D_1$, $y_1 \in D_2$ with $y_1 \in \overline{O^+(x)}$, and $y_2 \in D_2$, $z \in D_3$ with $z \in \overline{O^+(y_2)}$. Since $y_1, y_2 \in D_2$, we know that $y_2 \in \overline{O^+(y_1)}$, and hence, using continuous dependence on initial values $z \in \overline{O^+(x)}$, i.e. $D_1 \prec D_3$.

(ii) For all $x \in D$, an invariant control set, we have $\overline{O^+(x)} \subset \overline{D}$, and hence invariant control sets are maximal elements. Now consider the time reversed system associated with (1):

$$(7) \qquad \dot{x}^* = -X_0(x^*) - \sum_{i=1}^{m} u_i(t)X_i(x^*).$$

The positive orbits $O^{*+}(x)$ of (7) are exactly the negative orbits $O^-(x)$ of (1). Hence the interior of the closed (i.e. invariant) control sets of (7) are the open control sets of (1), and this proves the second assertion of (ii).

(iii) It was proved in Colonius and Kliemann (1989) that under our assumptions the control system (1) on a compact manifold M has at least one invariant control set D with int $D \neq \phi$. Furthermore for each $x \in M$ there exists an invariant control set $D \subset \overline{O^+(x)}$. Hence in this case the invariant control sets are exactly the maximal elements of $\prec$. Using time reversal, one sees that the open control sets are exactly the minimal elements of $\prec$. □

The next example shows that for noncompact M there need not exist invariant control sets, and maximal elements of $\prec$ need not be closed.

Example 3.12. Consider the control system in $\mathbb{R}^1$

$$(8) \qquad \dot{x} = X_0(x) + uX_1(x) = 2 + u(x^2 - 1)$$

with $U = [A, B] \subset [0, \infty)$. If $B < 2$, then (8) has no control set and $\lim_{t\to\infty} \varphi(t, x, u) = \infty$ for all $x \in \mathbb{R}$, all $u \in \mathcal{U}$.

If $A < 2$, $B \geq 2$, then (8) has a unique control set $D = \left[-\sqrt{1-\frac{2}{B}}, \sqrt{1-\frac{2}{B}}\right)$, which is neither open nor closed.

If $A \geq 2$, then (8) has two control sets $D_1 = \left(\sqrt{1-\frac{2}{A}}, \sqrt{1-\frac{2}{B}}\right)$, and $D_2 = \left[-\sqrt{1-\frac{2}{B}}, -\sqrt{1-\frac{2}{A}}\right]$. We have $D_1 \prec D_2$, and D_1 is open, D_2 is closed.

For chain control sets it is convenient to define the order via the lifts to $\mathcal{U}\times M$: Let $\mathcal{E}_1$ and $\mathcal{E}_2$ be two maximal invariant chain transitive sets of $(\mathcal{U}\times M, \phi)$, define

$$(9) \qquad \mathcal{E}_1 \prec \mathcal{E}_2 \text{ if there exists } (u, x) \in \mathcal{U} \times M \text{ sucht that } \omega^*(u, x) \subset \mathcal{E}_1 \text{ and } \omega(u, x) \subset \mathcal{E}_2.$$

where $\omega^*(u, x) = \{(v, y) \in \mathcal{U} \times M;$ there exists $t_k \to -\infty$ with $\phi(t_k, u, x) \to (v, y)\}$ is the α-limit set of (u, x).

Lemma 3.13. (i) $\prec$ *defined in* (9) *is an order on the chain recurrent components of*
$(\mathcal{U} \times M, \phi)$, *and hence on the chain control sets of* (1).

(ii) *Any finite collection* $\{\mathcal{E}_1, \ldots, \mathcal{E}_n\}$ *of chain recurrent components with the order* $\prec$ *defines a Morse decomposition of* $(\mathcal{U} \times M, \phi)$.

Proof. (i) According to Conley (1978, Section II.6.2), the chain recurrent set $\mathcal{CR}$ can be written as $\mathcal{CR} = \cap \{A \cup A^*,\ A$ is an attractor of $(\mathcal{U} \times M, \phi)$ and A^* its complementary repeller$\}$, and hence (9) defines an order on the components of the chain recurrent set.

(ii) Again using Conley (1978, Section II.7), any finite decomposition into disjoint invariant sets with the order $\prec$ defines a Morse decomposition. □

At this moment, we have primarily two areas of application in mind for the ergodic theory of stochastic and control flows: the theory of Lyapunov exponents and stochastic bifurcation theory. We will now characterize more precisely the control sets and chain control sets that come up in these areas.

Lyapunov exponents are the exponential growth rates of the linearized system. We consider here only the simple case, where the control system (1) has a rest point $x^0 \in M$, i.e. $X_j(x^0) = 0$ for $j = 0, \ldots, m$. Linearization around x^0 yields locally in a neighborhood of x^0

$$\dot{y} = A_0(x^0)y + \sum_{i=1}^{m} u_i(t)A_i(x^0)y \qquad \text{in } \mathbf{R}^d \quad (d = \dim M) \tag{10}$$

with $A_j(x^0) := X_{j*}(x^0)$, the linearization of X_j at x^0 for $j = 0, \ldots, m$. The Lyapunov exponents of (1) at x^0 are then defined by

$$\lambda(y_0, u) = \limsup_{t \to \infty} \frac{1}{t} \log |\varphi(t, y_0, u)|, \qquad y_0 \neq 0$$

where $\varphi(t, y_0, u)$ denotes the solution of (10). Since $\lambda(\alpha y_0, u) = \lambda(y_0, u)$ for all $\alpha \in \mathbf{R}$, $\alpha \neq 0$, it suffices to consider (10) on the projective space $\mathbf{P}^{d-1}$ in $\mathbf{R}^d$:

$$\dot{s} = h_0(s) + \sum_{i=1}^{m} u_i(t)h_i(s) \tag{11}$$

with $h_j(s) := \left(A_j(x^0) + s^T A_j(x^0)s \cdot Id\right) s$, the projected vector field on $\mathbf{P}^{d-1}$ for $j = 0, \ldots, m$. We will assume again the nondegeneracy condition for (11) on $\mathbf{P}^{d-1}$, i.e.

$$\dim \mathcal{LA}\left\{h_0 + \sum_{i=1}^{m} u_i h_i,\ (u_i) \in U\right\}(s) = d - 1 \quad \text{for all } s \in \mathbf{P}^{d-1}.$$

The system (10) is a bilinear control system in $\mathbf{R}^d$, its associated semigroup is given by

$$\mathcal{S} = \{\exp t_n B_n \cdot \ldots \cdot \exp t_1 B_1,\ B_i \in N,\ t_i \geq 0,\ i = 1, \ldots, n \in \mathbf{N}\} \subset G\ell(d, \mathbf{R}),$$

with $N := \left\{A_0(x^0) + \sum_{i=1}^{m} u_i A_i(x^0),\ (u_i) \in U\right\}$, the possible right hand sides of (10) for piecewise constant controls. $\mathcal{S}$ acts on $\mathbf{P}^{d-1}$ in a natural way via $s \mapsto \frac{1}{|gs|} gs$ for $s \in \mathbf{P}^{d-1}$, $g \in \mathcal{S}$, and the differential equation (11) corresponds to this action. (Compare Colonius and Kliemann (1990^b) for details of the entire set up.) For the control sets of (11) under the assumption (H) we obtain:

Theorem 3.14. (i) *There are k control sets D_i with int $D_i \neq \phi$, $i = 1, \ldots, k$, $1 \leq k \leq d$, called the main control sets.*

(ii) *The order, defined by (6) on the main control sets, is linear. We enumerate these sets by $D_1 \prec D_2 \prec \cdots \prec D_k$.*

(iii) *D_k is closed and $D_k = \bigcap_{s \in \mathbf{P}} \overline{\mathcal{O}^+(s)}$, D_1 is open and $\overline{D}_1 = \bigcap_{s \in \mathbf{P}} \overline{\mathcal{O}^-(s)}$.*

(iv) *For every $g \in$ int $\mathcal{S}$ and every $\lambda \in$ spec g there is a main control set D_i such that the corresponding generalized eigenspace $E(g, \lambda)$ satisfies $\mathbf{P}E(g, \lambda) \subset$ int D_i, where $\mathbf{P}E$ is the projection of $E \subset \mathbf{R}^d$ onto $\mathbf{P}^{d-1}$. Vice versa, the interior of the main control sets consists exactly of those elements $s \in \mathbf{P}$, which are eigenvectors for a (real) eigenvalue of some $g \in$ int $\mathcal{S}$.*

(v) *For every $g \in \mathcal{S}$ and every $\lambda \in$ spec g there is a main control set D_i with $\mathbf{P}E(g, \lambda) \cap \overline{D}_i \neq \phi$, and vice versa for every main control set D_i and every $g \in \mathcal{S}$ there exists a $\lambda \in$ spec g with $\mathbf{P}E(g, \lambda) \cap \overline{D}_i \neq \phi$.*

The proof of this result is given in Colonius and Kliemann (1990^b, Theorem 3.10).

For the chain control sets of (11) the corresponding result is

Theorem 3.15. (i) *There are ℓ chain control sets E_j, $j = 1, \ldots, \ell$, $1 \leq \ell \leq d$.*

(ii) *Every chain control set contains a main control set, in particular int $E_j \neq \phi$ for $j = 1, \ldots, \ell$ and $1 \leq \ell \leq k \leq d$.*

(iii) *The order defined by (9) on the (lifted) chain control sets is linear, and we write $E_1 < E_2 < \cdots < E_\ell$.*

(Compare Colonius and Kliemann (1990^b, Theorem 5.5) for a proof.) Note that the situations, described in Remark 3.9(ii) and (iii) can occur for systems of the type (11) as well. This can be seen from Example 3.10, which is the projection of the following bilinear system in $\mathbf{R}^2$ onto $\mathbf{P}^1$:

$$\dot{x} = \begin{pmatrix} 0 & 1 \\ -1 & -2b \end{pmatrix} x + u(t) \begin{pmatrix} 0 & 0 \\ -1 & 0 \end{pmatrix} x$$

has a projection onto $\mathbf{P}^1$ in polar coordinates $s = \begin{pmatrix} \cos\varphi \\ \sin\varphi \end{pmatrix}$, $\varphi \in \mathbf{R} \bmod \pi$

$$\dot{\varphi} = -\sin^2\varphi + (b^2 - 1)\cos^2\varphi - u(t)\cos^2\varphi.$$

For $b^2 - 1 = a$ and $u \in [A, a]$ this is Example 3.10. In particular we have in this case $k = 2$, $\ell = 1$, and there are points $s \in \mathbf{P}^1 \setminus \bigcup_{i=1}^{2} \overline{D}_i$ that are not in the (generalized) eigenspace of any $g \in \mathcal{S}$, but in int $E = \mathbf{P}^1$.

The next class of examples deals with situations arising in (stochastic) codimension one bifurcations, compare Arnold and Boxler (1990) for the general set up. Here one considers a family of one-dimensional systems (replacing the noise by controls)

(12) $$\dot{x} = X_\alpha(x) + u(t)Y_\alpha(x),$$

where X_α, Y_α are smooth vector fields on $\mathbf{R}$ or $\mathbf{S}^1$ and $\alpha \in I$ is the bifurcation parameter. We will analyze the case with compact $U \subset \mathbf{R}$, i.e. with bounded noise. The problem is to determine the invariant probability measures of (12), hence, first of all, the control sets and chain control sets need to be determined for each $\alpha \in I$. We will therefore consider the following system

(13) $$\dot{x} = X(x) + u(t)Y(x) \qquad \text{on } M = \mathbf{R}^1 \text{ or } \mathbf{S}^1$$

under the following assumptions:

(14)
- (i) X and Y are smooth vector fields on M,
- (ii) $u(t) \in U \subset \mathbf{R}$, U a compact interval,
- (iii) for each $u \in U$ there exists at most a finite number of zeros of $X + uY$.

These assumptions are typical for codimension one bifurcation diagrams with bounded noise, and cover also the case, where the bifurcation parameter α itself is noisy and appears linearly in (12), i.e. systems of the form $\dot{x} = X(x) + \alpha_t Y(x)$, with α_t a stochastic process with values in $U \subset \mathbf{R}$.

A general procedure for finding the control sets of one dimensional control systems was described in Kliemann (1980, Section II.6). Here we summarize these results and extend them to chain control sets. For each $u \in U$ define $S(u) = \{x \in M;\ X(x) + uY(x) = 0\}$, the rest points of (13) corresponding to

u. Denote $S = \bigcup_{u \in U} S(u)$ and

$$\begin{aligned}
SA &= \{s \in S;\ X(s) + uY(s) = 0 \text{ for all } u \in U\}, \\
SB &= \{s \in S;\ X(s) + uY(s) \geq 0 \text{ for all } u \in U, \text{ and there exist } u_1, u_2 \in U \\
&\qquad \text{with } X(s) + u_1 Y(s) = 0 \text{ and } X(s) + u_2 Y(s) > 0\}, \\
SC &= \{s \in S;\ X(s) + uY(s) \leq 0 \text{ for all } u \in U, \text{ and there exist } u_1, u_2 \in U \\
&\qquad \text{with } X(s) + u_1 Y(s) = 0 \text{ and } X(s) + u_2 Y(s) < 0\}, \\
SD &= \{s \in S;\ \text{there exist } u_1, u_2 \in U \text{ with } X(s) + u_1 Y(s) > 0 \\
&\qquad \text{and } X(s) + u_2 Y(s) < 0\}.
\end{aligned}$$

The control sets of (13) are intervals, possibly consisting of only one point. The points in $SA \cup SB \cup SC$ are the boundary points of these intervals, the elements of SD are the interior points. More precisely, all control sets for $M = \mathbb{R}$ can be found as follows:

Let $s_0 \in SA$, then four cases are possible:

(a) There exists an interval $(s_0 - \varepsilon,\ s_0 + \varepsilon)$ for some $\varepsilon > 0$, such that for all $p \in (s_0 - \varepsilon,\ s_0 + \varepsilon) \setminus \{s_0\}$ we have $p \notin S$, then $\{s_0\}$ is a one point invariant control set, and no other control set intersects this interval.

(b) There exists a (maximal) interval of the form $(s_0,\ s_0 + \varepsilon)$ (or $(s_0 - \varepsilon,\ s_0)$) for some $\varepsilon > 0$, such that all points in this interval are in SB (or in SC), then each point p in the interval is a one point control set $\{p\}$ (not invariant), and $\{s_0\}$ is an invariant control set.

(c) There exists an interval $(s_0,\ s_0 + \varepsilon)$ for some $\varepsilon > 0$, such that for all $p \in (s_0,\ s_0 + \varepsilon)$ one has $p \in SD$. Take $(s_0,\ s_0 + \varepsilon)$ as the maximal interval with: $p \in (s_0,\ s_0 + \varepsilon)$ implies $p \in SD$, and define $s_1 = s_0 + \varepsilon$. (Set $s_1 = +\infty$ if $p \in SD$ for all $p > s_0$.) Then there exists a control set of the form

$$\begin{array}{lll}
(s_0, s_1), & \text{if } s_1 \in SA, & \text{(invariant)}, \\
(s_0, s_1), & \text{if } s_1 \in SB, & \text{(not invariant)}, \\
(s_0, s_1], & \text{if } s_1 \in SC, & \text{(invariant)}, \\
(s_0, \infty), & \text{if } s_1 = \infty, & \text{(invariant)},
\end{array}$$

and $\{s_0\}$ is a one point invariant control set.

(d) There exists an interval $(s_0 - \varepsilon,\ s_0)$ for some $\varepsilon > 0$, such that for all $p \in (s_0 - \varepsilon,\ s_0)$ one has $p \in SD$. Take $(s_0 - \varepsilon,\ s_0)$ as the maximal interval with: $p \in (s_0 - \varepsilon,\ s_0)$ implies $p \in SD$, and define $s_1 = s_0 - \varepsilon$.

(Set $s_1 = -\infty$ if $p \in SD$ for all $p < s_0$.) Then there is a control set of the form

$$\begin{array}{lll} (s_1, s_0), & \text{if } s_1 \in SA, & \text{(invariant)}, \\ [s_1, s_0), & \text{if } s_1 \in SB, & \text{(invariant)}, \\ (s_1, s_0), & \text{if } s_1 \in SC, & \text{(not invariant)}, \\ (-\infty, s_0), & \text{if } s_1 = -\infty, & \text{(invariant)}, \end{array}$$

and $\{s_0\}$ is a one point invariant control set.

(Note that $SA = \phi$, if Assumption (H) holds for (13).)

If $s_0 \in SB$, then the following cases can occur:

(e) If there exists an interval of the form $(s_0 - \varepsilon,\ s_0 + \varepsilon)$ for some $\varepsilon > 0$, such that for all $p \in (s_0 - \varepsilon,\ s_0 + \varepsilon) \setminus \{s_0\}$ one has $p \notin S$, then $\{s_0\}$ is a one point (not invariant) control set.

(f) If there exists a (maximal) interval of the form $[s_0,\ s_0 + \varepsilon]$ or $[s_0 - \varepsilon,\ s_0]$ for some $\varepsilon > 0$, such that all points in the interval are in SB, then each point p in the interval is a one point (not invariant) control set $\{p\}$.

(g) If there exists an interval $(s_0,\ s_0 + \varepsilon)$ for some $\varepsilon > 0$, such that all points $p \in (s_0,\ s_0 + \varepsilon)$ are in SD, let $s_1 = s_0 + \varepsilon$ with $(s_0,\ s_0 + \varepsilon)$ the maximal interval as above. Then there exists a control set of the form

$$\begin{array}{lll} [s_0, s_1), & \text{if } s_1 \in SA \cup SB, & \text{(invariant if } s_1 \in SA\text{, not invariant otherwise)}, \\ [s_0, s_1], & \text{if } s_1 \in SC, & \text{(invariant)}, \\ [s_0, \infty), & \text{if } s_1 = \infty, & \text{(invariant)}. \end{array}$$

Similarly, if an interval $(s_0 - \varepsilon,\ s_0)$ exists with points in SD.

If $s_0 \in SC$, the cases are completely analogous to (e)–(g) above.

Finally, if $SD = M$, then M is the (unique, invariant) control set. Using Assumption (14(iii)) we see, that SA cannot contain infinitely many points. Hence after finitely many steps of the type (a)–(g) (and similarly for SC), all control sets of (13) in $\mathbb{R}$ are described. The general principle is: points in SA and intervals of points in SB and SC lead to one-point control sets, which are invariant iff the point is in SA. All other control sets are intervals with nonvoid interior, where the lower boundary belongs to the set, if the point is in SB, and similarly for the upper boundary, if this point belongs to SC. These intervals are invariant, iff the boundary points belong to the control sets or are in SA.

If M is compact, the same principles as above apply, except that, letting $M \simeq \mathbf{S}^1$, one has to consider the intervals mod 2π, when parametrizing $\mathbf{S}^1$ through the angle in $[0, 2\pi)$.

For the chain control sets of the control system (13) we obtain the following characterization: Define $\widehat{D} = \cup\{\overline{D},\ D \text{ is a control set of (13)}\}$, then by (14(iii)), $\widehat{D} = I_1\dot{\cup}\ldots\dot{\cup}I_n$, the finite disjoint union of closed 'intervals', constructed as above.

We call an interval I of $\widehat{D}$ isolated, if there exists an open neighborhood N of I such that for all $y \in N \setminus I$ we have for all $u \in U$: $X(y) + uY(y) < 0$ for $y < I$, and $X(y) + uY(y) > 0$ for $y > I$, (or $X(y) + uY(y) > 0$ for $y < I$, and $X(y) + uY(y) < 0$ for $y > I$).

Theorem 3.16. (i) *If $M = \mathbf{R}$, then the I_i, $i = 1, \ldots, n$ are exactly the chain control sets of* (13).

(ii) *If M is compact, then*

(a) *if $\widehat{D} = I_1$, then M is the chain control set,*

(b) *if $n \geq 2$, then the I_i, $i = 1, \ldots, n$ are exactly the chain control sets of* (13) *iff there is at least one isolated interval in $\widehat{D}$, otherwise M is the chain control set.*

(iii) *If $M = \mathbf{R}$ or if the system* (13) *has more than one chain control set, then for a control set D the set $\overline{D}$ is a chain control set iff there exists an open neighborhood N of $\overline{D}$, which intersects with no other control set. If M is compact and* (13) *has only one chain control set, then the closure of a control set D is a chain control set iff SA consists of at most one point and* (13) *is completely controllable in $M \setminus SA$.*

Proof. Note first of all that, by the construction of control sets above $\widehat{D} = S$, i.e. $x \in M \setminus \widehat{D}$ means for all $u \in U$ either $X(x) + uY(x) > 0$ or $X(x) + uY(x) < 0$. Recall also that by Lemma 3.8 chain control sets are pairwise disjoint, connected and closed.

(i) It is clear from the definition of chain control sets and from Lemma 3.8 that the intervals I_i, $i = 1, \ldots, n$ are contained in chain control sets. Let I be such an interval, and assume that there exists a chain control set $E \supsetneqq I$. Then there is a point $x \in E \setminus I$, say w.l.o.g. $x < p$ for all $p \in I$, and $x \notin \widehat{D}$. Then, by the remark above, there is an open neighborhood $N(x)$ of x such that $N(x) \cap \widehat{D} = \phi$ and for all $y \in N(x)$, all $u \in U$ we have either (a) $X(y)+uY(y) > 0$ or (b) $X(y)+uY(y) < 0$. In case (a) there cannot exist an (ε, T)-chain from I to y for ε small enough, and T large enough, in case (b) there is no (ε, T)-chain from y to I for small ε and large T. I.e. $E = I$ and this proves (i).

(ii) If M is compact, then the system (13) has at least one closed, invariant control set (this part of Lemma 3.11 holds without Assumption (H)), i.e. $\widehat{D} \neq \phi$, and there exists at least one chain control set, compare Remark 3.9(i).

(a) If $\widehat{D}$ is one 'interval', then there is again one chain control set $E \supset \widehat{D}$. But for all $u \in U$, all $y \in M \setminus \widehat{D}$ we have either $X(y) + uY(y) > 0$ or $X(y)+uY(y) < 0$, and hence there is $y_0 \in M\setminus\widehat{D}$ and $u_0 \in U$ with $\lim_{t\to-\infty} \varphi(t, y_0, u_0) \subset \widehat{D}$, and $\lim_{t\to+\infty} \varphi(t, y_0, u_0) \subset \widehat{D}$, where $\varphi(t, y_0, u_0)$ denotes the solution of (13) corresponding to the constant control $u(t) \equiv u_0$. Thus $E = M$.

(b) If all intervals $I_1 \dots I_n$ are not isolated, then it is easy to see that there exist $y_1, \dots, y_n \in M\setminus\widehat{D}$, and $u_1, \dots, u_n \in U$ such that y_i lies in the gap between I_i and I_{i+1} for $i = 1, \dots, n$, y_n lies in between I_n and I_1, and $\lim_{t\to-\infty} \varphi(t, y_i, u_i) \subset I_i$, $\lim_{t\to+\infty} \varphi(t, y_i, u_i) \subset I_{i+1}$, $\lim_{t\to-\infty} \varphi(t, y_n, u_n) \subset I_n$, and $\lim_{t\to+\infty} \varphi(t, y_n, u_n) \subset I_1$. Hence M is the chain control set.

If one interval I of $\widehat{D}$ is isolated, then there is an open neighborhood N of I such that either for all $y \in N \setminus I$ and all $u \in \mathcal{U}$ $\lim_{t\to-\infty} \varphi(t, y, u) \subset I$ and $\lim_{t\to+\infty} \varphi(t, y, u) \subset \widehat{D} \setminus I$, or for all such y and u $\lim_{t\to-\infty} \varphi(t, y, u) \subset \widehat{D} \setminus I$ and $\lim_{t\to+\infty} \varphi(t, y, u) \subset I$. Furthermore, because the right hand side of (13) depends continuously on u, it cannot happen that for some $z \in M \setminus \widehat{D}$ there exist $u_1 \in \mathcal{U}$, $u_2 \in \mathcal{U}$ with $\lim_{t\to-\infty} \varphi(t, z, u_1) \subset I$, $\lim_{t\to+\infty} \varphi(t, z, u_2) \subset I$. Hence, arguing as in (i), in this case the $I_1 \dots I_n$ are exactly the chain control sets of (13).

(iii) If the $I_1 \dots I_n$ are the chain control sets of (13), then the result follows directly from the definition. If this is not the case, then according to (i) and (ii), M is compact and the chain control set. Hence for $\overline{D} = M$ we need that (13) is completely controllable with the possible exception of one point in SA.

□

Remark 3.17. We have not assumed Hypothesis (H) for the system (13), because the situation, where $x \in M$ is a rest point for all $u \in U$, occurs frequently in bifurcation diagrams. If, however, we can assume (H), then the following simplifications hold: $SA = \phi$, and for the determination of the control sets we only have to go through the steps (e)–(g) for points in SB and SC. Theorem 3.16(iii) reads in this case: If M is compact and (13) has only one chain control set, then the closure of a control set D is a chain control set iff $D = M$.

Remark 3.18. Consider the projected linear system (11) on $\mathbb{P}^1$, i.e. $d = 2$, under Assumption (H). There are at most two main control sets, one closed, say C, and one open, denoted by C^-. In this situation we obtain for the chain control sets: $\mathbb{P}^1$ is the chain control set iff either

(i) the system is completely controllable, i.e. $C \cap C^- \neq \phi$, or
(ii) $C \cap \overline{C^-} \neq \phi$, or
(iii) there exists $u \in U$ such that $h_0(s) + \sum_{i=1}^{m} u_i h_i(s) = 0$ for all $s \in \mathbb{P}^1$.

In all other cases there are exactly two chain control sets, namely, C and $\overline{C^-}$.

4. Invariant Measures of Control Flows and Stochastic Flows

In this section we discuss existence and supports of invariant probability measures of control flows and stochastic flows. The situation for stochastic flows is distinguished by the following two facts: first of all, here an invariant measure P on the underlying shift space Ω is given a priori, and we are looking for invariant measures of the flow, whose marginal on Ω is P. Secondly one has to take measurability questions into account. We will first study the case of control flows, and then specialize the results to stochastic flows.

Definition 4.1. Let (S, Ψ) be a dynamical system. A probability measure μ on S is called *Ψ-invariant*, if $\Psi_t \mu = \mu$ for all $t \in \mathbb{R}$. We denote this set by M_Ψ.

Lemma 4.2. *Consider the control flow $(\mathcal{U} \times M, \phi)$ defined in (2.2). A probability measure μ on $\mathcal{U} \times M$ is ϕ-invariant iff μ is of the form $\mu(du, dx) = \mu_u(dx)\rho(du)$, where ρ is a θ- invariant measure on $\mathcal{U}$, and $\varphi(t, \cdot, u)\mu_u = \mu_{\theta_t u}$ for all $t \in \mathbb{R}$, where $\varphi(t, x, u)$ denotes again the solution of the control equation.*

For a proof see Crauel (1986, Lemma 2) or Colonius and Kliemann (1990[a], Proposition 5.2).

If a measure μ is Ψ-invariant for $t \geq 0$, then invertibility of Ψ_t implies $\mu = (\Psi_t)^{-1}\Psi_t\mu = \Psi_{-t}\mu$, i.e. μ is Ψ-invariant for all $t \in \mathbb{R}$.

Invariant measures of the control flow can be constructed via the Krylov-Bogolyubov device, i.e. for $(u, x) \in \mathcal{U} \times M$ consider the Cesaro limits for sequences $t_k \to \infty$

$$(1) \qquad \lim_{t_k \to \infty} \frac{1}{t_k} \int_0^{t_k} F(\theta_\tau(u), \varphi(\tau, x, u))\, d\tau = \int_{\mathcal{U} \times M} F(v, y)\, d\mu_{u,x}$$

for all $F \in C(\mathcal{U} \times M, \mathbb{R})$, the continuous functions from $\mathcal{U} \times M$ into $\mathbb{R}$. Note that in general $\mu_{u,x}$ is not unique for $(u, x) \in \mathcal{U} \times M$. The following properties of the probability measures $\mu_{u,x}$ are well known, see e.g. Mañé (1987, Chapter II.6):

(a) M_θ and M_ϕ are nonempty, if U and M are compact.

(b) Define

$$\Sigma_\phi^e = \{(u,x) \in \mathcal{U} \times M;\ \text{the measure } \mu_{u,x} \text{ defined in (1) is independent of the sequence } t_k \text{ and ergodic}\}$$

$$\Sigma_\phi^s = \{(u,x) \in \Sigma_\phi^e;\ (u,x) \in \operatorname{supp} \mu_{u,x}\},$$

then for M, U compact, $\Sigma_\phi^s \neq \phi$ and Σ_ϕ^s has total measure with respect to M_ϕ, i.e. $\mu\left(\Sigma_\phi^s\right)^c = 0$ for all $\mu \in M_\phi$. (Here supp μ denotes the support of μ, and A^c denotes the complement of the set A.)

(c) Each $\mu \in M_\phi$ has an ergodic decomposition: Every $F \in L^1(\mathcal{U} \times M,\ \mu)$ is $\mu_{u,x}$-integrable for μ-almost all $(u,x) \in \Sigma_\phi^s$ and $\int \left(\int F\, d\mu_{u,x}\right) d\mu = \int F\, d\mu$.

Hence, if we want to characterize the possible support of some $\mu \in M_\phi$, it suffices to characterize the set Σ_ϕ^s.

Define for a control set $D \subset M$

$$\mathcal{D}^+ = c\ell\{(u,x) \in \mathcal{U} \times M;\ \varphi(t,x,u) \in D \text{ for all } t \in \mathbb{R}\}.$$

If int $D \neq \phi$, then $\mathcal{D}^+ \supset \mathcal{D}$, with $\mathcal{D}$ as defined in (3) from Section 3. However, the lift $\mathcal{D}^+$ is nonvoid even if int $D = \phi$, as can be seen from Assertion (ii) in the following theorem.

Theorem 4.3.

(i) *For all $(u,x) \in \mathcal{U} \times M$ and all $\mu_{u,x}$ as in (1) we have supp $\mu_{u,x} \subset \omega(u,x) \subset \mathcal{E}$, the lift of some chain control set $E \subset M$ of (2.1) to $\mathcal{U} \times M$.*

(ii) *For all $(u,x) \in \mathcal{U} \times M$ and all $\mu_{u,x}$ as in (1) there exists $\Gamma \subset$ supp $\mu_{u,x}$ with $\mu_{u,x}\Gamma = 1$, such that for all $(v,y) \in \Gamma$ there is a control set $D \subset M$ with $\varphi(t,y,v) \in D \cap \pi_M\omega(u,x)$, for all $t \in \mathbb{R}$.*

(iii) *If $(u,x) \in \Sigma_\phi^e$, then there exists a (unique) control set $D \subset M$ such that for $\mu_{u,x}$-almost all (v,y) we have $\varphi(t,y,v) \in D$ for all $t \in \mathbb{R}$, i.e. supp $\mu_{u,x} \subset \mathcal{D}^+$, with $\mathcal{D}^+$ defined above. Also, if $\mu \in M_\phi$ is ergodic, then supp $\mu \subset \mathcal{D}^+$ for some control set D.*

(iv) *Vice versa, if $D \subset M$ is a control set with int $D \neq \phi$, then for each $x \in$ int D there exists $u \in \mathcal{U}$ such that $(u,x) \in \Sigma_\phi^s$.*

(v) *Assume that int $D \neq \phi$ for all control sets $D \subset M$. Then*

$$c\ell \cup \{D;\ D \text{ is a control set}\} = \pi_M c\ell \cup \{\operatorname{supp} \mu_{u,x};\ (u,x) \in \Sigma_\phi^s\}$$
$$= \pi_M c\ell \cup \{\operatorname{supp} \mu;\ \mu \in M_\phi\}.$$

The proof of this theorem can be found in Colonius and Kliemann (1990[a], Lemma 5.3, Proof of Theorem 5.5, and Corollary 5.7).

This theorem says in particular that for all $\mu \in M_\phi$ one has supp $\mu \subset \mathcal{E}$, where E is some chain control set, and supp $\mu \subset \{\mathcal{D}^+;\ D \subset E \text{ is a control set}\}$. If μ is ergodic, then supp $\mu \subset \mathcal{D}^+$ for some control set D. Note that an analogue of (iv) for chain control sets is not true, compare e.g. Example 3.10.

We now turn to the existence of invariant probability measures for control flows. Starting from $x \in M$, we know by Theorem 4.3 that for $x \notin \cup\{\overline{D};\ D \text{ is a control set}\}$ there is no Krylov-Bogolyubov measure for x and any $u \in \mathcal{U}$, (i.e. $(u,x) \in$ supp $\mu_{u,x}$ cannot hold). On the other hand, for $x \in$ int D, D some control set, there always exists a $u \in \mathcal{U}$ such that $(u,x) \in \Sigma_\phi^e$: Just take a periodic u, which leads to a periodic trajectory in int D through x. Furthermore, it is easy to construct examples for dim $M \geq 2$, such that for some $x \in \partial D$ we have that $x \notin \pi_M \Sigma_\phi^e$.

What is more important, however, by the characterization in Lemma 4.2, is to construct the measures μ_u on M for $u \in \mathcal{U}$. We proceed in the following way: Define for a control set $D \subset M$ and a control function $u \in \mathcal{U}$

$$D_u^+ = \{x \in \overline{D};\ \varphi(t,x,u) \in \overline{D} \text{ for all } t \geq 0\}. \tag{2}$$

Note that $D_u^+ \neq \phi$ iff $D_u = \{y \in \overline{D};\ \varphi(t,y,u) \in \overline{D} \text{ for all } t \in \mathbf{R}\} \neq \phi$.

Theorem 4.4. *Assume that M is compact, or more generally that $\widehat{D} = \cup\{D;\ D \text{ is a control set}\}$ is bounded. Then the following holds:*

(i) *Let $u \in \mathcal{U}$, and let $D \subset M$ be some control set. Then there exists an invariant measure $\mu_{u,x}$ of the form (1) with supp $\mu_{u,x} \subset \mathcal{D}^+$ iff $D_u^+ \neq \phi$.*

(ii) *Let μ be a Φ-invariant measure with decomposition $\mu = \mu_u \rho$ according to Lemma 4.2. Then for all control sets D with supp $\mu \cap \mathcal{D}^+ \neq \phi$ there exists $u \in \mathcal{U}$ such that $D_u^+ \neq \phi$. Conversely, if $\rho \in M_\theta$ and D is a control set with $D_u^+ \neq \phi$ for ρ-almost all $u \in \mathcal{U}$, then there is $\mu \in M_\phi$ with $\mu = \mu_u \rho$ and supp $\mu \subset \mathcal{D}^+$.*

(iii) *If $\mu \in M_\phi$ in (ii) is ergodic, then the control set D with supp $\mu \cap \mathcal{D}^+ \neq \phi$ is unique and $D_u^+ \neq \phi$ for ρ-almost all $u \in \mathcal{U}$. If $\rho \in M_\theta$ in (ii) is ergodic, and if $D_u^+ \neq \phi$ for ρ-almost all $u \in \mathcal{U}$, then $\mu \in M_\phi$.*

(iv) *Suppose $\rho \in M_\theta$ is ergodic and $\mu \in M_\phi$ can be desintegrated as $\mu = \mu_u \rho$. Then for any control set D with $\mu(\mathcal{D}^+) > 0$ we have that $D_u^+ \neq \phi$ for ρ-almost all $u \in \mathcal{U}$.*

Proof.

(i) One direction is obvious, the other one follows from Theorem 4.3(i), because $\mathcal{D}^+$ is ϕ-invariant.

(ii) Let $\mu \in M_\phi$ with $\mu = \mu_u \rho$, and let supp $\mu \cap \mathcal{D}^+ \neq \phi$. Then there exists $(u, x) \in \mathcal{D}^+$ and hence $x \in D_u^+ \neq \phi$.

Conversely, let $\rho \in M_\theta$ and $D_u^+ \neq \phi$ for ρ-almost all $u \in \mathcal{U}$. The map $u \mapsto D_u^+$ from $\mathcal{U}$ into M is a set valued map with compact values. It is measurable, because the set $\{u \in \mathcal{U};\ D_u^+ \cap A \neq \phi\}$ is closed for every closed set A. Hence by a selection theorem due to von Neumann-Aumann-Castaing (see e.g. Warga (1972, Theorems I.7.4 and I.7.7)) there exists a measurable selection $u \mapsto x(u) \in D_u^+$. Consider the measure $\hat{\mu} = \delta_{x(u)}\rho$ on $\mathcal{U} \times M$, where $\delta_{x(u)}$ denotes the Dirac measure at the point $x(u)$. Apply the Krylov-Bogolyubov construction to this measure $\hat{\mu}$ (instead of $\delta_{u,x}$ as in (1) above) in order to obtain a ϕ-invariant measure μ as

$$\lim_{t_k \to \infty} \frac{1}{t_k} \int_0^{t_k} \int_{\mathcal{U} \times M} F(\theta_\tau(u), \varphi(\tau, x, u))\, \hat{\mu}(d(u, x))\, d\tau = \int_{\mathcal{U} \times M} F(v, u) d\mu$$

for all $F \in C(\mathcal{U} \times M, \mathbf{R})$. Clearly supp $\mu \subset \mathcal{D}^+$ and $\mu = \mu_u \rho$, because the $\mathcal{U}$- component of $\hat{\mu}$ coincides with the θ-invariant measure ρ.

(iii) If μ is ergodic, then supp $\mu \subset \mathcal{D}^+$ for some control set D by Colonius and Kliemann (1990[a], Theorem 5.5(iii)). Hence D is unique and $D_u^+ \neq \phi$ for ρ-almost all $u \in \mathcal{U}$.

If $\rho \in M_\theta$ is ergodic, and if $D_u^+ \neq \phi$ for ρ-almost all $u \in \mathcal{U}$, then (ii) yields a ϕ-invariant measure $\tilde{\mu}$ with supp $\tilde{\mu} \subset \mathcal{D}^+$. Now any measure $\mu_{u,x}$ appearing in an ergodic decomposition of $\tilde{\mu}$ (see e.g. Mañé (1987, Theorem II.6.4)) is ϕ-invariant and ergodic with supp $\mu_{u,x} \subset \mathcal{D}^+$.

(iv) Suppose $\mu(\mathcal{D}^+) > 0$. Then the set $\{u \in \mathcal{U};\ D_u^+ \neq \phi\}$ is θ-invariant and has positive ρ-measure. By ergodicity of ρ, this set has ρ- measure 1.

□

Remark. If $\mu \in M_\phi$ is ergodic, then the proof of (iii) shows in particular that supp $\mu \subset \mathcal{D}^+$ for some control set D, compare Theorem 4.3(iii).

The crucial question for the existence of ϕ-invariant measures is therefore: Is $D_u^+ \neq \phi$. We will next present a general result for this problem and then analyze the systems (3.11) and (3.13) in more detail.

Proposition 4.5. *Consider the control flow* (2.2) *under* Assumption (H). *Then a control set* $C \subset M$ *is invariant iff* $C_u^+ = C$ *for all* $u \in \mathcal{U}$.

Proof. This result follows directly from the definitions, because for invariant control sets C we have $\overline{O^+(x)} = \overline{C}$ for all $x \in C$, and under (H) invariant control sets are closed.

□

For variant control sets $D \subset M$ the situation is different: We may have $D_u^+ \neq \phi$ for all $u \in \mathcal{U}$, or $D_u^+ = \phi$ for an open subset of $\mathcal{U}$, as the following analysis shows.

Consider the projected bilinear control system (3.11) on $\mathbf{P}^{d-1}$, always under the Assumption (H).

Proposition 4.6.

(i) *Let $D \subset \mathbf{P}^{d-1}$ be a main control set, i.e. int $D \neq \phi$, and let $g \in$ int $\mathcal{S}$. Denote by $u(g)$ a control corresponding to g. Then*

$$D_{u(g)} := \{x \in \overline{D};\ \varphi(t,x,u) \in \overline{D} \text{ for all } t \in \mathbf{R}\} = \mathbf{P} \bigoplus_{\lambda} E(g,\lambda),$$

where the sum is taken over all $\lambda \in$ spec g with $\mathbf{P}E(g,\lambda) \subset$ int D. (Recall that $E(g,\lambda)$ is the (generalized) eigenspace of g for the eigenvalue λ, and $\mathbf{P}E(g,\lambda)$ denotes its projection onto $\mathbf{P}^{d-1}$.)

(ii) *For a chain control set $E \subset \mathbf{P}^{d-1}$ of (3.11) denote by $\mathcal{E}$ its lift to $\mathcal{U} \times \mathbf{R}^d$. Then*

$$\mathcal{E} = cl\,\{(u(g),x) \subset \mathcal{U} \times \mathbf{R}^d;\ g \in \text{int } \mathcal{S} \text{ and } \mathbf{P}x \in \oplus D_{u(g)}, \text{ where the sum is taken over all main control sets } D \subset E\}.$$

(iii) *Let D be a main control set, then for all $x \in \overline{D}$ there exists $u \in \mathcal{U}$ with $\varphi(t,x,u) \in \overline{D}$ for all $t \geq 0$.*

(iv) *Let D be a main control set, then $D_u^+ \neq \phi$ for all $u \in \mathcal{U}$.*

Proof.

(i) Theorem 3.13 in Colonius and Kliemann (1990[b]).

(ii) Theorem 5.6 in Colonius and Kliemann (1990[b]).

(iii) The assertion is valid for all $x \in$ int D, and hence for all $x \in \overline{D}$ by compactness of $\mathcal{U}$.

(iv) By (i) this result is true for all $u(g)$ with $g \in$ int $\mathcal{S}$. But these controls are dense in $\mathcal{U}$ (see Colonius and Kliemann (1990[a], Lemma 2.2)), and the result follows from the compactness of $\overline{D}$.

□

Corollary 4.7. *Given a θ-invariant measure ρ on $\mathcal{U}$, then for each main control set D of (3.11) there exists a ϕ-invariant measure $\mu = \mu_u \rho$ with supp $\mu \subset \mathcal{D}^+$.*

Proof. This follows from Proposition 4.6(iv) and Theorem 4.4(ii).

□

The situation for the general one-dimensional nonlinear system (3.13) is more complicated. We will discuss this case again without Assumption (H).

Proposition 4.8. *Assume that the system* (3.13) *satisfies the hypothesis* (3.14) *and that* $\bigcup\{D;\ D$ *is a control set*$\}$ *is bounded. Then*

(i) C *is an invariant control set of* (3.13) *iff* $C_u^+ = \overline{C}$ *for all* $u \in \mathcal{U}$.

(ii) *Let* D *be a variant control set with int* $D \neq \phi$. *Assume: For all* $u \in U$ *there exists* $x \in \overline{D}$ *with* $X(x)+uY(x) = 0$. *Then* $D_u^+ \neq \phi$ *for all* $u \in \mathcal{U}$.

(iii) *For all other control sets* $D \subset M$ *there exists an open set* $\mathcal{V} \subset \mathcal{U}$ *such that* $D_u^+ = \phi$ *for all* $u \in \mathcal{V}$.

Proof. We will use again the notations introduced in Section 3 following (14).

(i) By definition of invariant control sets C we have $\overline{\mathcal{O}^+(x)} = \overline{C}$ for all $x \in C$. Without Assumption (H) C need not be closed and need not have nonvoid interior. If $C = \{x\}$ is a one point set, then $x \in SA$ and the result is obvious. If int $C \neq \phi$, then the boundary points are either in SA, or in SB (for the lower boundary), or in SC (for the upper boundary). Hence by inspection of the vector fields on the boundary, one sees that the assertion holds.

(ii) We will prove only the case, where for each $u \in U$ there exists a unique $x_u \in \overline{D}$ such that $X(x_u)+uY(x_u) = 0$, and x_u is an unstable rest point for this vector field. All other cases are similar, because each variant control set with nonvoid interior contains an interval of unstable rest points and by Assumption 14(iii) the construction below can be carried out in this interval.

Let $u \in \mathcal{U}$ be constant, then using the assumptions $A_u(t) := \{\varphi(-t,y,u),\ y \in \overline{D}\} \subset \overline{D}$ is, for all $t > 0$, a closed interval in $\overline{D}$ containing x_u. Consider now $u \in \mathcal{U}$, piecewise constant and periodic, i.e. there exists a time interval $[0,T]$ and a partition $0 = t_0 < t_1 < \cdots < t_n = T$ such that $u(\tau) = u_i$ if $\tau \in [t_{i-1},t_i)$. For $i = 1,\ldots,n$ consider the solution map $\varphi(-t_i,\cdot,u_i)\colon \overline{D} \to \overline{D}$, and note that im $\varphi(-t_i,\cdot,u_i) = A_{u_i}(t_i) \subset \overline{D}$. Then the map $\varphi(-T,\cdot,u) = \varphi(-t_1,\cdot,u_1) \circ \cdots \circ \varphi(-t_n,\cdot,u_n)$ is continuous (even a diffeomorphism) and maps the compact interval $\overline{D}$ into itself. Therefore $\varphi(-T,\cdot,u)$ has a fixed point $x_p \in \overline{D}$. By construction the solution $\varphi(t,x_p,u)$ is periodic and contained in $\overline{D}$ for all $t \geq 0$, hence $x_p \in D_u^+$. To complete the proof, note that the set of piecewise constant, periodic controls is dense in $\mathcal{U}$ (Colonius and Kliemann (1990[a], Lemma 2.2), and compactness yields $D_u^+ \neq \phi$ for all $u \in \mathcal{U}$.

(iii) By assumption there exists $u \in U$ such that $D_u^+ = \phi$, i.e. for all $x \in \overline{D}$ there is $T_x > 0$ with $\varphi(T_x, x, u) \notin \overline{D}$. Using compactness of $\overline{D}$, one sees that there is a universal $T > 0$ such that $\varphi(T, x, u) \notin \overline{D}$ for all $x \in \overline{D}$. Now continuity of the map $v \mapsto \varphi(t, x, v)$ from $\mathcal{U}$ into M implies the assertion.

□

Corollary 4.9. *Suppose that system* (3.13) *satisfies* (3.14) *and that* $\cup\{D;\ D$ *is a control set*$\}$ *is bounded. Then:*

(i) *Given a* θ*-invariant measure* ρ *on* $\mathcal{U}$ *and a control set* $D \subset M$ *of the kind described in* Proposition 4.8(i) *or* (ii). *Then there exists a* ϕ*-invariant measure* $\mu = \mu_u \rho$ *with supp* $\mu \subset \mathcal{D}^+$.

(ii) *Let* $D \subset M$ *be a control set of the kind described in* Proposition 4.8(iii). *Then there exists an ergodic* θ*-invariant measure* ρ*, such that there is no* ϕ*-invariant measure* μ *with marginal* ρ *on* $\mathcal{U}$ *and supp* $\mu \subset \mathcal{D}^+$.

Proof.

(i) Follows directly from Proposition 4.8(i) and (ii).

(ii) By the proof of Proposition 4.8(iii), for this kind of control sets D there exists a $u_0 \in U$ such that $X(x) + uY(x) > 0$, or < 0, for all $x \in \overline{D}$. Consider the constant control function $u_0(t) \equiv u_0$, then the Dirac measure δ_{u_0} is θ-invariant and ergodic. But $D_{u_0}^+ = \phi$, and no ϕ-invariant measure with the desired properties can exist.

□

The remainder of this paper is devoted to the study of invariant measures of stochastic flows, as described in Section 2. In order to use the results above directly, we will restrict ourselves here to flows associated with random differential equations of the following form:

Let $\{\xi_t,\ t \in \mathbf{R}$ or $t \in \mathbf{R}^+\}$ be a stationary, ergodic process taking values in $U \subset \mathbf{R}^m$, where U is compact and convex, and let $\mathcal{B}$ be the Borel σ- algebra of U. Denote by (Ω, θ) the trajectory space of measurable functions with values in U, with the shift θ (for $t \in \mathbf{R}$ or for $t \in \mathbf{R}^+$). Then there is a θ-invariant, ergodic measure P on Ω. Let $\zeta\colon \mathbf{R} \times \Omega \to U$ be the evaluation map $\zeta(t, w) = w(t)$, then $\{\zeta_t,\ t \in \mathbf{R}$ or $t \in \mathbf{R}^+\}$ is a stationary, ergodic process over $(\Omega, \mathcal{F}, P, \theta)$ where $\mathcal{F}$

is generated by the cylinder sets. Consider the random differential equation

$$(3) \qquad \dot{x} = X_0(x) + \sum_{i=1}^{m} \zeta_i(t) X_i(x) \text{ on a smooth manifold } M,$$

where $X_0, \ldots, X_m$ are smooth vectorfields on M. Then (3) defines a stochastic flow $(\Omega \times M, \phi)$, as described in Section 2. We continue to assume condition (H).

Remark 4.10. (One and two-sided stochastic flows, compare Crauel (1990)) A flow, defined for $t \in \mathbf{R}$, is called two-sided, if $t \in \mathbf{R}^+$, it is called one-sided. A two-sided flow can always be restricted to a one-sided: E.g. let $\mathcal{F}' = \sigma\{\phi_t,\ t \geq 0\}$ be the σ-algebra generated by ϕ for positive time, then the restriction to $(\Omega \times M, \mathcal{F}', \widetilde{P})$ and $t \geq 0$ is a one sided flow, where $\widetilde{P}$ is the measure induced by ϕ on $\Omega \times M$. Vice versa, if $\phi = (\theta, \varphi)$ is a one-sided flow, and if θ is invertible, then $\varphi(-t, x, w) = \varphi^{-1}(t, x, \theta_{-t} w)$ is a two-sided extension. Note that in our situation θ is always invertible.

For stochastic flows we are interested in ϕ-invariant measures μ, whose marginal on Ω is the given measure P, i.e. $\mu = \mu_w P$. For a σ-algebra $\tilde{\mathcal{F}} \subset \mathcal{F}$ we denote by $\mathbf{E}\{\mu \,|\, \tilde{\mathcal{F}}\}(w)$ the desintegration of μ restricted to $\tilde{\mathcal{F}} \otimes \mathcal{B}$ with respect to $P|_{\tilde{\mathcal{F}}}$. The measure μ is ϕ- invariant iff $\mathbf{E}\{\varphi(t, \cdot, w)\mu_w \,|\, \theta_t^{-1}\mathcal{F}\}(w) = \mu_{\theta_t w}$, which for θ_t invertible (i.e. in particular for the two-sided situation) reduces to $\varphi(t, \cdot, w)\mu_w = \mu_{\theta_t w}$, see Lemma 4.2. We denote by $M_\phi(P)$ the two-sided ϕ-invariant measures with marginal P on Ω, and by $M_\phi^+(P)$ the one-sided ones with marginal $P|_{\mathcal{F}'}$. Note that measures in $M_\phi(P)$ can always be restricted to measures in $M_\phi^+(P)$, while one-sided invariant measures can be extended if θ is invertible: Let $\mu^+ \in M_\phi^+(P)$ with desintegration $\mu^+ = \mu_w^+ P$, then $\varphi^{-1}(-t, w)\mu^+_{\theta_{-t} w}$ converges P-almost surely for $t \to \infty$ to a measure μ_w, such that $\mu_w P$ is ϕ-invariant for all $t \in \mathbf{R}$, compare e.g. Crauel (1990, Remark 2.3).

We have the following general result about the support of ϕ- invariant measures:

Theorem 4.11. *Consider the two-sided stochastic flow induced by the random differential equation* (3). *Let $\mu \in M_\phi(P)$ or $\mu \in M_\phi^+(P)$ be given. Then* $\operatorname{supp} \mu \subset cl \cup \{\mathcal{D}^+;\ D \text{ is a control set}\}$, *and for every control set D with $\mu(\mathcal{D}^+) > 0$ it holds that $D_u^+ \neq \phi$ P-almost surely.*

Proof. This follows from ergodicity of P and Theorem 4.4(iv).

□

For the existence of invariant measures for a two-sided stochastic flow the results given before still hold, compare Theorem 4.4, Proposition 4.5, Corollary 4.7, and Corollary 4.9(i). From Theorem 4.4(iv) one obtains in this context:

Corollary 4.12. *Let $D \subset M$ be a control set of the kind described in* Proposition 4.8(iii). *Denote $V := \{u \in U;$ there is no $x \in \overline{D}$ with $X(x) + uY(x) = 0\}$, and by $\mathcal{V}$ the corresponding trajectory space. If $P\{supp\ P \cap \mathcal{V}\} > 0$, then there exists no $\mu \in M_\phi^+(P)$ (or $\mu \in M_\phi(P)$) with supp $\mu \subset \mathcal{D}^+$.*

For the existence of ϕ-invariant measures the sets D_w^+ play a crucial role. In general, these sets can only be determined, if the entire trajectory $\{w(t),\ t \geq 0\}$ is known. (An exception are the invariant control sets C, for which we have $C_w^+ = C$ for all $w \in \Omega$.) This fact reduces the possible invariant measures, if we require certain measurability conditions, as in the case of Markov processes. We will discuss the Markovian case next, and use the following set up, see Crauel (1990):

If the stochastic process $\{\zeta_t,\ t \in \mathbf{R}$ or $t \in \mathbf{R}^+\}$ in (3) is a (time homogeneous) Markov process, then we call $\{\phi_t,\ t \in \mathbf{R}$ or $t \in \mathbf{R}^+\}$ a Markovian stochastic flow. In order that the pair process $\{(\zeta_t, \varphi_t),\ t \geq 0\}$ becomes a Markov process, we need a condition for the initial variable of φ_t: For a random variable $\eta\colon \Omega \to M$ the process $\{(\zeta_t, \varphi(t, \eta, w)),\ t \geq 0\}$ is a Markov process iff $\{\zeta_t,\ t \geq 0\}$ is a Markov process with respect to the enlarged family of σ-algebras $\mathcal{F}^\eta_{\leq t} := \sigma\{\eta, \mathcal{F}_{\leq t}\}$ for $t \geq 0$, compare Arnold and Kliemann (1983, Lemma 2.1) and Crauel (1990, Lemma 3.4). Here $\mathcal{F}_{\leq t} := \sigma\{\zeta_\tau,\ 0 \leq \tau \leq t\}$ is the σ- algebra generated by the Markov process $\{\zeta_\tau,\ \tau \geq 0\}$ between 0 and t.

The problem of a stationary (and ergodic) Markov solution (ζ_t, φ_t) of (3) can be formulated in this context as follows: Let ϕ be a Markovian stochastic flow, then $Q(t, (p, x), A) := P\{(\zeta_t(\cdot), \varphi(t, x, \cdot)) \in A \mid \zeta_0 = p\}$ defines a family of Markov transition probabilities on $U \times M$ for $t \geq 0$, where $A \in \mathcal{B}_U \times \mathcal{B}_M$, compare Bunke (1972, Satz 6.1). A probability measure μ on $\Omega \times M$ (with marginal P on Ω) is called a Markov measure, if the pair process $\{(\zeta_t(\cdot), \varphi(t, \cdot, \cdot)),\ t \geq 0\}$ is a (time homogeneous) Markov process with transition probabilities Q, and initial distribution $\pi_M\mu$ on M. The relation between this approach for Markovian stochastic flows and the usual definition of initial distributions via the Markov semigroup $\{Q_t,\ t \geq 0\}$, defined by the Markov transition probabilities Q, is as follows: For a Markov measure μ on $\Omega \times M$ satisfying $\mu. = \mathbf{E}\{\mu. \mid \mathcal{F}_{<\infty}\}$ define a family of probability measures on U by $\nu_u = \nu_{\zeta_0(w)} = \mu_w$, which gives (the desintegration of) an initial distribution ν on $U \times M$ for $\{Q_t,\ t \geq 0\}$. Vice versa, given such a ν, define $\mu_w = \nu_{\zeta_0(w)}$, yielding a Markov measure μ on $\Omega \times M$. Hence this correspondence is one-to-one, see again Crauel (1990, Section 5.2.1).

The following result characterizes the Markov measures of (3) among all probability measures on $\Omega \times M$ with marginal P on Ω:

Proposition 4.13.

(i) *If $\mu = \mu.P$ satisfies $\mu. = \mathbf{E}\{\mu. \,|\, \mathcal{F}_{\leq 0}\}$, then μ is a Markov measure.*

(ii) *If μ is a Markov measure, then $\mathbf{E}\{\mu. \,|\, \mathcal{F}_{\leq \infty}\} = \mathbf{E}\{\mu. \,|\, \mathcal{F}_{\leq 0}\}$ P-a.s.*

For a proof see Crauel (1990, Theorem 4.4).

In this paper we are primarily interested in ϕ-invariant measures and their support. From the discussion above we obtain immediately for invariant Markov measures of (3):

(a) Every invariant measure of the Markov semigroup $\{Q_t,\ t \geq 0\}$ corresponds to a (unique) invariant Markov measure of (3) for $t \geq 0$, with $\mu. = \mathbf{E}\{\mu. \,|\, \mathcal{F}_{<\infty}\}$.

If $\{\xi_t,\ t \geq 0\}$ is a diffusion process with generator $\mathcal{L}$, then $\{(\xi_t, \varphi_t),\ t \geq 0\}$ is a diffusion process with generator $\mathcal{L} + Y$, where Y is the right hand side of (3). In this case, the Q_t invariant measures μ are the solutions of the Fokker-Planck equation $(\mathcal{L} + Y)^* \mu = 0$, where $*$ denotes the adjoint operator.

(b) If M is compact, then an invariant Markov measure exists.

(c) The invariant Markov measures form a convex, $\{\phi_t,\ t \geq 0\}$ invariant subset of $M_\phi(P)$ (see Crauel (1990, Lemma 5.1)).

For invariant Markov measures of product type we obtain from our previous results:

Proposition 4.14.

(i) *Let $\mu = P \times \lambda$ be an invariant Markov measure for the one-sided flow $\{\phi_t,\ t \geq 0\}$ of a 'colored noise system', i.e. the $(\zeta_i,\ i = 1, \ldots, m)$ are the solution of a (Stratonovič type) stochastic differential equation in $\mathbb{R}^m$. Assume that supp $P = \mathcal{U}$. Then supp $\lambda = C$ for some invariant control set C of the associated control system.*

(ii) *Suppose that either* (H) *holds and $C \subset M$ is compact, or for the system* (3.13) *that the assumptions of* Proposition 4.8 *are met. Then for every stationary Markov process $\{\zeta_t,\ t \geq 0\}$ there exists an invariant Markov measure μ of* (3) *with supp $\mu \subset C^+$, the lift of C to $\Omega \times M$.*

Proof.

(i) By the discussion after Remark 4.10, $\mu = P \times \lambda$ is invariant iff P-a.s. for all $t \geq 0$ $\mathbf{E}\{\varphi(t, \cdot, w)\lambda \,|\, \theta_t^{-1}\mathcal{F}_{\geq 0}\} = \lambda$. Now $\varphi(t, \cdot, \cdot)$ and $\theta_t^{-1}\mathcal{F}_{\geq 0} = \sigma\{\zeta_{t+s} - \zeta_t \,|\, s \geq 0\}$ are independent for $t \geq 0$, and hence μ is invariant iff $\lambda = \mathbf{E}\{\varphi(t, \cdot, \cdot)\lambda\} = \int \varphi(t, \cdot, w) dP$ for all $t \geq 0$. If D is a variant control set, then there exist $u \in \mathcal{U}$ and $x \in D$ such that $\varphi(t,x,u) \notin \overline{D}$ for some $t > 0$. Hence there are open neighborhoods $N(u)$ and $N(x)$ such that $\varphi(t,y,v) \notin \overline{D}$ for all $v \in N(u)$, all $y \in N(x) \cap \overline{D}$. Since, by assumption supp $P = \mathcal{U}$, we have that $\overline{D} \cap \text{supp } \lambda = \phi$, which proves (i).

(ii) follows directly from Proposition 4.5, and from Proposition 4.8(i) for systems of the type (3.13) (without Assumption (H)).

□

Remark 4.15.

(i) The assumption "supp $P = \mathcal{U}$" applies e.g. to stochastic processes $\{\zeta_t, t \geq 0\}$ with continuous or cadlag trajectories, if supp P covers all these functions. Hence Proposition 4.14 applies in particular to noise processes $\{\zeta_t,\ t \geq 0\}$, which are stationary, nondegenerate diffusion processes, compare Kunita (1978). Hence this proposition generalizes results from Arnold, Kliemann and Oeljeklaus (1986) and Kliemann (1987).

(ii) If "supp $P = \mathcal{U}$" does not hold, then the conclusion (i) need not be true, as Example 4.16 below shows. In other words , if $\{\zeta_t,\ t \geq 0\}$ takes values in $V \subset U$, one should use $\Omega = \mathcal{V}$, the trajectory space over V, and the corresponding control system to check for supports of invariant Markov measures of product type.

(iii) At this moment we do not know, whether the result of Proposition 4.14(i) holds for Markov measures, which are not of product type, nor whether in (ii) a product type measure always exists.

The next example shows that a stochastic flow, defined by (3), can have ϕ-invariant measures, which are not Markov measures, and that without the assumption "supp $P = \mathcal{U}$" Proposition 4.14(i) need not hold.

Example 4.16. Consider again the system from Example 3.10

$$(4) \qquad \dot{x} = \sin^2 x + a\cos^2 x - \zeta_t \cos^2 x, \qquad x \in \mathbf{R} \bmod 2\pi,\ a > 0,$$

with $U = [A,a] \subset \mathbf{R}$, $A < a$. If $\{\zeta_t,\ t \geq 0\}$ is a stationary process in the set up as above, then there exist ϕ-invariant measures μ_1 and μ_2 with supp $\mu_i \subset \mathcal{D}_i^+$, $i = 1,2$ by Corollary 4.7. (Here D_i are the main control sets as in Example 3.10.)

By Proposition 4.14(i) the measure μ_2 cannot be an invariant Markov measure of product type. Furthermore, Remark 4.2 in Arnold and Kliemann (1983) shows that there need not be any invariant Markov measure for a Markovian stochastic system, which, nevertheless, may possess a ϕ-invariant probability.

Now fix $b \in (A, a)$, and consider the Markov process $\zeta_t \equiv b$. Then the system (4) has invariant Markov measures $\mu_i = P \times \delta_{x_i}$, $i = 1, \ldots, 4$, where x_i are the four rest points corresponding to $-\sin^2 x + a\cos^2 x - b\cos^2 x = 0$. More generally, if ζ_t has values in $V \subset U$, then the invariant control sets of the control system corresponding to V determine the possible supports of invariant Markov measures of product type.

Remark 4.17. (On uniqueness of invariant measures) In general the invariant measures in $\mathcal{D}^+$, the lift of some control set, need not be unique — this is also true for Markov measures and measures over invariant control sets. Consider e.g. the class of systems given by (3.11), with more than one main control set. Then, according to Corollary 4.7, for each θ-invariant measure P and over each main control set there exists a ϕ-invariant measure. In particular, for each $u \in \mathcal{U}$ there exist invariant Krylov-Bogolyubov measures in each $\mathcal{D}^+$. On the other hand, for each $x_0 \in \operatorname{int} D$, D a main control set, there exist different periodic functions $u \in \mathcal{U}$ and different periodic solutions $\varphi(t, x_0, u)$ of (3.11) in int D. Hence there are different ϕ- invariant measures, all of which have x_0 in their support projected onto $\mathbf{P}^{d-1}$.

Consider now the Markov situation of Proposition 4.14(i), i.e. supp $P = \mathcal{U}$ and $\mu = P \times \lambda$ is an invariant Markov measure for (3). Then supp $\lambda = \overline{C}$ for some invariant control set C, and λ is unique on C, if λ has a lower semicontinuous density with respect to the Lebesgue measure on M, compare Arnold and Kliemann (1987^a). This is true, in particular, if $\{(\zeta_t, \varphi_t),\ t \geq 0\}$ is a hypoelliptic diffusion process, compare Kliemann (1987) and Arnold and Kliemann (1987^a).

5. References

[1] Arnold, L., H. Crauel, J.-P. Eckmann (eds.) (1991). "Proceedings of the Conference on Lyapunov Exponents", Oberwolfach, May 1990, Springer.

[2] Arnold, L., P. Boxler (1989). *Eigenvalues, bifurcation and center manifolds in the presence of noise.* In: C. Dafermos (ed.) EQUADIFF '87, M. Dekker.

[3] Arnold, L., P. Boxler (1990). *Stochastic bifurcation: Instructive examples in dimension one.* This volume.

[4] Arnold, L., W. Kliemann (1983). *Qualitative theory of stochastic systems.* In: Probabilistic Analysis and Related Topics, Vol. 3 (A. T. Bharucha-Reid, ed.), Academic Press, 1–79.

[5] Arnold, L., W. Kliemann (1987^a). *On unique ergodicity for degenerate diffusions.* Stochastics **21**, 41–61.

[6] Arnold, L., W. Kliemann (1987[b]). *Large deviations of linear stochastic differential equations.* In: Stochastic Differential Systems (J. H. Engelbert and W. Schmidt, eds.), Lecture Notes in Control and Information Sciences No. 96, Springer, 117–151.

[7] Arnold, L., W. Kliemann, E. Oeljeklaus (1986). *Lyapunov exponents of linear stochastic systems.* In: Arnold, Wihstutz (1986), 85–125.

[8] Arnold, L., V. Wihstutz (eds.) (1986). Lyapunov Exponents. Lecture Notes in Mathematics No. 1186, Springer.

[9] Azencott, R. (1980). *Grandes déviations et application.* In: Ecole d'Eté de Probabilités de Saint Flour VIII 1978 (P. L. Hennequin, ed.). Lecture Notes in Mathematics No. 714, Springer, 1–176.

[10] Baxendale, P. (1980). *Wiener processes on manifolds of maps.* Proc. Royal Soc. Edinbourgh **87A**, 127- -152.

[11] Baxendale, P. (1986[a]). *The Lyapunov spectrum of a stochastic flow of diffeomorphisms.* In: Arnold, Wihstutz (1986), 322–337.

[12] Baxendale, P. (1986[b]). *Asymptotic behavior of stochastic flows of diffeomorphisms.* Proceedings of 15th SPA, Nagoya 1985. Lecture Notes in Mathematics No. 1203, Springer.

[13] Baxendale, P. (1990). In: Arnold (ed.) (1990).

[14] Baxendale, P., D. W. Stroock (1988). *Large deviations and stochastic flows of diffeomorphisms.* Probab. Th. Rel. Fields **80**, 169–215.

[15] Bhatia, N. P., G. P. Szegö (1970). Stability Theory of Dynamical Systems. Springer.

[16] Boothby, W. M. (1975). An Introduction to Differentiable Manifolds and Riemannian Geometry. Academic Press.

[17] Boxler, P. (1989). *A stochastic version of center manifold theory.* Probab. Th. Rel. Fields **83**, 509- -545.

[18] Bunke, H. (1972). Gewöhnliche Differentialgleichungen mit zufälligen Parametern. Akademie Verlag.

[19] Brockett, R. W. (1973). *Lie algebras and Lie groups in control theory.* In: Geometric Methods in Systems Theory (R. W. Brockett, D. Q. Mayne, eds.) Reidel, 43–82.

[20] Carverhill, A. (1985[a]). *Flows of stochastic dynamical systems: ergodic theory.* Stochastics **14**, 273–317.

[21] Carverhill, A. (1985[b]). *A formula for the Lyapunov numbers of a stochastic flow. Applications to a perturbation theorem.* Stochastics **14**, 209–226.

[22] Carverhill, A. P., M. J. Chappell, K. D. Elworthy (1986). *Characteristic exponents for stochastic flows.* In: Lecture Notes in Mathematics No. 1158, Springer, 52–80.

[23] Clark, J. M. C. (1973) *An introduction to stochastic differential equations on manifolds.* In: Geometric Methods in Systems Theory (R. W. Brockett, D. Q. Mayne, eds.) Reidel, 131–149.

[24] Colonius, F., W. Kliemann (1989). *Infinite time optimal control and periodicity.* Appl. Math. Opt. **20**, 113–130.

[25] Colonius, F., W. Kliemann (1990[a]). *Some aspects of control systems as dynamical systems.* Submitted.

[26] Colonius, F., W. Kliemann (1990[b]). ***Linear control semigroups acting on projective space.*** Submitted.
[27] Colonius, F., W. Kliemann (1990[c]). ***Maximal and minimal Lyapunov exponents of bilinear control systems.*** To appear in J. Diff. Equations.
[28] Colonius, F., W. Kliemann (1990[d]). ***Stability radii and Lyapunov exponents.*** In: Control of Uncertain Systems (D. Hinrichsen, B. Mårtensson, eds.), Birkhäuser, 19–55.
[29] Colonius, F., W. Kliemann (1990[e]). ***Stabilization of uncertain linear system.*** To appear in: Modeling and Control of Uncertain Systems (G. DiMasi, A. Gombani, A. Kurzhanski, eds.), Birkhäuser.
[30] Colonius, F., W. Kliemann (1991[a]). ***Lyapunov exponents of control flows.*** In: Arnold et al. (eds.) (1991).
[31] Colonius, F., W. Kliemann (1991[b]). ***The Lyapunov spectrum of bilinear control systems.*** In preparation.
[32] Conley, C. (1978). ***Isolated Invariant Sets and the Morse Index.*** Regional Conference Series in Mathematics No. 38. American Mathematical Society.
[33] Crauel, H. (1986). ***Lyapunov exponents and invariant measures of stochastic systems on manifolds.*** In: L. Arnold, V. Wihstutz (eds.).
[34] Crauel, H. (1987). ***Random Dynamical Systems: Positivity of Lyapunov Exponents and Markov Systems.*** Ph.D. dissertation. Bremen, FRG.
[35] Crauel, H. (1990). ***Markov measures for random dynamical systems.*** Preprint, Bremen, FRG.
[36] Crauel, H. (1991). In: L. Arnold et al. (ed.) (1991).
[37] Dahlke, S. (1989). ***Invariante Mannigfaltigkeiten für Produkte zufälliger Diffeomorphismen.*** Ph.D. dissertation, Bremen, FRG.
[38] Elliott, D. L. (1973). ***Diffusions on a manifold arising from controllable systems.*** In: Geometric Methods in Systems Theory (R. W. Brockett, D. Q. Mayne eds.), Reidel, 285–294.
[39] Elworthy, K. D. (1978). ***Stochastic dynamical systems and their flows.*** In: Stochastic Analysis (A. Friedman, M. Pinsky, eds.), Academic Press, 79–95.
[40] Emery, M. (1990). Stochastic Calculus in Manifolds. Springer.
[41] Ehrhardt, M., W. Kliemann (1982). ***Controllability of linear stochastic systems.*** Systems and Control Letters **2**, 145–153.
[42] Ethier, S. N., T. G. Kurtz (1986). Markov Processes. Wiley.
[43] Freidlin, M. I., A. D. Wentzell (1984). Random Perturbations of Dynamical Systems. Springer.
[43] Friedman, A. (1975). Stochastic Differential Equations and Applications, Vol. I and II (1976). Academic Press.
[44] Hasminskii, R. Z. (1980). Stochastic Stability of Differential Equations. Sijthoff and Nordhoff. (Russian edition 1969).
[45] Hijab, O. (1987). Stabilization of Control Systems. Springer.
[46] Hörmander, L. (1967). ***Hypoelliptic second order differential equations.*** Acta Math. **119**, 147–171.
[47] Ikeda, N., S. Watanabe (1981). Stochastic Differential Equations and Diffusion Processes. North Holland.
[48] Isidori, A. (1989). Nonlinear Control Theory. Springer. 2nd ed.

[49] Kalman, R. E., R. S. Bucy (1961). *New results in linear filtering and prediction theory.* ASME Trans. **80**, Part D, 95–108.
[50] Kifer, Y. (1986). Ergodic Theory of Random Transformations. Birkhäuser.
[51] Kifer, Y. (1988). Random Perturbations of Dynamical Systems. Birkhäuser.
[52] Kliemann, W. (1980). *Qualitative Theory of Nonlinear Stochastic Systems* (in German). Ph. D. dissertation. Bremen, FRG.
[53] Kliemann, W. (1987). *Recurrence and invariant measures for degenerate diffusions.* Ann. Prob. **15**, 690–707.
[54] Kolmogorov, A. N. (1931). *Über die analytischen Methoden in der Wahrscheinlichkeitsrechnung.* Math. Ann. **104**, 415–458.
[55] Kunita, H. (1971). *Asymptotic behavior of the filtering errors of Markov processes.* J. Multivar. Ana. **1**, 365–393.
[56] Kunita, H. (1974). Diffusion Processes and Control Systems. Lecture Notes University of Paris VI.
[57] Kunita, H. (1978). *Supports of diffusion processes and control systems.* In: Proc. Intern. Symp. Stochastic Diff. Equ. (K. Itō, ed.), Wiley, 163–185.
[58] Kunita, H. (1984). *Stochastic differential equations and stochastic flows of diffeomorphisms.* In: Ecole d'Eté de Probabilités de Saint Flour XII, 1982 (P. L. Hennequin, ed.) Lecture Notes in Mathematics No. 1097, Springer, 143–303.
[59] Kunita, H. (1990). Stochastic Flows and Stochastic Differential Equations. To appear.
[60] Kwakernaak, H., R. Sivan (1972). Linear Optimal Control. Wiley.
[61] Ledrappier, F., L.-S. Young (1985). *The metric entropy of diffeomorphisms I.* Ann. of Math. **122**, 509–539.
[62] Ledrappier, F., L.-S. Young (1988). *Entropy formula for random transformations.* Probab. Th. Rel. Fields **80**, 217–240.
[63] Le Jan, Y. (1986). *Equilibre statistique pour les produits de difféomorphismes aléatoires indépendants.* C. R. Acad. Sc. Paris **302**, Serie I, No. 9, 351–354.
[64] Mañé, R. (1987). Ergodic Theory and Differentiable Dynamics. Springer.
[65] Nemytskii, V. V., V. V. Stepanov (1960). Qualitative Theory of Differential Equations. Princeton University Press. (Russian edition 1949).
[66] Nijmeijer, H., A. J. van der Schaft (1990). Nonlinear Dynamical Control Systems. Springer.
[67] Oseledeč, V. I. (1968). *A multiplicative ergodic theory. Lyaponov characteristic numbers for dynamical systems.* Trans. Moscow Math. Soc. **19**, 197–231.
[68] Rozanow, Y. A. (1967). Stationary Random Processes. Holden-Day.
[69] Stroock, D. W., S. R. S. Varadhan (1972). *On the support of diffusion processes with applications to the strong maximum principle.* Proc. 6th Berkeley Symp. Math. Stat. Probab. **3**, 333–359.
[70] Sussmann, H. J., V. Jurdjevic (1972). *Controllability of nonlinear systems.* J. Diff. Equs. **12**, 95–116.
[71] Varsan, C. (1982). *Controllability of Nonlinear Stochastic Control Systems I, II.* preprint, National Institute for Scientific and Technical Creation, Bucharest, Romania.

[72] Willems, J. L., J. C. Willems (1983). *Robust stabilization of uncertain systems*. SIAM J. Control Optim. **21**, 352–374.

[73] Wong, E., M. Zakai (1969). *Riemann Stieltjes approximations of stochastic integrals*. Z. Wtheorie verw. Gebiete **12**, 87–97.

[74] Wonham, W. M. (1979). Linear Multivariable Control: A Geometric Approach. Springer.

[75] Zabczyk, J. (1981). *Controllability of linear stochastic systems*. Systems and Control Letters **1**, 25–31.

[76] Zeeman, E. C. (1988). *Stability of dynamical systems*. Nonlinearity **1**, 115–155.

FRITZ COLONIUS
UNIVERSITÄT AUGSBURG
AND
WOLFGANG KLIEMANN
IOWA STATE UNIVERSITY
AMES, IOWA 50010 USA

Stochastic bifurcation: instructive examples in dimension one

by

Ludwig Arnold and Petra Boxler

Summary. A general framework of a stochastic version of bifurcation theory is proposed. The concepts are exemplified by one dimensional examples which are perturbed versions of deterministic differential equations exhibiting the elementary bifurcation scenarios. As explosion in finite time is possible, local stochastic dynamical systems have to be introduced.

1. Deterministic bifurcation theory

Many systems of physical interest contain parameters in their defining equations. If these parameters are varied, the solutions may change their qualitative behavior for certain parameter values. These qualitative changes are called *bifurcations*, and the parameter values at which they occur are called *bifurcation points*. The intuitive idea of bifurcation may be formalized by the concept of *structural stability*. A bifurcation point would then just be a parameter value at which our given family of equations is *not* structurally stable (Guckenheimer and Holmes [15], p. 119).

Bifurcation theory is at the heart of the theory of deterministic dynamical systems (see e.g. Guckenheimer and Holmes [15], V. I. Arnold [5], V. I. Arnold (ed.) [6] and Ruelle [20]). It tells us that the most simple bifurcations of equilibria in dimension one appear in the following differential equations which depend on a parameter $\alpha \in \mathbb{R}$:

$$\dot{x} = \alpha - x^2 \qquad \text{(saddle node)} \qquad (1.1a)$$

$$\dot{x} = \alpha x - x^2 \qquad \text{(transcritical)} \qquad (1.1b)$$

$$\dot{x} = \alpha x - x^3 \qquad \text{(pitchfork)} \qquad (1.1c)$$

The corresponding *bifurcation diagrams* below show that bifurcation is related to the appearance of new solutions and a change in the stability behavior. Solid/dotted lines correspond to stable unstable steady states:

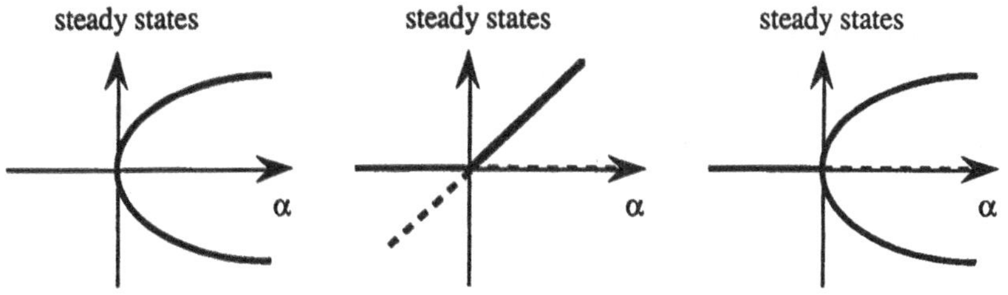

(a) saddle node (b) transcritical (c) pitchfork

The cases (b) and (c) have the additional property that there is a reference steady state $x = 0$ for all $\alpha = 0$. The bifurcation point $\alpha = 0$ could thus also be described by the property that for each α in a neighborhood of $\alpha = 0$ there are nontrivial steady states which converge to $x = 0$ as $\alpha \to 0$. This latter property is also used as a (static) definition of a bifurcation point (cf. Chow and Hale [12], p.2). It implies that the Jacobian of the mapping is necessarily singular at $\alpha = 0$.

The disadvantage of the examples (1.1) is that they only define *local* flows as solutions might (and will) explode in finite time. This complicates the study of their perturbed versions, as will be seen in section 3.

2. Stochastic bifurcation theory

Stochastic bifurcation theory should of course study qualitative changes in parametrized families of stochastic dynamical systems. The problem is how to formalize this intuitive idea.

Physicists, chemists and biologists have for a long time been interested in (and in fact have experimentally observed) qualitative changes in systems perturbed by noise. For a systematic study see Horsthemke and Lefever [16]. In the white noise case they have examined the invariant probability density (steady state solution of the Fokker-Planck equation) and have e.g. observed a change from a one-peak to a two-peak density.

As this is certainly a qualitative change one could call it a stochastic bifurcation, and the point at which it occurs a stochastic bifurcation point.

This concept can be made rigorous using ideas of Zeeman [22], [23]. He calls two probability densities p, q *equivalent* if there are diffeomorphisms α, β such that $p \circ \alpha = \beta \circ q$. This gives rise to a notion of structural stability and thus of bifurcation of densities which captures the above observations.

We have been arguing against this concept for some time (Arnold [1], Arnold and Boxler [2]). Besides being restricted to the white noise (i.e. Markovian) case it has the basic drawback of being a *static* definition: The invariant density describes the asymptotic proportion of time the system spends in a volume element. It is thus not related to the stability of the system since this is determined by the behavior of two neighboring orbits. There are examples of parametrized families of stochastic differential equations which change from 'exponentially stable' to 'exponentially unstable' whereas the Fokker-Planck equation is independent of the parameter (see Baxendale [7]). Furthermore, invariant densities may 'disappear' at certain parameter values, as the notion of an invariant measure as a solution of the Fokker-Planck equation is too narrow (cf. Crauel [14]). At the end of his paper [22] Zeeman himself says: "It seems a pity to have to represent a dynamical system by a static picture."

We have therefore advocated another, more dynamic concept of stochastic bifurcation which is based on the theory of stochastic flows and on multiplicative ergodic theory (see Arnold and Boxler [2], Boxler [10], [11]) and which we are now going to describe briefly.

Many random and stochastic differential and difference equations on a manifold M, which will often be $\mathbb{R}^d$, generate through their solutions a *stochastic dynamical system*

(SDS) φ in the following sense (see Boxler [8], Kunita [17]).

(i) $(\Omega, \mathcal{F}, P)$ is a probability space, and $(\vartheta_t)_{t\in T}$, $T = \mathbb{Z}$ or $\mathbb{R}$, is a flow of ergodic P-preserving maps of Ω, i.e. $\vartheta_0 = \mathrm{id}$, $\vartheta_{t+s} = \vartheta_t \circ \vartheta_s$ for all $t, s \in T$.

(ii) For P-a.a. ω and for all $t \in T$,

$$\begin{aligned} \varphi(t,\omega): M &\rightarrow M, \\ x &\rightarrow \varphi(t,\omega)x \end{aligned}$$

is a diffeomorphism of M which is continuous in (t,x) and measurable in (t,ω,x).

(iii) For P-a.a. ω the family $(\varphi(t,\omega))_{t\in T}$ of diffeomorphisms is a *cocycle* in the following sense:

$$\varphi(t+s,\omega) = \varphi(t,\vartheta_s\omega) \circ \varphi(s,\omega) \quad \text{for all } t, s \in T.$$

If φ is an SDS then

$$\begin{aligned} \Theta_t: \Omega\times M &\rightarrow \Omega\times M, \\ (\omega,x) &\rightarrow (\vartheta_t,\varphi(t,\omega)x) \end{aligned}$$

is a flow, called *skew product flow* induced by φ.

A probability measure μ on $\Omega\times M$ is said to be *invariant* for the SDS φ if

(i) its marginal on Ω is P,

(ii) it satisfies $\Theta_t\mu = \mu$ for all $t \in T$.

Using (i) and the disintegration $\mu(d\omega,dx) = \mu_\omega(dx)P(dx)$, (ii) is equivalent to $\varphi(t,\omega)\mu_\omega = \mu_{\vartheta_t\omega}$ for all $t \in T$, P-a.s.

The examples below support our belief that the proper level to discuss stochastic bifurcation is the level of invariant (and more general) measures: New invariant measures are expected to bifurcate from reference measures at certain points. More specifically: Let φ_α, $\alpha \in \mathbb{R}$, be a family of SDS with invariant measures μ_α. Then $(\alpha_0, \mu_{\alpha_0})$ is called a *stochastic bifurcation point* if for each α in a neighborhood of $\alpha = \alpha_0$, there is an invariant $\nu_\alpha \neq \mu_{\alpha_0}$ with $\nu_\alpha \Rightarrow \mu_{\alpha_0}$ $(\alpha \rightarrow \alpha_0)$.

A crucial role in a future stochastic bifurcation theory along these lines will be played by Oseledec's *Multiplicative Ergodic Theorem* (MET) (see Oseledec [18], for the version used here see Boxler [8]):

Let φ be an SDS with ergodic invariant measure μ and assume that $\log^+ \sup_{0\le t\le 1} \|T\varphi(t,\omega,x)^{\pm 1}\| \in L^1(\Omega\times M,\mu)$ where $T\varphi(t,\omega,x): T_xM \rightarrow T_{\varphi(t,\omega)x}M$ is the derivative (linearization) of $\varphi(t,\omega)$ at $x \in M$. Then there are r real numbers $\lambda_1 > \ldots > \lambda_r$ (*Lyapunov exponents*) with multiplicities d_i, $\sum_{i=1}^{r} d_i = d$, such that for an invariant set Γ

$\subset \Omega \times M$ with $\mu(\Gamma) = 1$ there are random subspaces (*Oseledec spaces*) $E_i(\omega,x)$ of T_xM such that for each $(\omega,x) \in \Gamma$:

(i) $T_xM = E_1(\omega,x) \oplus \ldots \oplus E_r(\omega,x)$, $\dim E_i(\omega,x) = d_i$,

$T\varphi(t,\omega)E_i(\omega,x) = E_i(\Theta_t(\omega,x))$.

(ii) $\lim_{t\to\pm\infty} \frac{1}{t} \log \| T\varphi(t,\omega,x)v \| = \lambda_i$ if and only if $v \in E_i(\omega,x)$, $v \neq 0$.

This theorem enables one to develop the linear stochastic theory in analogy to the deterministic one as it provides the stochastic versions of eigenvalues and eigenspaces. Boxler [11] recently proved that at a stochastic bifurcation point there is necessarily a vanishing Lyapunov exponent.

A lot of work is still to be done to arrive at a full-fledged stochastic bifurcation theory although there exists already a stochastic center manifold theory (Boxler [8], [9]), and a stochastic normal form theory is on its way (Xu [21]).

3. Local stochastic dynamical systems

The perturbed versions of the models (1.1) will, as in the deterministic case, explode in finite (random) time. Before these models can be studied in detail we will thus have to 'localize' our concept of an SDS.

A theory of stochastic flows of local diffeomorphisms was developed by Kunita [17], Sect. IV.7. However, this theory is not yet in the 'cocycle formulation' we need here to apply multiplicative ergodic theory (see also Boxler [8], Section 6, for another, but equivalent local theory).

We define a *local stochastic dynamical system* (local SDS) φ as follows:

(i) same as (i) for an SDS,

(ii) For P-a.a. ω and for all $t \in T$, $\varphi(t,\omega): D_t(\omega) \to R_t(\omega)$ is a (local) diffeomorphism of the random (open) domain $D_t(\omega) \subset M$ onto the random range $R_t(\omega) = \varphi(t,\omega)D_t(\omega) \subset M$ (thus an open set) where $D_t(\omega) \downarrow$ ($|t| \uparrow \infty$), and $\varphi(0,\omega) = \mathrm{id}|_M$.

(iii) For P-a.a. ω the family $(\varphi(t,\omega))_{t \in T}$ of local diffeomorphisms is a local cocycle in the following sense:
Let $s, t \in T$ and $x \in D_s(\omega)$. Then $x \in D_{t+s}(\omega)$ if and only if $\varphi(t,\omega)x \in D_t(\vartheta_s\omega)$, and in that case:

$$\varphi(t+s,\omega)x = \varphi(t,\vartheta_s\omega)\varphi(s,\omega)x. \tag{3.1}$$

Putting $t = -s$ in (iii) and using $D_0(\omega) = M$ we find that if $x \in D_s(\omega)$ then $\varphi(s,\omega)x \in D_s(\vartheta_s\omega)$. In other words, $R_t(\omega) = D_{-t}(\vartheta_t\omega)$ for all $t \in T$, and (3.1) reads

$$\varphi(t,\omega)^{-1} = \varphi(-t,\vartheta_t\omega): D_{-t}(\vartheta_t\omega) \to D_t(\omega). \tag{3.2}$$

Examples

a) Random ODE's $\dot{x} = f(\vartheta_t\omega,x)$ with smooth r.h.s. generate a local SDS by classical pathwise ODE theory.

b) Stochastic differential equations $dx = X_0(x)dt + \sum_{i=1}^{m} X_i(x) \circ dW_i$ with smooth vector fields $X_0, \ldots, X_m$ generate a local SDS (see Kunita [17], IV.7.). Notice that here $T = \mathbb{R}$ (and not $\mathbb{R}^+$ as in classical stochastic analysis).

Remarks. 1) For random or stochastic differential equations on $M = \mathbb{R}$, the uniqueness of the solution implies that $D_t(\omega)$ (and thus $R_t(\omega)$) is an open interval.
2) A local SDS is global iff $D_t(\omega) \to R_t(\omega) = M$ for all $t \in T$, P-a.s.

Since we want to apply the MET we look for those points $x \in M$ for which the linearization $T\varphi(t,\omega,x)$ exists for *all* $t \in T$. $T\varphi(t,\omega,x)$ being defined for each $x \in D_t(\omega)$ these are the points in the set

$$D(\omega) = \bigcap_{t \in T} D_t(\omega).$$

Hence this is the set of those initial values whose orbits live 'from eternity to eternity'.

Lemma. $D(\omega)$ *is invariant in the sense that* $\varphi(t,\omega)D(\omega) = D(\vartheta_t\omega)$ *for all* $t \in T$.

Proof.
$$\begin{aligned} x \in D(\omega) &\Leftrightarrow x \in D_s(\omega) \text{ for all } s \\ &\Leftrightarrow x \in D_{s+t}(\omega) \text{ for all } s \\ &\Leftrightarrow \varphi(t,\omega)x \in D_s(\vartheta_t\omega) \text{ for all } s \text{ (by (iii))} \\ &\Leftrightarrow \varphi(t,\omega)x \in D(\vartheta_t\omega) \end{aligned}$$
∎

In order to apply the MET as in the global case to obtain Lyapunov exponents and Oseledec spaces (and all ensuing things) we finally need an *invariant* measure $\mu(d\omega,dx) = \mu_\omega(dx)P(d\omega)$ for the local SDS φ. It has to satisfy $\mu_\omega(D(\omega)) = 1$ P-a.s., otherwise it would lose mass due to explosion. Then $\varphi(t,\omega)\mu_\omega$ is again a probability measure with $\varphi(t,\omega)\mu_\omega(D(\vartheta_t\omega)) = 1$ P-a.s.

4. The transcritical case

We now perturb the parameter α in $\dot{x} = \alpha x - x^2$ by a zero mean stationary stochastic process ξ_t (which can be white noise), i.e. we look at

$$\dot{x} = (\alpha + \sigma\xi_t)x - x^2$$

where $\sigma > 0$ is a strength parameter. We will restrict our development to the white noise case as all expressions become cleaner. Thus we are going to investigate the local SDS generated by the Stratonovich equation

$$dx = (\alpha x - x^2)dt + \sigma x \circ dW. \tag{4.1}$$

The transformation $y = \frac{1}{x}$ turns (4.1) into the linear equation

$$dy = (-\alpha y + 1)dt - \sigma y \circ dW$$

whose solution is

$$y(t) = e^{-\alpha t - \sigma W_t}\left(y_0 + \int_0^t e^{\alpha s + \sigma W_s}\, ds\right).$$

Consequently, the local SDS φ_α generated by (4.1) is given by

$$\begin{aligned} 0 &\to \varphi_\alpha(t,\omega)0 = 0, \\ 0 \neq x &\to \varphi_\alpha(t,\omega)x = \frac{e^{\alpha t + \sigma W_t(\omega)}}{\frac{1}{x} + \int_0^t e^{\alpha s + \sigma W_s(\omega)}\, ds} \end{aligned} \tag{4.2}$$

We can immediately read off (4.2) the following facts:

Theorem. *For the local SDS* $\varphi_\alpha(t,\omega)\colon D_t^\alpha(\omega) \to R_t^\alpha(\omega)$ *generated by (4.1),*

$$D_t^\alpha(\omega) = \begin{cases} \mathbb{R} & \text{if } t = 0, \\ (d_t^\alpha(\omega), \infty) & \text{if } t > 0, \\ (-\infty, d_t^\alpha(\omega)) & \text{if } t < 0, \end{cases}$$

where

$$d_t^\alpha(\omega) = -\frac{1}{\int_0^t e^{\alpha s + \sigma W_s(\omega)}\, ds}.$$

Thus

$$R_t^\alpha(\omega) = D_{-t}^\alpha(\vartheta_t\omega) = \begin{cases} \mathbb{R} & \text{if } t = 0, \\ (-\infty, r_t^\alpha(\omega)) & \text{if } t > 0, \\ (r_t^\alpha(\omega), \infty) & \text{if } t < 0, \end{cases}$$

where

$$r_t^\alpha(\omega) = \frac{e^{\alpha t + \sigma W_t(\omega)}}{\int_0^t e^{\alpha s + \sigma W_s(\omega)}\, ds}.$$

Using $W_s(\vartheta_t\omega) = W_{t+s}(\omega) - W_t(\omega)$ one checks that $r_t^\alpha(\omega) = d_{-t}^\alpha(\vartheta_t\omega)$. Further studies depend on the parameter α:

Case $\alpha < 0$:

Since $e^{\alpha t+\sigma W_t} = e^{\alpha t(1+o(1))}$ we have $\int_0^{\infty} e^{\alpha s+\sigma W_s} ds < \infty$ and $\int_0^{-\infty} e^{\alpha s+\sigma W_s} ds = -\infty$.
Consequently $D^\alpha(\omega) = \bigcap_{t\in\mathbb{R}} D_t^\alpha(\omega) = [d^\alpha(\omega), 0]$ where

$$d^\alpha(\omega) = -\frac{1}{\int_0^{\infty} e^{\alpha s+\sigma W_s} ds}.$$

The boundary points of $D^\alpha(\omega)$ constitute two ergodic invariant measures (in fact: stationary solutions)

$$\mu_\omega^\alpha = \delta_0, \quad \nu_\omega^\alpha = \delta_{d^\alpha(\omega)}.$$

Lemma. *μ^α and ν^α are the only ergodic invariant measures for φ_α.*

Proof. Suppose $\rho \neq \mu^\alpha, \nu^\alpha$ is invariant. Then $\rho_\omega(\text{int } D^\alpha(\omega)) = 1$. For each $f \in \mathcal{C}^b(\mathbb{R})$,

$$\varphi_\alpha(t,\omega)\rho_\omega(f) = \int_{\mathbb{R}} f(\varphi_\alpha(t,\omega)x)\, \rho_\omega(dx) = \rho_{\vartheta_t\omega}(f).$$

But for $t \to \infty$, $\varphi_\alpha(t,\omega)x \to 0$ for all $x \in \text{int } D^\alpha(\omega)$, so that the l.h.s. converges to $f(0)$ while the r.h.s. is a stationary process entailing $\rho_\omega(f) = f(0) = \delta_0(f)$. This, however, was excluded. ∎

We now apply the MET. The linearized flow $T\varphi_\alpha(t,\omega,x)v = v_t$ satisfies the linearization of (4.1):

$$dv_t = (\alpha - 2\varphi_\alpha(t,\omega)x)v_t\, dt + \sigma v_t \circ dW_t.$$

Thus

$$T\varphi_\alpha(t,\omega,x)v = v \exp\left(\alpha t - 2\int_0^t \varphi_\alpha(s,\omega)x\, ds + \sigma W_t\right)$$

and

$$\lim_{t\to\infty} \frac{1}{t} \log \| T\varphi_\alpha(t,\omega,x)v \| = \alpha - 2 \lim_{t\to\infty} \frac{1}{t} \int_0^t \varphi_\alpha(s,\omega)x\, ds.$$

For $\mu_\omega^\alpha = \delta_0$ we insert $x = 0$ and obtain

$$\lambda(\mu^\alpha) = \alpha < 0.$$

For $\nu_\omega^\alpha = \delta_{d^\alpha(\omega)}$ we insert $x = d^\alpha(\omega)$, take into account that $\varphi_\alpha(t,\omega)d^\alpha(\omega) = d^\alpha(\vartheta_t\omega)$ by invariance and use the ergodic theorem to get

$$\lambda(\nu^\alpha) = \alpha - 2\, E d^\alpha(\omega).$$

This looks hopeless at first sight but note that

$$d^{\alpha}(\vartheta_t\omega) = -\frac{e^{\alpha t+\sigma W_t}}{\int_t^{\infty} e^{\alpha s+\sigma W_s}\,ds} = \frac{\Psi'(t)}{\Psi(t)},$$

so that

$$Ed^{\alpha}(\omega) = \lim_{t\to\infty}\frac{1}{t}\int_0^t d^{\alpha}(\vartheta_s\omega)\,ds = \lim_{t\to\infty}\frac{1}{t}\log\int_t^{\infty} e^{\alpha s+\sigma W_s}\,ds = \alpha$$

and in turn

$$\lambda(v^{\alpha}) = -\alpha > 0.$$

Since it is easily checked that for all $x > d^{\alpha}(\omega)$, $\varphi_{\alpha}(t,\omega)x \to 0$ $(t\to\infty)$ with exponential rate α, we immediately obtain the stochastic stable manifold consisting of those y for which for $t\to\infty$, $|\varphi_{\alpha}(t,\omega)y - \varphi_{\alpha}(t,\omega)x| \to 0$ (exponentially fast):

$$M_s^{\alpha}(\omega,x) = (d^{\alpha}(\omega),\infty) \qquad \text{for all } x > d^{\alpha}(\omega).$$

Similarly for the unstable manifold:

$$M_u^{\alpha}(\omega,x) = (-\infty,0) \qquad \text{for all } x < 0.$$

Case $\alpha > 0$:

Here $D^{\alpha}(\omega) = [0, e^{\alpha}(\omega)]$, where

$$e^{\alpha}(\omega) = \frac{1}{\int_{-\infty}^{0} e^{\alpha s+\sigma W_s}\,ds} < \infty.$$

The only invariant measures are

$$\mu_{\omega}^{\alpha} = \delta_0 \text{ with } \lambda(\mu^{\alpha}) = \alpha > 0,\ M_u^{\alpha}(\omega,x) = (-\infty, e^{\alpha}(\omega)) \text{ for all } x < e^{\alpha}(\omega),$$

$$v_{\omega}^{\alpha} = \delta_{e^{\alpha}(\omega)} \text{ with } \lambda(v^{\alpha}) = -\alpha < 0,\ M_s^{\alpha}(\omega,x) = (0,\infty) \text{ for all } x > 0.$$

Case $\alpha = 0$:

We claim that $D^0(\omega) = \{0\}$. This follows from:

Lemma. $\displaystyle\int_{-\infty}^{0} e^{\sigma W_s}\,ds = \int_0^{\infty} e^{\sigma W_s}\,ds = \infty.$

Proof. Put $N(\omega) = \{t \geq 0: W_t(\omega) \geq 0\}$. Then

$$\int_0^\infty e^{\sigma W_s(\omega)}\, ds \geq \int_0^\infty 1_{N(\omega)}\, ds = \text{Leb}\, N(\omega).$$

But by the arc sin law, $\limsup_{t \to \infty} \frac{\text{Leb}\,\{s \leq t: W_s \geq 0\}}{t} = 1$ P-a.s., so that Leb $N(\omega) = \infty$ P-a.s. ■

There is thus a unique invariant measure $\mu^0_\omega = \delta_0$ with $\lambda(\mu^0) = 0$.

In order to make sure that there is a stochastic bifurcation at $\alpha = 0$ we finally have to check that for $\alpha \to 0$, $\nu^\alpha \Rightarrow \delta_0 \times P$. But this follows by monotone convergence and the last lemma because $\lim_{\alpha \uparrow 0} d^\alpha(\omega) = 0 = \lim_{\alpha \downarrow 0} e^\alpha(\omega)$.

The final result is thus:

Theorem. *For the local SDS* φ_α *generated by equation (4.1),* $\alpha = 0$ *is a stochastic bifurcation point. The stochastic bifurcation diagram describes a scenario which will be called a* stochastic transcritical bifurcation. *It looks as follows:*

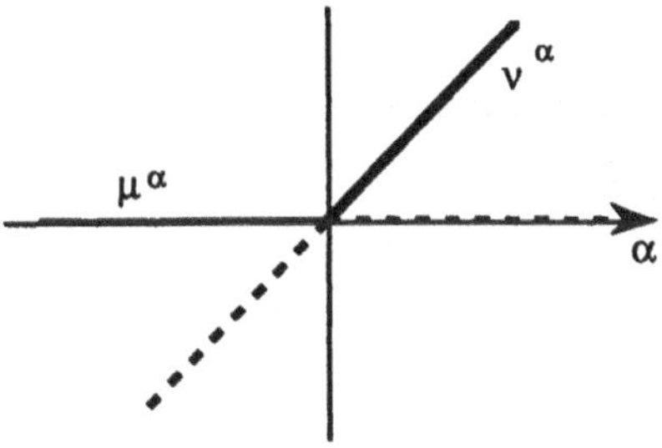

The straight lines for ν^α are justified by $Ed^\alpha(\omega) = Ee^\alpha(\omega) = \alpha$.

Remark. For the real noise case just replace W_t by $\int_0^t \xi_s\, ds$.

5. The pitchfork case

As things are similar to the transcritical case, our presentation will be very brief.

We consider the perturbed version of (1.1c):

$$\dot{x} = (\alpha + \sigma\xi_t)x - x^3$$

and stick again to the white noise case

$$dx = (\alpha x - x^3)\, dt + \sigma x \circ dW. \tag{5.1}$$

The transformation $y = \frac{1}{x^2}$ turns (5.1) into a solvable linear equation yielding the local SDS φ_α

$$0 \to \varphi_\alpha(t,\omega)0 = 0,$$

$$0 \neq x \to \varphi_\alpha(t,\omega)x = \operatorname{sign}(x)\frac{e^{\alpha t+\sigma W_t}}{\sqrt{\frac{1}{x^2} + 2\int_0^t e^{2\alpha s+2\sigma W_s}\,ds}}. \tag{5.2}$$

Inspection of (5.2) immediately gives:

Theorem. *For the local SDS* $\varphi_\alpha(t,\omega)$: $D_t^\alpha(\omega) \to R_t^\alpha(\omega)$ *generated by (5.1),*

$$D_t^\alpha(\omega) = \begin{cases} \mathbb{R} & \text{if } t \geq 0, \\ (-d_t^\alpha(\omega), d_t^\alpha(\omega)) & \text{if } t < 0, \end{cases}$$

where

$$d_t^\alpha(\omega) = \frac{1}{\sqrt{2\int_t^0 e^{2\alpha s+2\sigma W_s}\,ds}}.$$

Thus

$$R_t^\alpha(\omega) = D_{-t}^\alpha(\vartheta_t\omega) = \begin{cases} (-r_t^\alpha(\omega), r_t^\alpha(\omega)) & \text{if } t > 0, \\ \mathbb{R} & \text{if } t \leq 0, \end{cases}$$

where

$$r_t^\alpha(\omega) = \frac{e^{\alpha t+\sigma W_t}}{\sqrt{2\int_0^t e^{2\alpha s+2\sigma W_s}\,ds}}.$$

Further studies depend on α:

Case $\alpha < 0$:

$D^\alpha(\omega) = \bigcap_{t\in\mathbb{R}} D_t^\alpha(\omega) = \{0\}$, and $\mu_\omega^\alpha = \delta_0$ is the unique invariant measure. The linearized flow is

$$T\varphi_\alpha(t,\omega,x)v = v\,\exp\left(\alpha t - 3\int_0^t (\varphi_\alpha(s,\omega)x)^2\,ds + \sigma W_t\right)$$

and

$$\lim_{t\to\infty} \frac{1}{t} \log \| T\varphi_\alpha(t,\omega,x)v \| = \alpha - 3 \lim_{t\to\infty} \frac{1}{t} \int_0^t (\varphi_\alpha(s,\omega)x)^2\,ds.$$

For x = 0 this gives

$$\lambda(\mu^\alpha) = \alpha < 0.$$

We have

$$M_s^\alpha(\omega,x) = \mathbb{R} \quad \text{for all } x \in \mathbb{R}.$$

Case $\alpha > 0$:

$D^\alpha(\omega) = [-d^\alpha(\omega), d^\alpha(\omega)]$, where

$$d^\alpha(\omega) = \frac{1}{\sqrt{2\int_{-\infty}^{0} e^{2\alpha s + 2\sigma W_s}\,ds}}.$$

We now have exactly three ergodic invariant measures (proof as in section 4):

$$\mu_\omega^\alpha = \delta_0 \text{ with } \lambda(\mu^\alpha) = \alpha > 0,$$

$$\nu_{1,\omega}^\alpha = \delta_{d^\alpha(\omega)} \text{ with } \lambda(\nu_1^\alpha) = -2\alpha < 0,$$

$$\nu_{2,\omega}^\alpha = \delta_{-d^\alpha(\omega)} \text{ with } \lambda(\nu_2^\alpha) = -2\alpha < 0.$$

The Lyapunov exponents are calculated by the same trick as in section 4. We have:

$$M_u^\alpha(\omega,x) = (-d^\alpha(\omega), d^\alpha(\omega)) \qquad \text{for } -d^\alpha(\omega) < x < d^\alpha(\omega).$$

$$M_s^\alpha(\omega,x) = (0,\infty) \qquad \text{for } x > 0.$$

$$M_s^\alpha(\omega,x) = (-\infty,0) \qquad \text{for } x < 0.$$

Case $\alpha = 0$:

Again, since $\int_{-\infty}^{0} e^{2\sigma W_s}\,ds = \infty$ for all $\sigma > 0$, $D^0(\omega) = \{0\}$, and $\mu_\omega^0 = \delta_0$ is the unique invariant measure with $\lambda(\mu^0) = 0$. The linearized flow

$$T\varphi_0(t,\omega,0)v = ve^{\sigma W_t}$$

is unstable, while for the nonlinear flow we obtain for all $x \neq 0$:

$$\varphi_0(t,\omega)x = \text{sign}(x)\frac{e^{\sigma W_t}}{\sqrt{\frac{1}{x^2}+2\int_0^t e^{2\sigma W_s}\,ds}} \to 0 \quad (t\to\infty).$$

This can be seen by checking that for $dx = -x^3\,dt + \sigma x \circ dW$, the points $\pm\infty$ are repelling and 0 is attracting (c.f. e.g. Prohorov and Rozanov [19], p. 268-269). The nonlinear flow is thus asymptotically stable for $\alpha = 0$, as is the deterministic case.

Since by monotone convergence $\lim_{\alpha\downarrow 0} d^\alpha(\omega) = 0$, i.e. $v_{1,2}^\alpha \Rightarrow \delta_0 \times P$ $(\alpha \downarrow 0)$, the final result is:

Theorem. *For the local SDS* φ_α *generated by equation (5.1),* $\alpha = 0$ *is a stochastic bifurcation point. The stochastic bifurcation diagram describes a scenario which will be called a* stochastic pitchfork bifurcation. *It looks as follows:*

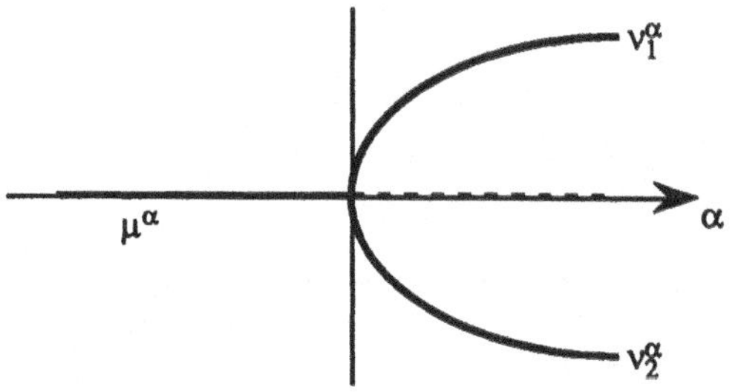

The parabola for $v_{1,2}^\alpha$ is justified by the fact that $E\,d^\alpha(\omega)^2 = \alpha$.

Remark. For the real noise case just replace W_t by $\int_0^t \xi_s\,ds$.

6. The saddle node case: global effects of noise

The perturbed version of (1.1a) is

$$\dot{x} = \alpha + \sigma\xi_t - x^2,$$

in its white noise version

$$dx = (\alpha - x^2)\,dt + \sigma \circ dW. \tag{6.1}$$

We put $x = \frac{\dot{u}}{u}$ and obtain, with $z = \begin{pmatrix} z_1 \\ z_2 \end{pmatrix} = \begin{pmatrix} u \\ \dot{u} \end{pmatrix}$,

$$dz = \begin{bmatrix} 0 & 1 \\ \alpha & 0 \end{bmatrix} z\,dt + \begin{bmatrix} 0 & 0 \\ \sigma & 0 \end{bmatrix} z \circ dW \tag{6.2}$$

which is nothing but the white noise version of

$$\ddot{u} - (\alpha + \sigma\xi_t)u = 0,$$

i.e. the undamped random linear oscillator. Explosion of (6.1) is equivalent to z_1 becoming 0 or (6.2), projected onto S^1,

$$d\gamma = (-\sin^2\gamma + \alpha\cos^2\gamma)\,dt + \sigma\cos^2\gamma \circ dW, \tag{6.3}$$

crossing the angles $\gamma = \pm\frac{\pi}{2}$, $\gamma = \arctan\frac{\dot{u}}{u}$.

Equations of type (6.2) and their projections are extremely well investigated, see e.g. Arnold, Oeljeklaus and Pardoux [4]. It follows that in our case (6.3) is recurrent, i.e. all solutions of (6.1) explode P-a.s. which means that $D^\alpha(\omega) = \emptyset$ for all $\alpha \in \mathbb{R}$.

Why does the saddle node bifurcation 'disappear' after perturbation? The reason is the global effect of white noise: It pushes the system around all possible scenarios for the different values of α, and the resulting stochastic behavior is a nontrivial 'mixture' of all deterministic scenarios. In the saddle node case white noise always pushes the system to $\alpha < 0$ where it explodes.

We can avoid this in the real noise case. Look at

$$\dot{x} = \alpha(1 + \xi_t) - x^2 \tag{6.4}$$

where ξ_t is a nice diffusion in an interval [a, b], $a > -1$. The analysis of the corresponding real noise system $\left(x = \frac{\dot{u}}{u},\ z = \begin{pmatrix} u \\ \dot{u} \end{pmatrix}\right)$

$$\ddot{u} - \alpha(1 + \xi_t)u = 0 \quad \text{or} \quad \dot{z} = \begin{bmatrix} 0 & 1 \\ \alpha(1+\xi_t) & 0 \end{bmatrix} z \tag{6.5}$$

shows that

$$D^\alpha(\omega) = \begin{cases} \emptyset & \text{if } \alpha < 0, \\ \{0\} & \text{if } \alpha = 0, \\ [d_-^\alpha(\omega), d_+^\alpha(\omega)] & \text{if } \alpha > 0, \end{cases}$$

where

$$d_\pm^\alpha(\omega) = \tan\gamma_\pm^\alpha(\omega), \quad -\frac{\pi}{2} < \gamma_-^\alpha(\omega) < 0 < \gamma_+^\alpha(\omega) < \frac{\pi}{2},$$

and $\gamma_\pm^\alpha(\omega)$ are the angles defining the two different Oseledec spaces of (6.5) (see Arnold, Kliemann, Oeljeklaus [3] and Crauel [13]). The two invariant measures of (6.4) for $\alpha > 0$ are

$$\mu_{\pm,\omega}^\alpha = \delta_{d_\pm^\alpha(\omega)} \text{ with } \lambda(\mu_{\pm,\omega}^\alpha) = -2\,E\tan\gamma_\pm^\alpha(\omega).$$

Since $d_\pm^\alpha(\omega) \to 0$ $(\alpha \downarrow 0)$, we would say that (6.4) exhibits a *stochastic saddle node bifurcation* with the familiar bifurcation diagram.

REFERENCES

[1] *Arnold, L.:* Lyapunov exponents of nonlinear stochastic systems. In: G. I. Schueller, F. Ziegler (eds.): Nonlinear Stochastic Dynamic Engineering Systems. Springer 1988.

[2] *Arnold, L., Boxler, P.:* Eigenvalues, bifurcation and center manifolds in the presence of noise. In: C. Dafermos et al. (eds.): EQUADIFF'87. M. Dekker 1989.

[3] *Arnold, L., Kliemann, W., Oeljeklaus, E.:* Lyapunov exponents of linear stochastic systems. Proceedings of a workshop Bremen 1984. Lecture Notes in Mathematics vol. 1186. Springer 1986.

[4] *Arnold, L., Oeljeklaus, E., Pardoux, E.:* Almost sure and moment stability for linear Ito equations. Proceedings of a workshop Bremen 1984. Lecture Notes in Mathematics vol. 1186. Springer 1986.

[5] *Arnold, V.I.:* Geometrical Methods in the Theory of Ordinary Differential Equations. Springer 1977.

[6] *Arnold, V. I.(ed.):* Dynamical Systmes V: Theory of Bifurcations and Catastrophes. Springer 1990.

[7] *Baxendale, P.:* Asymptotic behaviour of stochastic flows of diffeomorphisms. Proceedings of 15th SPA, Nagoya 1985. Lecture Notes in Mathematics, Springer 1985.

[8] *Boxler, P.:* A stochastic version of center manifold theory. Probab. Th. Rel. Fields 83 (1989), 509 - 545.

[9] *Boxler, P.:* Center manifolds for randomly perturbed dynamical systems. Report No. 212, Institut für Dynamische Systeme, Universität Bremen 1989.

[10] *Boxler, P.:* Lyapunov exponents indicate stability and detect stochastic bifurcations. In: P. Krée, W. Wedig (ed.): Effective Stochastic Analysis I. Springer 1990 (to appear).

[11] *Boxler, P.:* A necessary condition for a stochastic bifurcation. In preparation.

[12] *Chow, S., Hale, J.:* Methods of Bifurcation Theory. Springer 1982.

[13] *Crauel, H.:* Lyapunov exponents and invariant measures of stochastic systems on manifolds. Proceedings of a workshop Bremen 1984. Lecture Notes in Mathematics vol. 1186. Springer 1986.

[14] *Crauel, H.:* Markov measures for random dynamical systems. Submitted.

[15] *Guckenheimer, J., Holmes, Ph.:* Nonlinear oscillations, dynamical systems and bifurcations of vector fields.Springer, 2nd ed. 1986.

[16] *Horsthemke, W., Lefever, R.:* Noise-induced transitions. Springer 1984.

[17] *Kunita, H.:* Stochastic flows and stochastic differential equations. To appear.

[18] *Oseledec,V. I.:* A multiplicative ergodic theorem. Lyapunov characteristic numbers for dynamical systems. Trans. Mosc. Math. Soc. 19 (1968), 197 - 231.

[19] *Prohorov, Y., Rozanov, Y.:* Probability Theory. Springer 1969.

[20] *Ruelle, D.:* Elements of Differentiable Dynamics and Bifurcation Theory. Academic Press 1989.

[21] *Xu, K.:* Stochastic normal forms. Ph.D. Thesis. Bremen 1990.

[22] *Zeeman, E. C.:* On the classification of dynamical systems. Bull. London Math Soc. 20 (1988), 545 - 557.

[23] *Zeeman, E. C.:* Stability of dynamical systems. Nonlinearity 1 (1988), 115 - 155.

Ludwig Arnold

Institut für Dynamische Systeme
Universität Bremen, Postfach 330 440
2800 Bremen 33
W-Germany

Petra Boxler

Department of Mathematics
University of Southern California
DRB 306, University Park
Los Angeles, CA 90089-1113
U.S.A.

on leave from:

Institut für Dynamische Systeme
Universität Bremen
Postfach 330 440
2800 Bremen 33
W-Germany

Lyapunov exponent and rotation number of the linear harmonic oscillator

Mark A. Pinsky

1. INTRODUCTION.

Many authors have studied the asymptotic behavior of solutions of the stochastic equation

$$x''(t) + x(t)(1 + \varepsilon N(t)) = 0$$

where $N(t)$ is a suitable "noise process" and ε is a small parameter. In case $N(t)$ is a mean-zero function of a diffusion process, the Lyapunov exponent and rotation number were obtained by Arnold, Papanicolaou and Wihstutz [1] by a perturbation method in the small parameter ε. We carried out a similar analysis in the case of a mean-zero function of a finite-state reversible Markov process [2], where the expansion coefficients are expressed in term of the eigenfunctions and eigenvalues of the noise generator. This approach has the advantage that we can study the sign of the coefficients. Meanwhile, Pardoux and Wihstutz [3] have studied the case of a white-noise process $N(t)$, which can be viewed as a limiting case of the "real-noise models" just mentioned. The convergence of the real-noise results to the white noise results was carried out by Pardoux [5].

It is the purpose of the present study to extend and unify the previous results, using the method of [1]. We obtain the two non-zero terms in the Lyapunov exponent and rotation number which agree with the white noise result in the central limit scaling.

The white noise driven harmonic oscillator is also discussed here separately and is singular from at least two points of view: (i) the white noise-driven harmonic oscillator has the largest Lyapunov exponent in comparison with real noise processes of the same variance (this is discussed in section 7); (ii) the rotation number for the white noise-driven oscillator is numerically greater than for the noiseless case, whereas the rotation number for the real-noise driven oscillator is numerically smaller than for the noiseless case. Thus the generic effect of noise upon rotation is somewhat obscure.

A novel feature of the present method is the systematic use of martingale methods to justify the estimates of the Lyapunov exponents, thus avoiding any discussions of the existence or smoothness of the invariant measure on the product space.

2. Specification of the process.

Let $\{\xi(t), t \geqslant 0\}$ be a non-degenerate reversible Markov process on a compact space M. This could be a diffusion on a manifold or a jump process or a finite set. The infinitesimal generator is denoted G and the invariant measure ν. Thus we have for any two $f_1, f_2 \in L^2(d\nu) \cap \mathcal{D}(G)$

$$\int_M f_1(x)(Gf_2)(x)\nu(dx) = \int_M (Gf_1)(x)f_2(x)\nu(dx).$$

In particular zero is a simple isolated eigenvalue with $G1 = 0$ and $\nu G = 0$.

In addition, we suppose given a real-valued function $F : M \to \mathbf{R}$ with $F(\xi) \not\equiv 0$ and

$$\int_M F(\xi)\nu(d\xi) = 0.$$

The Fredholm alternative yields a unique solution of the equation $GH = F$ with $\int_M H(\xi)\nu(d\xi) = 0$, denoted $H = G^{-1}F$.

Finally, we suppose given a 2×2 matrix $B = \begin{pmatrix} b_{11} & b_{12} \\ b_{21} & b_{22} \end{pmatrix}$. The stochastic oscillator process is defined as the solution of the system

$$X_t' = \begin{pmatrix} 0 & \beta \\ -\beta & 0 \end{pmatrix} X_t + \varepsilon F(\xi_t) \begin{pmatrix} b_{11} & b_{12} \\ b_{21} & b_{22} \end{pmatrix} X_t \qquad (\beta > 0). \tag{2.1}$$

This is a Markov process on $\mathbf{R}^2 \times M$ with infinitesimal generator

$$L_X = G + \beta\left(x_2\frac{\partial}{\partial x_1} - x_1\frac{\partial}{\partial x_2}\right) + \varepsilon(BX \cdot \nabla)F(\xi). \tag{2.2}$$

A system of "logarithmic polar coordinates" is defined by $x_1 = e^\rho \cos\theta$, $x_2 = e^\rho \sin\theta$. In this notation the stochastic system (2.1) is written as

$$\begin{aligned} \dot\theta_t &= -\beta + \varepsilon F(\xi_t)h(\theta_t) \\ \dot\rho_t &= \varepsilon F(\xi_t)q(\theta_t) \end{aligned} \tag{2.3}$$

where

$$q(\theta) = \langle Bx, x\rangle = \cos\theta(b_{11}\cos\theta + b_{12}\sin\theta) + \sin\theta(b_{21}\cos\theta + b_{22}\sin\theta)$$
$$h(\theta) = \langle Bx, x^{\perp}\rangle = -\sin\theta(b_{11}\cos\theta + b_{12}\sin\theta) + \cos\theta(b_{21}\cos\theta + b_{22}\sin\theta).$$

It is readily verified that we have

$$q'(\theta) = 2h(\theta) + (b_{12} - b_{21})$$
$$h'(\theta) = -2q(\theta) + (b_{11} + b_{22}).$$

where the prime denotes $\partial/\partial\theta$. In this notation the infinitesimal generator is written

$$L_{\rho,\theta,\xi} = G - \beta\frac{\partial}{\partial\theta} + \varepsilon F(\xi)\left(h(\theta)\frac{\partial}{\partial\theta} + q(\theta)\frac{\partial}{\partial\rho}\right). \tag{2.4}$$

To rigorously define the process, we let (θ_t, ρ_t) be the solution of (2.3), viewed as a process on $\mathbf{R}\times\mathbf{R}$. Defining X_t by $x_1 = e^{\rho}\cos\theta$, $x_2 = e^{\rho}\sin\theta$, it is readily verified that equation (2.1) is satisfied.

3. Perturbation method.

To determine the Lyapunov exponent and rotation number, it suffices to find bounded functions J, K and real numbers λ, α so that when we define

$$\mathcal{J}(\rho,\theta,\xi) \doteq \rho + J(\rho,\theta,\xi), \qquad \mathcal{K}(\rho,\theta,\xi) \doteq \theta + K(\rho,\theta,\xi)$$

then we have $L_{\rho,\theta,\xi}\mathcal{J} = \lambda$ and $L_{\rho,\theta,\xi}\mathcal{K} = \alpha$. Indeed, by applying Itô's formula, we then have

$$\rho(t) + J(\rho(t),\theta(t),\xi(t)) = \lambda t + M_1(t)$$
$$\theta(t) + K(\rho(t),\theta(t),\xi(t)) = \alpha t + M_2(t)$$

where $M_1(t)$ and $M_2(t)$ are martingale processes with bounded quadratic variation. Dividing by t and taking the limit, we then have

$$\lim_{t\to\infty}\rho(t)/t = \lambda, \quad \lim_{t\to\infty}\theta(t)/t = \alpha.$$

Therefore, the Lyapunov exponent and rotation number can equivalently be characterized in terms of the Fredholm alternative; for example the Lyapunov exponent is the unique number λ such that there exists a bounded solution J of the equation $L_{\rho,\theta,\xi}J = L_{\rho,\theta,\xi}\rho - \lambda$; similarly for the rotation number.

In practice it is impossible to solve the equations $L\mathcal{J} = \lambda$ and $L\mathcal{K} = \alpha$ in the indicated forms. Therefore, we develop a perturbative method to obtain an approximate solution. When $\varepsilon = 0$ the generator of the stochastic oscillator process is $L = G - \beta\frac{\partial}{\partial\theta}$ and we have the *exact* solutions $\mathcal{J} = \rho$, $\mathcal{K} = \theta$ with $\lambda = 0$, $\alpha = -\beta$. In order to develop an effective perturbation theory, we look for an approximate solution in the form

$$\mathcal{J}^{\varepsilon} = \rho + \varepsilon J_1(\theta,\xi) + \varepsilon^2 J_2((\theta,\xi) + \cdots + \varepsilon^N J_N(\theta,\xi)$$
$$\mathcal{K}^{\varepsilon} = \theta + \varepsilon K_1(\theta,\xi) + \varepsilon^2 K_2(\theta,\xi) + \cdots + \varepsilon^N K_N(\theta,\xi).$$

Inserting these into the generator, we have

$$\begin{aligned}
L_{\rho,\theta,\xi}\mathcal{J} &= \varepsilon\left[\left(G - \beta\frac{\partial}{\partial\theta}\right)J_1 + F(\xi)q(\theta)\right]\\
&\quad + \varepsilon^2\left[\left(G - \beta\frac{\partial}{\partial\theta}\right)J_2 + F(\xi)h(\theta)\frac{\partial}{\partial\theta}J_1\right]\\
&\quad + \ldots + \varepsilon^N\left[\left(G - \beta\frac{\partial}{\partial\theta}\right)J_N + F(\xi)h(\theta)\frac{\partial}{\partial\theta}J_{N-1}\right]\\
&\quad + \varepsilon^{N+1}F(\xi)h(\theta)\frac{\partial}{\partial\theta}J_N\\
L_{\rho,\theta,\xi}\mathcal{K} &= -\beta\frac{\partial}{\partial\theta} + \varepsilon\left[\left(G - \beta\frac{\partial}{\partial\theta}\right)K_1 + F(\xi)h(\theta)\right]\\
&\quad + \varepsilon^2\left[\left(G - \beta\frac{\partial}{\partial\theta}\right)K_2 + F(\xi)h(\theta)\frac{\partial}{\partial\theta}K_1\right]\\
&\quad + \ldots + \varepsilon^N\left[\left(G - \beta\frac{\partial}{\partial\theta}\right)K_N + F(\xi)h(\theta)\frac{\partial}{\partial\theta}K_{N-1}\right]\\
&\quad + \varepsilon^{N+1}F(\xi)h(\theta)\frac{\partial}{\partial\theta}K_N.
\end{aligned}$$

The correctors $J_1, J_2, \ldots, J_N, K_2, \ldots, K_N$ are determined by solving the equations

$$\begin{aligned}
\left(G-\beta\frac{\partial}{\partial\theta}\right) J_1 + F(\xi)q(\theta) &= \lambda_1 \\
\left(G-\beta\frac{\partial}{\partial\theta}\right) J_2 + F(\xi)h(\theta)\frac{\partial}{\partial\theta} J_1 &= \lambda_2 \\
&\vdots \\
\left(G-\beta\frac{\partial}{\partial\theta}\right) J_N + F(\xi)h(\theta)\frac{\partial}{\partial\theta} J_{N-1} &= \lambda_N \\
\left(G-\beta\frac{\partial}{\partial\theta}\right) K_1 + F(\xi)h(\theta) &= \alpha_1 \\
\left(G-\beta\frac{\partial}{\partial\theta}\right) K_2 + F(\xi)h(\theta)\frac{\partial}{\partial\theta} K_1 &= \alpha_2 \\
&\vdots \\
\left(G-\beta\frac{\partial}{\partial\theta}\right) K_N + F(\xi)h(\theta)\frac{\partial}{\partial\theta} K_{N-1} &= \alpha_N.
\end{aligned}$$

If these equations can be solved for bounded functions $J_1, \ldots, K_N$ then we have

$$\begin{aligned}
L_{\rho,\theta,\xi}\mathcal{J} &= \varepsilon\lambda_1 + \cdots + \varepsilon^N\lambda_N + O(\varepsilon^{N+1}) \qquad (\varepsilon \downarrow 0) \\
L_{\rho,\theta,\xi}\mathcal{K} &= \varepsilon\alpha_1 + \cdots + \varepsilon^N\alpha_N + O(\varepsilon^{N+1}) \qquad (\varepsilon \downarrow 0).
\end{aligned}$$

Applying Itô's formula and the assumed boundedness, we have

$$\begin{aligned}
\lim_{t\to\infty} \rho(t)/t &= \varepsilon\lambda_1 + \cdots + \varepsilon^N\lambda_N + O(\varepsilon^{N+1}) \qquad (\varepsilon \downarrow 0) \\
\lim_{t\to\infty} \theta(t)/t &= \varepsilon\alpha_1 + \cdots + \varepsilon^N\alpha_N + O(\varepsilon^{N+1}) \qquad (\varepsilon \downarrow 0).
\end{aligned}$$

It remains to find these functions.

4. Perturbative computation of the rotation number.

Now we will solve the equations

$$\left(G-\beta\frac{\partial}{\partial\theta}\right) K_1 + F(\xi)h(\theta) = \alpha_1 \tag{4.1}$$

$$\left(G-\beta\frac{\partial}{\partial\theta}\right) K_2 + F(\xi)h(\theta)\frac{\partial}{\partial\theta} K_1 = \alpha_2 \tag{4.2}$$

on the space $(-\pi,\pi)\times M$ with the measure $(2\pi)^{-1}d\theta\times\nu$ where ν is the invariant measure of the noise process $(\xi(t))$. The notation $<\cdot>$ denotes the average on $(-\pi,\pi)\times M$ with respect to the measure $(2\pi)^{-1}d\theta v(d\xi)$, while the notation $(\cdot,\cdot)$

denotes the Hilbert space inner product. If (4.1) is to have a solution, then we may integrate and use the mean-zero property of $F(\xi)$ to deduce $\alpha_1 = 0$. This equation can be solved in terms of the spectral measure of G by

$$\begin{aligned} K_1 &= -\left(G - \beta\frac{\partial}{\partial\theta}\right)^{-1} F(\xi)h(\theta) \\ &= \iint_{[0,\infty)\times\mathbf{R}} \frac{1}{\lambda + i\mu} d(E^{(1)}_\lambda F, F)(E^{(2)}_\mu h, h) \end{aligned}$$

Here we have used the functional calculus for functions of commuting self- adjoint operators, in the weak sense; $E^{(1)}_\lambda$ is the spectral measure of the self-adjoint operator $-G$ and $E^{(\ )}_\mu$ is the spectral measure of the self-adjoint operator $\frac{\beta}{i}\frac{\partial}{\partial\theta}$ leading to

$$\begin{aligned} \alpha_2 &= \langle F(\xi)h(\theta)\frac{\partial}{\partial\theta}K_1\rangle \\ &= \langle F(\xi)h'(\theta) \left(G - \beta\frac{\partial}{\partial\theta}\right)^{-1} F(\xi)h(\theta)\rangle \\ &= \iint_{[0,\infty)\times\mathbf{R}} \frac{1}{\lambda + i\mu} d(E^{(1)}_\lambda F, F)(E^{(2)}_\mu h, h'). \end{aligned}$$

The spectral measure $E^{(2)}_\mu$ is concentrated on the lattice $\beta Z = [0, \pm\beta, \pm 2\beta, \dots]$. In case $B = \begin{pmatrix} 0 & 0 \\ 1 & 0 \end{pmatrix}$ we have $h(\theta) = \cos^2\theta$ and $\int_{\mathbf{R}} \frac{1}{\lambda + i\mu} d(E^{(2)}_\mu h, h') = \frac{1}{2}\frac{\beta}{\lambda^2 + 4\beta^2}$, leading to

$$\alpha_2 = \frac{\beta}{2}\int_0^\infty \frac{d(E^{(1)}_\lambda F, F)}{\lambda^2 + 4\beta^2} > 0.$$

This leads to the asymptotic estimate

$$\alpha(\varepsilon) = -\beta + \varepsilon^2\alpha_2 + O(\varepsilon^3)$$

and we again retrieve the result that real noise retards rotation.

5. Perturbative computation of the Lyapunov exponent.

We need to solve the equations

$$\left(G - \beta\frac{\partial}{\partial\theta}\right) J_1 + F(\xi)q(\theta) = \lambda_1 \tag{5.1}$$

$$\left(G - \beta\frac{\partial}{\partial\theta}\right) J_2 + F(\xi)h(\theta)\partial J_1/\partial\theta = \lambda_2 \tag{5.2}$$

on the product space $(-\pi, \pi) \times M$. As before, equation (5.1) has a solution iff $\lambda_1 = 0$. The solution is given by

$$J_1(\theta, \xi) = -\left(G - \beta\frac{\partial}{\partial\theta}\right)^{-1} F(\xi)q(\theta).$$

Averaging (5.2) gives

$$\begin{aligned}\lambda_2 &= \langle F(\xi)h(\theta)\frac{\partial J_1}{\partial\theta}\rangle \\ &= \langle F(\xi)h'(\theta)\left(G-\beta\frac{\partial}{\partial\theta}\right)^{-1}F(\xi)q(\theta)\rangle \\ &= \iint_{\mathbf{R}\times\mathbf{R}}\frac{1}{\lambda+i\mu}\,d(E_\lambda^{(1)}F,F)d(E_\mu^{(2)}q,h').\end{aligned}$$

In case $B=\begin{pmatrix}0&0\\1&0\end{pmatrix}$ then $q(\theta)=\cos\theta\sin\theta$, $h(\theta)=\cos^2\theta$ and

$$\int_{\mathbf{R}}\frac{1}{\lambda+i\mu}\,d(E_\mu^{(2)}q,h')=\frac{1}{4}\frac{\lambda}{\lambda^2+4\beta^2}$$

with the result

$$\lambda_2\equiv\frac{1}{4}\int_0^\infty\frac{\lambda}{\lambda^2+4\beta^2}\,d(E_\lambda^{(1)}F,F)>0$$

and the expansion

$$\lambda(\varepsilon)=\varepsilon^2\lambda_2+O(\varepsilon^3).$$

6. White-noise driven oscillator.

In this section we give the asymptotic analysis of the Lyapunov exponent and rotation number in case of a white-noise driven harmonic oscillator. This is the solution of the Itô system of equations

$$\begin{aligned}dx_1 &= \beta x_2dt+\varepsilon(Bx)_1\circ dw\\ dx_2 &= -\beta x_1dt+\varepsilon(Bx)_2\circ dw\end{aligned}$$

where $\{w(t),\ t\geqslant 0\}$ is a Wiener process with mean zero and variance σ^2t. The infinitesimal generator is

$$L_{x_1x_2}=\beta(x_2\partial/\partial x_1-x_1\partial/\partial x_2)+\frac{1}{2}\varepsilon^2\sigma^2(Bx\cdot\nabla)^2.$$

Equivalently we may specify the process in polar coordinates by the Stratonovich equations

$$\begin{aligned}d\theta &= -\beta dt+\varepsilon h(\theta)\circ dw\\ d\rho &= \varepsilon q(\theta)\circ dw\end{aligned}$$

which has infinitesimal generator

$$L_{\theta\rho} = -\beta\partial/\partial\theta + \frac{1}{2}(\varepsilon\sigma)^2\left(h(\theta)\frac{\partial}{\partial\theta} + q(\theta)\frac{\partial}{\partial\rho}\right)^2.$$

The asymptotics of the Lyapunov exponent and rotation number are obtained by looking for the solution in the form

$$\mathcal{J}^\varepsilon(\rho,\theta) = \rho + \varepsilon^2 J_2(\theta) + \cdots + \varepsilon^{2N} J_{2N}(\theta)$$
$$\mathcal{K}^\varepsilon(\theta) = \theta + \varepsilon^2 K_2(\theta) + \cdots + \varepsilon^{2N} K_{2N}(\theta).$$

To obtain an approximate solution of the equations $L\mathcal{J} = \lambda$, $L\mathcal{K} = \alpha$ we are led to the perturbation equations

$$\begin{aligned}
-\beta\partial J_2/\partial\theta + \frac{1}{2}\sigma^2 h(\theta)q'(\theta) \quad &= \lambda_2\\
-\beta\partial J_4/\partial\theta + \frac{1}{2}\sigma^2(h(\theta)\partial/\partial\theta))^2 J_2 \quad &= \lambda_4\\
&\vdots\\
-\beta\partial J_{2N}/\partial\theta + \frac{1}{2}\sigma^2(h(\theta)\partial/\partial\theta)^2 J_{2N-2} \quad &= \lambda_{2N}\\
-\beta\partial K_2/\partial\theta + \frac{1}{2}\sigma^2(h'(\theta)h(\theta)) \quad &= \alpha_2\\
-\beta\partial K_4/\partial\theta + \frac{1}{2}\sigma^2(h(\theta)\partial/\partial\theta)^2 K_2 \quad &= \alpha_4\\
&\vdots\\
-\beta\partial K_{2N}/\partial\theta + \frac{1}{2}\sigma^2(h(\theta)\partial/\partial\theta)^2 K_{2N-2} \quad &= \alpha_{2N}.
\end{aligned}$$

This leads to the value $\lambda_2 = \frac{1}{2}\sigma^2\langle h(\theta)q'(\theta)\rangle$, where the angle bracket denotes the average on the circle : $<\Psi>= \frac{1}{2\pi}\int_0^{2\pi}\Psi(\theta)d\theta$. In case $B = \begin{pmatrix} 0 & 0 \\ 1 & 0 \end{pmatrix}$, then $\lambda_2^W = \sigma^2/8$ and $J_2(\theta) = \frac{1}{4}\sin 2\theta + (1/8)\sin 4\theta$. The next correction term is $\lambda_4 = \frac{1}{2}\sigma^2\langle(\cos^2\theta\partial/\partial\theta)^2 J_2\rangle$. But $\cos^2\theta$ is even and its derivative is odd, so that $\lambda_4 = 0$ and we have the expansion

$$\lambda(\varepsilon) = (\varepsilon^2\sigma^2)/8 + O(\varepsilon^6) \qquad (\varepsilon \downarrow 0).$$

To deal with the rotation number, we must solve for $\alpha_2 = -\frac{1}{2}\langle\cos^3\theta\sin\theta\rangle = 0$, with $K_2(\theta) = (\sigma^2/4)\cos^4\theta$. The next correction is therefore given by $\alpha_4 = (\frac{1}{2}\sigma^2)\langle(\cos^2\theta\partial/\partial\theta)^2 K_2\rangle$. But $\cos^2\theta\partial K_2/\partial\theta = -\frac{1}{2}\sigma^2\cos^5\theta\sin\theta$,

$$(\cos^2\theta\partial/\partial\theta)(\cos^2\theta\partial K_2/\partial\theta) = -\frac{1}{2}\sigma^2(\cos^8\theta - 5\sin^2\theta\cos^6\theta)$$

leading to the evaluation

$$\begin{aligned}\alpha_4 &= \left(\frac{1}{2}\sigma^2\right)\left(-\frac{1}{2}\sigma^2\right)[\langle\cos^8\theta\rangle - 5\langle\sin^2\theta\cos^6\theta\rangle] \\ &= -2(\sigma^2/2)^2[35/128 - 5(5/128)] \\ &= -5\sigma^4/128\end{aligned}$$

and the expansion

$$\alpha(\varepsilon) = -1 - \varepsilon^4(5\sigma^4/128) + O(\varepsilon^6).$$

7. Extremal property of white noise.

If we consider a class of stationary Markov noise processes with the same variance, we can assert that white noise gives a strictly larger Lyapunov exponent than all other competing (small) noise processes. The extremal characterization is a consequence of the following inequality, which pertains to general self-adjoint operators on a Hilbert space. The formulation and proof are due to Allen Devinatz [6].

Let H_1 and H_2 be Hilbert spaces, A and B self-adjoint operators in H_1 and H_2 respectively so that A^{-1} exists. Let dE_A and dE_B be the canonical spectral measures of A and B respectively.

The operators $A_1 = A\otimes I_2$ and $B_1 = I_1\otimes B$ are permuting self-adjoint operators in the tensor product $H = H_1 \otimes H_2$, in the sense that their canonical spectral measures $dE_A\otimes I_2$ and $I_1\otimes dE_B$ permute. Then $A_1 + iB_1$ is a normal operator in H and its canonical spectral measure in the plane is $dE_A \otimes dE_B$. Further, since A^{-1} exists, $dE_A\otimes dE_B$ has no atom at $(0,0)$, so that $(A_1+iB_1)^{-1}$ exists, (possibly unbounded).

THEOREM. *If $a \in H_1$, $b \in H_2$ and $a \in |A^{-1}|, a\otimes b \in (A_1 + iB_1)^{-1}$ then*

$$|((A_1+iB_1)^{-1}a\otimes b, a\otimes b)| \leqslant (|A^{-1}|a,a)_1(b,b)_2.$$

Equality occurs if and only if $Bb = 0$. Here $|C|$ denotes the non-negative square root of C^*C if C is a densely defined operator on H_1.

PROOF: The functional calculus for commuting self-adjoint operators gives the weak-sense equation

$$(A_1+iB_1) = \iint_{\mathbf{R}\times\mathbf{R}} (\lambda + i\mu)dE_A(\lambda)dE_B(\mu)$$

so that

$$(A_1 + iB_1)^{-1} = \iint_{\mathbf{R}\times\mathbf{R}} 1/(\lambda + i\mu) dE_A(\lambda) dE_B(\mu).$$

Now if Δ_1 and Δ_2 are in the Borel field of $\mathbf{R}$, then

$$((E_A \otimes E_B)(\Delta_1 \times \Delta_2) a \otimes b, a \otimes b) = (E_A(\Delta_1)a, a)_1 (E_B(\Delta_2)b, b)_2.$$

Hence, by the usual construction, $((dE_A \otimes dE_B) a \otimes b, a \otimes b)$ is a product measure. Consequently,

$$\begin{aligned}
|((A_1 + iB_1)^{-1} a \otimes b, a \otimes b)| &= \left| \iint_{\mathbf{R}\times\mathbf{R}} 1/(\lambda + i\mu)(dE_A(\lambda) dE_B(\mu) a \otimes b, a \otimes b) \right| \\
&\leqslant \iint_{\mathbf{R}\times\mathbf{R}} (1/|\lambda|)(dE_A(\lambda) \otimes dE_B(\mu) a \otimes b, a \otimes b) \\
&= \int_{\mathbf{R}} (1/|\lambda|) d(E_A(\lambda)a, a) \int_{\mathbf{R}} d(E_B(\mu)b, b) \\
&= (|A^{-1}|a, a)_1 (b, b)_2.
\end{aligned}$$

If equality occurs here, then the spectral measure $(dE_B(\mu), b, b)$ must be concentrated at $\mu = 0$, which is equivalent to the statement that $Bb = 0$. Of course, if $Bb = 0$ then $(A_1 + iB_1)^{-1} a \otimes b = (A^{-1}a) \otimes b$ and the equality applies above. The proof is complete.

To apply these ideas to the Lyapunov exponent, we recall that for the white-noise driven process we have $\lambda(\varepsilon) = \varepsilon^2 \lambda_2^w + O(\varepsilon^3)$ where $\lambda_2^w = \frac{\sigma^2}{2} \langle hq' \rangle_{S^1}$ and σ^2 is the variance of the driving Wiener process.

For a real-noise driven process we have a corresponding expansion $\lambda(\varepsilon) = \varepsilon^2 \lambda_2^R + O(\varepsilon^3)$ with λ_2^R given in the preceding sections. The limiting variance of the real-noise process is given by ([4],p. 256)

$$\operatorname{Var} \int_0^t F(\xi(s)) ds \sim \sum\nolimits^2 t$$

where $\frac{1}{2} \sum^2 = -\langle G^{-1}F, F \rangle$. To obtain a corresponding white-noise process we choose a Wiener process with variance $\sigma^2 = \sum^2$. For the corresponding harmonic oscillator, we have

$$\lambda_2^w = \frac{1}{2} \langle hq' \rangle \sigma^2 = -\langle hq' \rangle (G^{-1}F, F)$$

where we again distinguish between the angle bracket notation for averages on the circle and the Hilbert-space inner product.

THEOREM. *Suppose that* $trB \doteq (b_{11} + b_{22}) = 0$. *Then* $\lambda_2^R < \lambda_2^w$.

PROOF: We have

$$\begin{aligned}
\lambda_2^R &= \langle (G - \beta\frac{\partial}{\partial\theta})^{-1} qFh'F \rangle_{M\times S^1} \\
&= -\langle (G - \beta\frac{\partial}{\partial\theta})^{-1} qF2qF \rangle_{M\times S^1} \\
&< -2(G^{-1}F, F)_M \langle q, q\rangle_{S^1} \\
&= (G^{-1}F, F)_M \langle h', q\rangle_{S^1} \\
&= \lambda_2^W
\end{aligned}$$

where we have integrated by parts twice and used the identity $h' = -2q + trB$.

REFERENCES

[1] L. Arnold, G. Papanicolaou and V. Wihstutz, *Asymptotic analysis of the Lyapunov exponent and rotation number of the random oscillator and applications*, SIAM Journal of Applied Mathematics, 46 (1986), 427–450.

[2] M. Pinsky, *Instability of the harmonic oscillator with small noise*, SIAM Journal of Applied Mathematics, 46 (1986), 451–463.

[3] E. Pardoux and V. Wihstutz, *Lyapunov exponent and rotation number of two-dimensional linear stochastic systems with small diffusion*, SIAM Journal of Applied Mathematics, 48 (1988) 442–457.

[4] M. Iosifescu, "Finite Markov Processes and their Applications," John Wiley and Sons, 1980.

[5] E. Pardoux, *Wide Band Limit of Lyapunov Exponents*, Lecture Notes in Control & Information Sciences 78 (1986) 305–315.

[6] A. Devinatz, private communication.

Mark A. Pinsky
Northwestern University
Evanston,IL 60208

The growth of energy of a free particle of small mass with multiplicative real noise

Volker Wihstutz

1. INTRODUCTION.

Aim. Considering a particle under the influence of multiplicative noise η_t, i.e. $m\ddot{y}(t) = \eta_t y(t)$, we know that the exponential growth rate λ_m of its energy $(y(t)^2 + \dot{y}(t)^2)^{1/2}$ is positive, if the mass m is finite. If m tends to zero, does this growth rate or Lyapunov exponent λ_m tend to infinity or stay bounded?

For a *free particle* under the influence of *white* noise $\eta_t = \dot{W}_t$, it is known that $\lambda_m \sim const.(1/m)^{2/3}$ as $m \to 0$ (in fact equality holds in this case, see [7]) ; while the question remains open for *real* noise (and continuous time). In case of discrete time Avron, Craig and Simon showed in [4] that for large m, $\lambda_m \sim \log(1/m)$. On the other hand, in [3] it was proved that in the presence of *real* noise (continuous time), λ_m is bounded, if there is a *restoring force* which is not too small, i.e. for $\eta_t = -\gamma + F^{\#}(\xi_t)$, ξ_t a nice diffusion.

The main purpose of this paper is to find the order of the Lyapunov exponent for the case of small γ and $\gamma = 0$ (free particle system) in presence of real noise. Once this is established, we can easily treat the Lyapunov exponents λ_σ and rotation numbers ρ_σ for *real* noise driven 2-dimensional systems

$$\dot{x} = [A + \sigma F(\xi_t)B]x$$

where A and B are constant 2×2 matrices, B having a double eigenvalue with degenerate eigenspace.

Set-up. We shall analyze

$$\ddot{y}(t) + \sigma F(\xi_t)y(t) = 0 \tag{1.1}$$

for large parameter σ. The noise ξ_t is assumed to be an ergodic stationary Markov process with compact state space M which is a smooth connected Riemannian manifold (with or without boundary) or a finite set. By ν we denote the invariant probability measure of ξ and by G its generator which we assume to be a nice self-adjoint operator $G^*\nu = G\nu = 0$. More precisely, we assume that G is a self-adjoint elliptic diffusion operator on M (M a manifold) as in Arnold/Papanicolaou/Wihstutz [3], or a symmetric matrix with non-zero entries representing the jump intensities of the noise on M (M a finite set) as in Pinsky [6]. $F : M \to \mathbf{R}$ is a non-constant function with mean zero with respect to ν (and smooth for M a manifold). We are interested in the order of the *Lyapunov exponent*

$$\lim_{t\to\infty} \frac{1}{t} \log(y_\sigma(t,\omega,x_0)^2 + \dot{y}_\sigma(t,\omega,x_0)^2)^{1/2}, \tag{1.2}$$

$x_0 = (y(0), \dot{y}(0))'$, and of the *rotation number*

$$\lim_{t\to\infty} \frac{1}{t} \tan^{-1}(\dot{y}_\sigma(t,\omega,x_0)/y_\sigma(t,\omega,x_0)), \tag{1.3}$$

in terms of σ, $\sigma \to \infty$. In system form, putting $x = (y, \dot{y})'$, (1.1) becomes

$$\dot{x} = \widetilde{A}_\sigma(\xi_t)x = Ax + \sigma F(\xi_t)Bx, \tag{1.4}$$

where $A = \begin{bmatrix} 0 & 1 \\ 0 & 0 \end{bmatrix}$ and $B = \begin{bmatrix} 0 & 0 \\ -1 & 0 \end{bmatrix}$ are nilpotent constant matrices. We define polar coordinates by

$$r = \|x\| = (x_1^2 + x_2^2)^{1/2}, \quad \varphi = \tan^{-1}(x_2/x_1).$$

Then the pair (ξ_t, φ_t) is a Markov process on the compact space $M \times \mathbf{P}$, where we identify the one-dimensional projective space $\mathbf{P}$ with the interval $[-\pi/2, \pi/2]$ or $[0, \pi]$ for the angle φ in an obvious way. (ξ_t, φ_t) is generated by

$$G + \widetilde{h}_\sigma(\xi, \varphi)\partial/\partial\varphi, \tag{1.5}$$

where

$$\widetilde{h}_\sigma(\xi, \varphi) = -(\sin^2\varphi + \sigma F(\xi)\cos^2\varphi) \tag{1.6}$$

is the projection of the vectorfield $x \to A_\sigma(\xi)x$ onto $\mathbf{P}$, that is to say that the angle φ obeys

$$\dot{\varphi}_t = \widetilde{h}_\sigma(\xi_t, \varphi_t). \tag{1.7}$$

The radius r satisfies $\dot{r}_t = q_\sigma(\xi_t, \varphi_t)r$, with

$$\widetilde{q}_\sigma(\xi,\varphi) = ((\cos\varphi, \sin\varphi) \quad ,$$
$$A_\sigma(\xi)(\cos\varphi, \sin\varphi)) = (1-\sigma F(\xi))\cos\varphi\sin\varphi$$

being the radial projection of the original linear vectorfield.

It is known that under our assumptions on G, the generator (1.5) is not too degenerate and there is a unique ergodic invariant probability measure $\widetilde{p}_\sigma(\xi,\varphi)d\xi d\varphi$ on $M \times \mathbf{P}$ with density $\widetilde{p}_\sigma$ (which is smooth in the diffusion case) and there are constants λ_σ and ρ_σ such that almost surely for all $x_0 \neq 0$, all $\sigma \geq 0$.

$$\lambda_\sigma = \iint \widetilde{q}_\sigma(\xi,\varphi)\widetilde{p}_\sigma(\xi,\varphi)d\xi d\varphi = \lim_{t\to\infty} \frac{1}{t}\log\|x_\sigma(t,\omega,x_0\| \geq 0, \tag{1.8}$$

and

$$\rho_\sigma = \iint \widetilde{h}_\sigma(\xi,\varphi)\widetilde{p}_\sigma(\xi,\varphi)d\varphi = \lim_{t\to\infty}\frac{1}{t}\varphi_\sigma(t,\omega,x_0). \tag{1.9}$$

(For the diffusion case, see e.g. Arnold/Kliemann/Oeljeklaus [1], here (1.5) is hypoelliptic; for the finite state Markov process, see e.g. Pinsky [6]).

In section 2, we will discuss systems of the form (1.4) with nilpotent A and nilpotent B and show that λ_σ and ρ_σ are of order $\sqrt{\sigma}$, if $\sigma \to \infty$. In section 3, we will extend our consideration to arbitrary matrices A and degenerate B having no eigenvectors in common with A (so that the generator of (ξ_t, φ_t) is hypoelliptic). We also will treat noise F with non-zero mean. The results will contain the special case of dichotomic noise which was considered in Arnold/Kloeden [2].

2. Systems with nilpotent matrices A and B.

Scaling. Lyapunov exponents and rotation numbers of nilpotent systems driven by *white* noise were investigated in Pinsky/Wihstutz [7] for both small and large intensity σ. We also could treat the situation with *small real* noise by reducing it to the case of small white noise using a homogenization procedure (see [8]). But for *large* σ, the white noise results do not provide any orientation. However, we learn from the white noise situation that in order to treat nilpotent systems one has to "balance" the systematic and noisy terms in the generator by means of appropriate scaling. It is not possible to "balance" all three terms in (1.5), i.e. the G term as well as the cosine and sine terms. But we can balance the systematic and noisy terms of the vectorfield h on $\mathbf{P}$ by transforming $\mathbf{R}^2$ linearly with $T = \begin{bmatrix} \sqrt{\sigma} & 0 \\ 0 & 1 \end{bmatrix}$. (Note that λ_σ and ρ_σ do not depend on the choice of the coordinate system.) Then $\widetilde{A}_\sigma(\xi)$ becomes

$$A_\sigma(\xi) = \sqrt{\sigma}[A + F(\xi)B],$$

the new projected vectorfields are

$$h_\sigma(\xi,\varphi) = -\sqrt{\sigma}H(\xi,\varphi)\ , \quad H(\xi,\varphi) = \sin^2\varphi + F(\xi)\cos^2\varphi, \tag{2.1}$$

$$q_\sigma(\xi,\varphi) = \sqrt{\sigma}Q(\xi,\varphi)\ , \quad Q(\xi,\varphi) = (1 - F(\xi))\cos\varphi\sin\varphi, \tag{2.2}$$

The new generator reads

$$L_\sigma = G + \sqrt{\sigma}L_0, \quad L_0 = (-H)\partial/\partial\varphi, \tag{2.3}$$

and if the density of the new invariant measure μ_σ is denoted by p_σ, then

$$\lambda_\sigma = \sqrt{\sigma}\iint Qp_\sigma d\varphi d\xi = \sqrt{\sigma}\langle Q, p_\sigma\rangle \geq 0, \tag{2.4}$$

$$\rho_\sigma = \sqrt{\sigma}\iint Hp_\sigma d\varphi d\xi = \sqrt{\sigma}\langle H, p_\sigma\rangle. \tag{2.5}$$

with p_σ solving the Fokker-Planck equation $L_\sigma^* p_\sigma = 0$. Since $\iint p_\sigma d\varphi d\xi = 1$ for all $\sigma \geq 0$, immediately, $\lambda_\sigma \leq \sqrt{\sigma}\max|Q|$, $|\rho_\sigma| \leq \sqrt{\sigma}\max|H|$ (the maximum taken over the compact space $M \times \mathbf{P}$). So the order is at most $\sqrt{\sigma}$, therefore at any rate less than the order $\sigma^{2/3}$ which holds in the white noise case. We will show that λ_σ, ρ_σ are actually of order $\sqrt{\sigma}$, starting with the Lyapunov exponent λ_σ.

Representation of the Lyaunov exponent as Fredholm alternative. Instead of solving the Fokker-Planck equation, we try to solve the adjoint problem $L_\sigma f_\sigma = q_\sigma - c$ where we hope to determine the "Fredholm alternative" c, which in view of (2.4), heuristically speaking, equals the Lyapunov exponent λ_σ. Although it will hardly be possible to solve this equation with our degenerate operator L_σ, the idea of the Fredholm alternative leads the way to an asymptotic expansion.

Suppose there is a real number λ_0 and a (possibly generalized) function $f_0(\xi,\varphi)$ such that

$$L_0 f_0 = Q - \lambda_0, \tag{2.6}$$

then, since $L_\sigma^* p_\sigma = 0$,

$$\lambda_\sigma = \sqrt{\sigma}\langle Q, p_\sigma\rangle = \sqrt{\sigma}\lambda_0 + \langle -Gf_0, p_\sigma\rangle. \tag{2.7}$$

Being again guided by the Fredholm alternative, one suspects that $\lambda_0 = \langle Q, \mu_0\rangle$ is the right choice for the real number, where μ_0 solves the zeroth order part of the Fokker-Planck equation

$$L_0^*\mu_0 = \frac{\partial}{\partial\varphi}(H\mu_0) = 0. \tag{2.8}$$

Solving $L_0^*\mu_0 = 0$. Other than for small noise intensity, for large σ we cannot orient ourselves at the spectrum associated with the undisturbed system $\dot{x} = Ax$. For small σ, $\lambda_\sigma + i\rho_\sigma$ can be viewed as perturbation of the "maximum" eigenvalue of A and $L_\sigma^*\mu_\sigma = 0$ as perturbation of $L_0^*\mu_0 = 0$, the solution of which is associated with the eigendirection corresponding to the maximum eigenvalue. So one can see immediately the "right" solution of the equation analogous to (2.8).

Our situation is somewhat more involved. Since L_0 is an operator with respect to the two-dimensional space $M \times \mathbf{P}$, in view of such degeneration, we expect solutions in the distributional sense rather than smooth functions. Of course, we will confine ourselves to positive distributions which are probability measures and (as limit of μ_σ's with marginal ν) have the marginal measure ν.

Let $\Psi(\xi,\varphi)$ denote the smooth test functions which are π-periodic with respect to φ and let $'$ denote the partial derivative $\partial/\partial\varphi$. Then for all Ψ

$$\begin{aligned} 0 = \langle L_0^*\mu_0, \Psi\rangle &= -\langle H\mu, \Psi'\rangle \\ &= \iint \Psi'(\xi,\varphi)H(\xi,\varphi)\mu_0(d\xi, d\varphi) \\ &= \iint \Psi'(\xi,\varphi)H(\xi,\varphi)\mu_0^\xi(d\varphi)\nu(d\xi), \end{aligned}$$

since μ_0 has the marginal ν. In particular, for all $\Psi(\xi,\varphi) = \Psi_1(\xi)\Psi_2(\varphi)$ with π-periodic Ψ_2,

$$0 = \iint \Psi_1(\xi)\left[\int_{-\pi/2}^{\pi/2} \Psi_2'(\varphi)H(\xi,\varphi)\mu^\xi(d\xi)\right]\nu(d\xi),$$

so $g(\xi) = \int_{-\pi/2}^{\pi/2} \Psi_2' H\mu^\xi(d\varphi)$ is orthogonal to all smooth functions Ψ_1, thus to all $L_2(\nu)$-functions, and therefore vanishing ν-almost surely. For (ν-almost) each fixed ξ, we can read $g(\xi) = 0$ as

$$(L^\xi)^*\mu^\xi = [H(\xi,\cdot)\mu^\xi(\varphi)]' = 0 \tag{2.9}$$

where $L^\xi = -H(\xi,\cdot)\partial/\partial\varphi$ is considered as an operator with respecct to $\varphi \in I := [-\pi/2,\pi/2]$ only. The function $H(\xi,\varphi) = \sin^2\varphi + F(\xi)\cos^2\varphi$ has no zeros, if $F(\xi) > 0$ and it vanishes for $F(\xi) \leq 0$ at

$$\phi^{\pm}(\xi) = \tan^{-1}\left(\pm\sqrt{-F(\xi)}\right). \tag{2.10}$$

Define $D = (M \times I)\backslash N$, where

$$N = \begin{cases} \emptyset & \text{if } F(\xi) > 0 \\ \{(\xi,\phi^{\pm}(\xi)) | \xi \in M\}, & \text{if } F(\xi) \leq 0, \end{cases}$$

that is to say that we take away from $M \times \mathbf{P}$ the switching curves $(\xi,\phi^{\pm}(\xi))$ where the vectorfield $H(\xi,\cdot)$ switches signs.

On the open set $D^\xi = D \cap [\xi = \text{const}]$ (open, if we identify $-\pi/2$ and $\pi/2$), $H(\xi,\varphi) \neq 0$; therefore is the first order differential operator L^ξ hypoelliptic and (2.9) has a smooth solution $p^\xi(\varphi)$ on D^ξ with $H(\xi,\varphi)p^\xi(\varphi) = c(\xi)$ or

$$p^\xi(\varphi) = c(\xi)/H(\xi,\varphi). \tag{2.11}$$

Let $F(\xi) \leq 0$. Then $1/H(\xi,\varphi)$ is unbounded on D^ξ, but

$$1 \geq \mu^\xi(D_\xi) = \int 1_{D^\xi}(\varphi)\frac{c(\xi)}{H(\xi,\varphi)}d\varphi;$$

which entails $c(\xi) = 0$, $\mu^\xi(D^\xi) = 0$ and $\sup \mu^\xi = \{\phi^+(\xi),\ \phi^-(\xi)\}$. Since μ^ξ is a probability measure, therefore on D^ξ

$$\mu^\xi(\varphi) = \alpha\delta(\varphi - \phi^+(\xi)) + (1-\alpha)\delta(\varphi - \phi^-(\xi)) \tag{2.12}$$

with $0 \leq \alpha \leq 1$ and δ the Dirac measure. For $F(\xi) > 0$, $D^\xi = [-\pi/2,\pi/2]$ and $c(\xi)/H(\xi,\varphi)$ is bounded and smooth. Normalizing μ^ξ to 1 yields $c(\xi) = \sqrt{F(\xi)}/\pi$. So for ν-almost all ξ, the solutions of (2.8) in the sense of probability measures with marginal ν are of the form

$$\mu_0 = \begin{cases} [\pi^{-1}\sqrt{F(\xi)}H^{-1}(\xi,\varphi)]\nu & \text{, if } F(\xi) > 0 \\ [\alpha\delta(\varphi - \phi^+(\xi)) + (1-\alpha)\delta(\varphi - \phi^-(\xi))]\nu & \text{, if } F(\xi) \leq 0 \end{cases} \tag{2.13}$$

with $0 \leq \alpha \leq 1$.

Computing λ_0. Since $\{\mu_\sigma, \sigma \to \infty\}$ is a tight family of measures on $M \times \mathbf{P}$, any sequence contains a subsequence which converges weakly to a measure μ_0 from (2.13) with σ suitably chosen, i.e. for $\sigma_n \to \infty$

$$\iint Q(\xi,\varphi)\mu_{\sigma_n}(d\xi,d\varphi) \to \iint Q(\xi,\varphi)\mu_0(d\xi.d\varphi) =: \lambda_0,$$

$$\sigma^{-1/2}\lambda_\sigma \to \lambda_0$$

and the order of λ_σ is $\sigma^{1/2}$, if $\lambda_0 > 0$. From (2.2) and (2.13) we obtain

$$\lambda_0 = \int_{F>0} (1 - F(\xi))\sqrt{F(\xi)}\frac{1}{\pi}\int_{-\pi/2}^{\pi/2} \cos\varphi \sin\varphi H^{-1}(\xi,\varphi)d\varphi\nu(d\varphi)$$
$$+ \int_{F\leq 0} (1 - F(\xi)) \int_{-\pi/2}^{\pi/2} \cos\varphi \sin\varphi[\alpha\delta(\varphi - \phi^+) + (1-\alpha)\delta(\varphi - \phi^-)]d\varphi\nu(d\xi).$$

The first term with $F > 0$ vanishes, since $H(\xi, \cdot)$ is even and $\cos\varphi \sin\varphi$ is odd and both functions are π-periodic. With (2.10) the second term, thus λ_0, becomes

$$\lambda_0 = \alpha E_\nu \left\{+\sqrt{[-F(\xi)]^+}\right\} + (1-\alpha)E_\nu \left\{-\sqrt{-F(\xi)]^+}\right\} \tag{2.14}$$

where the expectation E_ν is taken with respect to ν and $0 \leq \alpha \leq 1$.

With regard to solutions of (2.8) we cannot expect less ambiguity. In order to obtain a well-defined constant λ_0 and to determine its sign, we consider the passage to the limit, $\mu_\sigma \to \mu_0$, hoping that the process will break the non-uniqueness by picking exactly one measure from (2.13) as limit, that is to say by choosing exactly one α from $[0, 1]$. First of all, since $\lambda_\sigma \geq 0$, this α has to be $\geq 1/2$, otherwise we would get $\lambda_0(\alpha) < 0$ and therefore negative Lyapunov exponents λ_σ for σ large. For the positivity of $\lambda_0(\alpha)$ it suffices that $\alpha > 1/2$ (if we assume, as we do, that $\nu\{F(\xi) < 0\} > 0$). We can show that, actually, $\alpha = 1$. This is true for $F(\xi) = 0$ as can be seen immediately from (2.13). For $F(\xi) < 0$ we use a large deviation argument and consider the point $z = (\xi, \phi^-(\xi))$ on the unstable branch of the switching curve and a small tube T around this branch in a neighborhood around z bounded away from the line $F(\xi) = 0$. We then can see that $\mu_\sigma(T) \to 0$ as $\sigma \to \infty$ by observing the sejourn time $\tau_\sigma(T)$ becoming shorter and shorter, since the "speed" in F-direction does not change with σ, while in φ-direction, for each fixed point not on the unstable branch, the system is driven away from this branch with a speed growing to infinity. The details will be published elsewhere.

So we have obtained

$$\mu_0 = \begin{cases} \pi^{-1}\sqrt{F(\xi)}H^{-1}(\xi,\varphi)\nu & \text{if } F(\xi) > 0 \\ \delta(\varphi - \phi^+(\xi))\nu & \text{if } F(\xi) \leq 0 \end{cases} \tag{2.15}$$

and

$$\lambda_0 = E_\nu \left\{+\sqrt{[-F(\xi)]^+}\right\} \tag{2.16}$$

which is positive e.g. for non-constant F with mean zero.

Computing ρ_0 for the rotation number. Similar reasoning holds for the rotation number.

$$\rho_\sigma = \iint H d\mu_\sigma, \quad \iint H d\mu_\sigma \to \iint H d\mu_0 =: \rho_0.$$

In the latter expression for ρ_0 the integral over non-positive F's vanishes, while the integral over positive F's yields

$$\rho_0 = -E_\nu \left\{ \sqrt{F[(\xi)]^+} \right\} \tag{2.17}$$

which is positive e.g. for non-constant F with mean zero.

Main result. Summarizing, we obtain the following theorem.

THEOREM. *Given the real noise driven system (1.1) or (1.4) with non-constant zero mean F. Then the top Lyapunov exponent λ_σ and rotation number ρ_σ are unbounded for $\sigma \to \infty$ and*

$$\lambda_\sigma \sim \sqrt{\sigma}\lambda_0, \quad \rho_\sigma \sim -\sqrt{\sigma}\rho_0 \quad (\sigma \to \infty) \tag{2.18}$$

with positive λ_0, ρ_0 given by (2.16) or (2.17) respectively. ∎

Remarks.

(i) We remark that the order of both λ_σ and ρ_σ is smaller than in the corresponding white noise case (where it is $\sigma^{2/3}$, see [7]) and larger than in case of discrete time (were it is $\log \sigma$, see [4]).

(ii) It is interesting to note that the contribution to the leading term comes either from $F < 0$ only, in case of λ_σ (for $F < 0$ (1.4) represents a hyperbolic system) or from $F > 0$ only, in case of ρ_σ (for $F > 0$ (1.4) is elliptic).

(iii) I conjecture that the next term in the expansion of λ_σ and ρ_σ is of order 0(1). However, I was not yet able to prove this for general Markovian noise ξ_t.

3. Real noise driven two-dimensional systems with arbitrary A and degenerate B.

With respect to the spectrum, $B = \begin{bmatrix} 0 & 0 \\ -1 & 0 \end{bmatrix}$ is a degenerate matrix in the sense of having a double eigenvalue with degenerate eigenspace of dimension one.

In this section we will extend the results from systems with nilpotent matrices A and B to two-dimensional systems

$$\dot{x} = [A + \sigma F(\xi_t)B]x \tag{3.1}$$

with a degenerate matrix B (in the above sense) and arbitrary A having no eigenvector in common with B. (The latter condition guarantees (1.8) and (1.9); in the diffusion case the associated pair (ξ_t, φ_t) has a hypelliptic generator.) We also drop the assumption that F has mean zero.

Noise with non-zero mean and bounded exponents. For deriving (2.16) and (2.17), we did not use that $E_\nu F = 0$. Let $F = \gamma + F^{\#}$, γ a real constant, $E_\nu F^{\#} = 0$. (Then, for example, (1.1) reads $m\ddot{y} + \sigma[\gamma + F^{\#}(\xi)]y = 0$ and γy can be interpreted as restoring force.) If γ is small such that $\gamma + F^{\#}$ takes both positive and negative values, Theorem 1 still holds. If γ is large enough, $\gamma \geq (|\min F^{\#}|, |\max F^{\#}|)$, then $F(M) \times [-\pi/2, \pi/2]$ does not contain any switching cure and both $[-F]^+ = [-(\gamma + F^{\#})]^+$ and the $\sqrt{\sigma}$-term of λ_σ vanish. In this case, equation (2.6), $L_0 f_0 = Q - \lambda_0$, has a smooth solution, therefore in (2.7)

$$\sup_\sigma |\langle -Gf_0, p_\sigma \rangle| \leq \max |Gf_0|$$

and λ_σ is bounded for $\sigma \to \infty$. This recovers the result in [3].

Comparison with white noise. By (2.16) boundedness of λ_σ can only occur if F is bounded. So we do not expect a bounded λ_σ, if $F_t^{\#} = \dot{W}_t$ is white noise. In this case (1.4) becomes

$$\begin{aligned} dx &= \begin{bmatrix} 0 & 1 \\ 0 & 0 \end{bmatrix} x dt + \sigma \begin{bmatrix} 0 & 0 \\ -1 & 0 \end{bmatrix} x \circ (\gamma dt + dW_t) \\ &= \begin{bmatrix} 0 & 1 \\ -\sigma\gamma & 0 \end{bmatrix} x dt + \sigma \begin{bmatrix} 0 & 0 \\ -1 & \partial \end{bmatrix} x \circ dW_t) \end{aligned}$$

with generator

$$(-\sin^2 \varphi)\partial/\partial\varphi + \sigma(-\gamma \cos^2 \varphi)\partial/\partial\varphi + \frac{1}{2}\sigma^2[(\cos^2 \varphi)\partial/\partial\varphi]^2.$$

By transformation with $T = \begin{bmatrix} \sigma^{2/3} & 0 \\ 0 & 1 \end{bmatrix}$ as in [7] we obtain the new generator

$$\begin{aligned} &\sigma^{2/3}\hat{L}_0 + \sigma^{1/3}\hat{L}_1 \\ &= \sigma^{2/3}[(-\sin^2 \varphi)\partial/\partial\varphi + (-\cos^2 \varphi\partial/\partial\varphi)^2] + \sigma^{1/3}[-\gamma \cos^2 \varphi\partial/\partial\varphi]; \end{aligned}$$

and the function $\hat{q}$ occurring in the representation of $\hat{\lambda}_\sigma$ analogous to (1.8), reads

$$\hat{q}_\sigma = \sigma^{2/3}\hat{Q}_0 + \sigma^{1/3}\hat{Q}_1 = \sigma^{2/3}\hat{Q}_0 + \sigma^{1/3}(-\gamma \cos\varphi \sin\varphi)$$

with some smooth π-periodic function $\hat{Q}_0$. This yields

$$\hat{\lambda}_\sigma = \sigma^{2/3}\hat{\lambda}_0 + \sigma^{1/3}\hat{\lambda}_1 + 0(1), \quad \hat{\lambda}_0 > 0, \tag{3.2}$$

rather than the exact formula $\hat{\lambda}_\sigma = \sigma^{2/3}\hat{\lambda}_0$ which occurs for $\gamma = 0$, see [7]. Here $\hat{\lambda}_1$ is obtained by the following algorithm: solve $\hat{L}_0\hat{p}_0 = 0$, compute $\hat{\lambda}_0 = \int \hat{Q}_0 p_0 d\varphi$, solve $\hat{L}_0\hat{f}_0 = \hat{Q}_0 - \hat{\lambda}_0$, then

$$\hat{\lambda}_1 = \int_{-\pi/2}^{\pi/2} (-\hat{L}_1\hat{f}_0 + \hat{Q}_1)\hat{p}_0 d\varphi$$

(All functions are smooth and π-periodic functions of φ.).

Arbitrary matrix A and degenerate B. Let B have a double eigenvalue with degenerate one-dimensional eigenspace. Then, without loss of generality, $B = \begin{bmatrix} b & 0 \\ b_{21} & b \end{bmatrix}$ with $b_{21} \neq 0$. If the pair (A, B) satisfies the hypoellipticity condition, that is if A and B have no common eigenvector, then $a_{12} \neq 0$; and again, without loss of generality, we can assume $a_{12} = 1$ (if necessary, after transformation with $\begin{bmatrix} 1/a_{12} & 0 \\ 0 & 1 \end{bmatrix}$ which leaves diagonal elements unchanged). So $A = \begin{bmatrix} a_{11} & 1 \\ a_{21} & a_{22} \end{bmatrix}$, $B = \begin{bmatrix} b & 0 \\ \bar{b}_{21} & b \end{bmatrix}$, $\bar{b}_{21} \neq 0$. Considering $A + FB$ and replacing, if necessary, F by $-\bar{b}_{21}^{-1}F$ and b by $-\bar{b}_{21}^{-1}b$, we can assume that $A + \sigma F(\xi)B$ has the form

$$\begin{bmatrix} a_{11} & 1 \\ a_{21} & a_{22} \end{bmatrix} + \sigma F(\xi) \begin{bmatrix} b & 0 \\ -1 & b \end{bmatrix}. \tag{3.3}$$

The σ-dependent linear transformation $T = \begin{bmatrix} \sqrt{\sigma} & 0 \\ 0 & 1 \end{bmatrix}$ then yields

$$\begin{aligned} &\begin{bmatrix} a_{11} & \sqrt{\sigma} \\ \frac{1}{\sqrt{\sigma}} a_{21} & a_{22} \end{bmatrix} + \sigma \cdot F(\xi) \begin{bmatrix} b & 0 \\ \frac{-1}{\sqrt{\sigma}} & b \end{bmatrix} \\ &= \sqrt{\sigma}\left(\begin{bmatrix} 0 & 1 \\ 0 & 0 \end{bmatrix} + F(\xi)\begin{bmatrix} 0 & 0 \\ -1 & 0 \end{bmatrix}\right) + \begin{bmatrix} a_{11} & 0 \\ 0 & a_{22} \end{bmatrix} + \frac{1}{\sqrt{\sigma}}\begin{bmatrix} 0 & 0 \\ a_{21} & 0 \end{bmatrix} \\ &\quad + \sigma F(\xi) \begin{bmatrix} b & 0 \\ 0 & b \end{bmatrix}. \end{aligned}$$

Hence

$$h_\sigma(\xi, \varphi) = -\sqrt{\sigma}H + (-a_{11} + a_{22})\cos\varphi \sin\varphi + \sqrt{\sigma}^{-1} a_{21} \cos^2\varphi + 0,$$

$$q_\sigma(\xi, \varphi) = \sqrt{\sigma}Q + a_{11}\cos^2 + a_{22}\sin^2\varphi + \sqrt{\sigma}^{-1} a_{21}\cos\varphi \sin\varphi + \sigma F(\xi) b,$$

which leads to the same invariant limit μ_0 as in the case of nilpotent A and B. Since all the extra terms with coefficients from A are smooth functions, the corresponding extra terms in $\int\int q_\sigma d\mu_\sigma$ and $\int\int h_\sigma d\mu_\sigma$ are at most of order 0(1). The extra term in q_σ which we have obtained by changing the trace of B does not effect ρ_σ, but yields an extra term for λ_σ : $\sigma b E_\nu F$. Summarizing, we obtain the following corollary to the theorem from section 2.

COROLLARY. *Given the two-dimensional real noise driven system*

$$\dot{x}^- = (A + \sigma F(\xi)B)x^-,$$

where B has the double eigenvalue b with degenerate one-dimensional eigenspace and A is arbitrary having no eigenvector with B in common. Then the top Lyapunov exponent λ_σ and the rotation number ρ_σ tend to infinity for $\sigma \to \infty$ and

$$\lambda_\sigma \sim \sigma b E_\nu F(\xi) + \sqrt{\sigma}\lambda_0 \qquad (3.4)$$

$$\rho\sigma \sim \qquad -\sqrt{\sigma}\rho_0,$$

where λ_0 and ρ_0 are given by (2.16), (2.17) and the expectation E_ν is taken with respect to the invariant measure ν of the noise ξ_t. ∎

Remarks.

(i) If B has the double eigenvalue 0 and/or F has mean zero, then $\lambda_\sigma \sim \sqrt{\sigma}$ for arbitrary A (having no eigenvector with B in common).

(ii) **Comparison with white noise.** We obtain $B = \begin{bmatrix} b & 0 \\ -1 & b \end{bmatrix}$ from $B = \begin{bmatrix} 0 & 0 \\ -1 & 0 \end{bmatrix}$ by changing its trace via adding bI. For a white noise driven system $dx = Ax + \sigma Bx \circ W_t$, such a change of the trace of the diffusion term has no effect on the Lyapunov exponent, since not only the function analog to h_σ remains unchanged (as it does for real noise) but also the function q^{white} which occurs in the representation of λ_σ analogous to (1.8).

REFERENCES

[1] L. Arnold, W. Kliemann, E. Oeljeklaus, *Lyapunov exponents of linear stochastic systems, in: Lyapunov Exponents* (L. Arnold, V. Wihstutz, eds.), Springer Lecture Notes in Mathematics no. 1186 (1986), 129–159.

[2] L. Arnold, P. Kloeden, *Lyapunov exponents and rotation numbers of two-dimensional systems with telegraphic noise*, SIAM J. Appl. Math. 49 (1989),1242-1274.

[3] L. Arnold, G. Papanicolaou, V. Wihstutz, *Asymptotic analysis of the Lyapunov exponent and rotation number of the random oscillator and applications*, SIAM J. Appl. Math. 46 (1986), 427–450.

[4] J. Avron, W. Craig, B. Simon, *Large coupling behavior of the Lyapunov exponent for tight binding one-dimensional random systems*, J. Phys. A: Math. Gen. 16 (1983), L209–L211.

[5] E. Pardoux, V. Wihstutz, *Lyapunov exponent of linear stochastic systems with large diffusion terms*, preprint 1989.

[6] M. Pinsky, *Instability of the harmonic oscillator with small noise*, SIAM J. Appl. Math. 46 (1986), 451–463.

[7] M. Pinsky, V. Wihstutz, *Lyapunov exponents of nilpotent Itô systems*, Stochastics 25 (1988), 43–57.

[8] ________, *Lyapunov exponents of real-noise driven nilpotent systems and harmonic oscillators*, Stochastics, to appear.

Volker Wihstutz
University of North Carolina at Charlotte
Charlotte, NC 28223

PART V

Iterated Function Systems

Iterated Function Systems and Multiplicative Ergodic Theory

Ludwig Arnold and Hans Crauel

Abstract

Given a random dynamical system of affine mappings of $\mathbf{R}^d$ whose linear part is hyperbolic, we prove that it has a unique invariant measure. The proof uses multiplicative ergodic theory. The result generalizes and sheds new light on results of Barnsley and Elton on Iterated Function Systems.

1 Introduction

Iterated function systems (IFS) have recently found considerable interest in the context of image generation and encoding, cf. Barnsley [3].
An iterated function system is a finite collection of maps $\mathcal{T} = \{T_i : X \to X \mid i = 1, \ldots, N\}$, X a complete metric space, such that each T_i is a contraction, i.e., there exists a constant $c < 1$ such that $d(T_i x, T_i y) \le c\, d(x, y)$ for all $x, y \in X$ and $i = 1, \ldots, N$.
Associated with an IFS $\mathcal{T}$ there is a unique compact set $A_{\mathcal{T}} \subset X$ with $A_{\mathcal{T}} = \bigcup_1^N T_i(A_{\mathcal{T}})$. In fact, the map $K \mapsto \bigcup_1^N T_i(K)$ is a contraction on the set of compact subsets of X with the Hausdorff metric (Hutchinson [14] Theorem 3.2.1, p. 728).

Another characterization of $A_{\mathcal{T}}$ is

$$A_{\mathcal{T}} = \text{closure}\{\text{fixed point of} \quad T_{i_k} \circ T_{i_{k-1}} \circ \ldots \circ T_{i_2} \circ T_{i_1} \mid 1 \le i_j \le N, k \in \mathbf{N}\};$$

the contractivity of the T_i's guarantees that arbitrary iterates of the form $T_{i_k} \circ T_{i_{k-1}} \circ \ldots \circ T_{i_2} \circ T_{i_1}$ are contractions which thus have unique fixed points (Hutchinson [14] Theorem 3.1.3, p. 724).
Put $\Omega = \{1, \ldots, N\}^{\mathbf{N}}$; each $\omega \in \Omega$ has a representation $\omega = \omega_0 \omega_1 \omega_2 \ldots$ The contractivity of the T_i's implies that for all sequences of indices $\omega \in \Omega$

$$\bar{x}(\omega) = \lim_{n \to \infty} T_{\omega_1} T_{\omega_2} T_{\omega_3} \ldots T_{\omega_n} x$$

exists and is independent of $x \in X$. The map $\omega \mapsto \bar{x}(\omega)$ is continuous and surjective onto $A_{\mathcal{T}}$ (Ω equipped with the product topology).

A third characterization of $A_{\mathcal{T}}$ uses probabilistic arguments. Let Ω be equipped with the product σ-algebra, and choose probabilities $p = (p_1, \ldots, p_N)$, $p_i > 0$, $\sum_1^N p_i = 1$. Let $P = p^{\mathbf{N}}$ be the product measure on Ω. Then P is invariant under the shift ϑ,

i. e., $P \circ \vartheta^{-1} = P$, where $\vartheta : \Omega \to \Omega$ is defined by $(\vartheta\omega)_i = \omega_{i+1}$.
Define a "random map"

$$T : \Omega \to \{T_1, \dots, T_N\} \subset X^X$$

$$\omega \mapsto T_{\omega_0}.$$

Then $\{T(\vartheta^n\omega) \mid n \in \mathbb{N}\}$ is an iid sequence of maps of X. Iteration of the random maps $T(\vartheta^n\omega)$ induces a Markov chain on X in the following way:
For each $x \in X$, the process

$$x_n(\omega) = T(\vartheta^{n-1}\omega) \circ \dots \circ T(\vartheta\omega) \circ T(\omega)x, \qquad n > 0, \quad x_0 = x$$

is a homogeneous Markov process with transition probabilities

$$Q(x, B) = \sum_1^N p_i \, 1_B(T_i(x)) = \sum_1^N p_i \, \delta_{T_i(x)}(B).$$

The transition probabilities induce an operator Q on (probability) measures by

$$Q\rho = \sum_1^N p_i \, (T_i\rho), \tag{1}$$

where $\rho \in Pr(X)$ (= probability measures on X), and $T\rho = \rho \circ T^{-1}$ for a measurable map $T : X \to X$. The following theorem is due to Hutchinson [14] Theorem 4.4.1, p. 733.

Theorem *Let $T = \{T_1, \dots, T_N\}$ be an IFS with probabilities $p_i > 0$, $i = 1, \dots, N$. Then the Markov operator Q (as defined in (1)) is a contraction on the set of probability measures on X with compact support, equipped with the metric*

$$d(\mu, \nu) = \sup\left\{ \int_X f \, d\mu - \int_X f \, d\nu \,\middle|\, \mathrm{Lip}(f) \le 1 \right\},$$

where $\mathrm{Lip}(f) = \sup\left\{ \frac{d(f(x), f(y))}{d(x,y)} \,\middle|\, x, y \in X, x \neq y \right\}$ *denotes the Lipschitz constant of* $f : X \to \mathbb{R}$.

In particular, the theorem implies that for any $\sigma \in Pr(X)$ with compact support $Q^n\sigma$ converges to ρ exponentially fast in this metric.

A measure $\rho \in Pr(X)$ is said to be *attractive* (for a Markov operator Q) if $Q^n\sigma$ converges to ρ for any $\sigma \in Pr(X)$. Clearly, an attractive measure ρ is invariant, i. e., $Q\rho = \rho$, and it is the unique measure with this property.

The relation between the attractor A_T of an IFS and the attractive measure for the Markov operator Q induced by some choice of probabilities p_i, $i = 1, \dots, N$, is given by

$$A_T = \operatorname{supp} \rho, \tag{2}$$

where $\operatorname{supp}\sigma$ is the support of σ, $\operatorname{supp}\sigma = \bigcap_{F \text{ closed}, \, \sigma(F)=1} F$. See Hutchinson [14] Theorem 4.4.4, p. 733 for a proof of (2).

Note that though (1) holds regardless of the choice of probabilities, the attractive measure ρ does depend on this choice. If only $p_i \geq 0$ is assumed, still $A_{\mathcal{T}} \subset \operatorname{supp} \rho$ holds.

A generalization of Hutchinson's existence result for an attractive measure is due to Barnsley and Elton [4] Theorem 1, p. 20:

Theorem *Let $\mathcal{T} = \{T_i : X \to X \mid i = 1, \ldots, N\}$ be maps (none of which need to be a contraction) and suppose there exist probabilities $p = (p_1, \ldots, p_N)$, $p_i \geq 0$, $\sum_1^N p_i = 1$ such that for some $c < 1$*

$$\prod_1^N d(T_i x, T_i y)^{p_i} \leq c\, d(x, y) \tag{3}$$

for all $x, y \in X$. Then there exists an attractive measure ρ for the Markov operator Q as defined in (1).

Furthermore, for P-almost all $\omega \in \Omega$ $(=\{1, \ldots, N\}^{\mathbb{N}})$ the limit

$$\bar{x}(\omega) = \lim_{n \to \infty} T_{\omega_1} T_{\omega_2} T_{\omega_3} \ldots T_{\omega_n} x$$

exists and is independent of $x \in X$. The random variable $\bar{x}$ is distributed like ρ, i. e., $P(\bar{x}(\omega) \in B) = \rho(B)$.

Note that (3) is equivalent to $E\left(\log \dfrac{d(T(\omega)x, T(\omega)y)}{d(x,y)}\right) < 0$ uniformly in $x, y \in X$. It is called *average contractivity.*

A collection of maps $\mathcal{T} = \{T_1, \ldots, T_N\}$ with probabilities $(p_1, \ldots, p_N)$ can be average contractive without any of the T_i's being a contraction. For instance, consider two affine maps on $X = \mathbb{R}^2$:

$$T_1(x) = \begin{pmatrix} r & 0 \\ 0 & s \end{pmatrix} x + b_1 \qquad \text{and} \qquad T_2(x) = \begin{pmatrix} s & 0 \\ 0 & r \end{pmatrix} x + b_2$$

with $r < s^{-1}$, which becomes average contractive with $p_1 = p_2 = \frac{1}{2}$.
Note that the support of the attractive measure ρ need not be compact under the conditions of Barnsley and Elton's theorem.

Existence of an attractive measure for a Markov operator Q has been established under somewhat different conditions by Dubins and Freedman [10] Theorem 4.4, p. 840. They assumed that the space X is compact, and that the T's (not necessarily finitely many) satisfy $P(\operatorname{Lip}(T) \leq 1) = 1$ and $P(\operatorname{Lip}(T) < 1) > 0$.

The aim of the present paper is to generalize the existence results on attractive measures for Markov operators. We prove existence of a unique invariant measure for *random dynamical systems* of affine maps on $X = \mathbb{R}^d$ under the assumption that all Lyapunov exponents of the linear part of the maps are nonzero. The unique invariant measure is attractive if and only if all exponents are negative.

2 Random Dynamical Systems

Suppose $(\Omega, \mathcal{F}, P)$ is an abstract probability space and $\vartheta_t : \Omega \to \Omega$, $t \in T$ is a *flow of P-preserving maps*, i. e., $\vartheta_0 = \mathrm{id}$ and $\vartheta_{t+s} = \vartheta_t \circ \vartheta_s$ for all $t, s \in T$, where $T = \mathbb{R}$ or $\mathbb{Z}$. Assume $\{\vartheta_t \mid t \in T\}$ to be ergodic and measurable, i. e., $(t, \omega) \mapsto \vartheta_t\omega$ is jointly measurable.

Let X be a Polish space, i. e., separable, metric, and complete. A *random dynamical system (RDS)* is a family of continuous maps

$$\varphi(t, \omega) : X \to X,$$

measurable in (t, ω), such that $\varphi(0, \omega) = \mathrm{id}$ for P-almost all ω and

$$\varphi(t + s, \omega) = \varphi(t, \vartheta_s\omega) \circ \varphi(s, \omega) \tag{4}$$

for all $t, s \in T$ P-almost surely.

Note that for discrete time ($T = \mathbb{Z}$) an RDS is a product of random maps of the form

$$\varphi(n, \omega) = \begin{cases} T(\vartheta_{n-1}\omega) \circ \ldots \circ T(\vartheta\omega) \circ T(\omega) & \text{for } n > 0 \\ \mathrm{id} & \text{for } n = 0 \\ T^{-1}(\vartheta_n\omega) \circ T^{-1}(\vartheta_{n+1}\omega) \circ \ldots \circ T^{-1}(\vartheta_{-1}\omega) & \text{for } n < 0, \end{cases} \tag{5}$$

where $T : \Omega \to C(X, X)$ is given by $T(\omega) = \varphi(1, \omega)$.

Vice versa, any stationary and ergodic sequence of maps $\{T(\vartheta_n\omega) \mid n \in \mathbb{Z}\}$ induces an RDS φ by (5). (For the sake of notational coherence we write ϑ_n instead of ϑ^n in the following.)

Suppose φ is an RDS. Then $\{\Theta_t \mid t \in T\}$, where

$$\Theta_t : X \times \Omega \to X \times \Omega$$

$$(x, \omega) \mapsto (\varphi(t, \omega)x, \vartheta_t\omega),$$

is called the *skew product flow* induced by φ.

Given an RDS φ, we define *stable* and *unstable sets (at (x, ω))* by

$$M^s(x, \omega) = \{y \in X \mid d(\varphi(t, \omega)x, \varphi(t, \omega)y) \to 0 \text{ for } t \to \infty\}, \tag{6}$$

$M^u(x, \omega)$ analogously with $t \to -\infty$ (d denoting the metric on X). The stable and unstable sets are random invariant sets in the sense

$$\varphi(t, \omega)M^{s,u}(x, \omega) = M^{s,u}(\Theta_t(x, \omega)) = M^{s,u}(\varphi(t, \omega)x, \vartheta_t\omega).$$

A probability measure $\mu \in Pr(X \times \Omega)$ is said to be an *invariant measure for* φ if μ has marginal P on Ω and if $\Theta_t\mu = \mu$ for all $t \in T$.

Henceforth all measures $\mu \in Pr(X \times \Omega)$ will be understood to have marginal P on Ω. Any such μ is uniquely (up to P nullsets) characterized by its disintegration $\omega \mapsto \mu_\omega$ with respect to μ, $d\mu(x, \omega) = d\mu_\omega(x)\, dP(\omega)$. Here we use separability of X.

It is straightforward to verify that μ is an invariant measure for an RDS φ if and only if

$$\varphi(t,\omega)\mu_\omega = \mu_{\vartheta_t\omega} \qquad P\text{-a.s.} \tag{7}$$

for all $t \in T$.

Only in rather particular cases an RDS φ induces a Markov process on X. For instance, for discrete time:

1. Lemma *Suppose φ is a discrete time RDS. Then $\{\varphi(n,\omega)x \mid n \in \mathbb{N}\}$ is a Markov process for all $x \in X$ if and only if the $\{T(\vartheta_n\omega) \mid n \in \mathbb{Z}\}$ is an independent family. The family of Markov processes $\{\varphi(n,\omega)x\}_{x\in X}$ is homogeneous if and only if $\{T(\vartheta_n\omega)\}_n$ are, in addition, identically distributed (hence iid).*

The proof of Lemma 1 follows from Theorem 1.1 in Kifer [17], p. 8.

Suppose the $\{T(\vartheta_n\omega)\}_n$ are iid. Putting $Q(x,B) := P(T(\omega)x \in B)$ defines transition probabilities on X. The associated *Markov operator* Q on X acts on measures $\rho \in Pr(X)$ by

$$Q\rho = \int T(\omega)\rho\, dP(\omega) = E(T(\cdot)\rho). \tag{8}$$

2. Lemma *Let φ be a discrete time RDS such that $\{T(\vartheta_n\omega) \mid n \in \mathbb{Z}\}$ is iid, where $T(\omega) = \varphi(1,\omega)$. Then the following two assertions are equivalent:*

(i) *μ is an invariant measure for φ with the additional property that the disintegration $\omega \mapsto \mu_\omega$ is measurable (mod P) with respect to $\sigma\{\omega \mapsto T(\vartheta_n\omega) \mid n < 0\}$ ("measurable with respect to the past").*

(ii) $\rho = \int \mu_\omega\, dP(\omega)$ *is an invariant measure for the Markov operator Q*

(more precisely, ρ is defined by $\rho(f) = \int(\int f\, d\mu_\omega)\, dP(\omega)$ for $f : X \to \mathbb{R}$ bounded and measurable).

Furthermore, if ρ is an invariant measure for the Markov operator Q then P-almost surely

$$\varphi(-n,\omega)^{-1}\rho \xrightarrow[n\to\infty]{\text{weak}^*} \mu_\omega, \tag{9}$$

where μ is an invariant measure for φ.

Lemma 2 is a particular case of Theorem 4.4 of Crauel [9]. Under the present conditions, (9) is due to Le Jan [19], Lemme 1, p. 112.

A measure μ whose disintegration $\omega \mapsto \mu_\omega$ is measurable with respect to the past is said to be a *Markov measure*.

2.1 RDS of affine transformations

From now on we shall restrict ourselves to RDS consisting of invertible affine maps of $\mathbb{R}^d$.

1. The affine group

The group $\mathcal{A}(d,\mathbb{R})$ of invertible affine transformations of $\mathbb{R}^d$ has the structure of a *semi-direct* (or *skew*) *product* of $\mathrm{Gl}(d,\mathbb{R})$ and $\mathbb{R}^d$. In particular, every element φ of $\mathcal{A}(d,\mathbb{R})$ is represented by $x \mapsto \Phi x + \psi$, where $\Phi \in \mathrm{Gl}(d,\mathbb{R})$ and $\psi \in \mathbb{R}^d$. If $\varphi_1, \varphi_2 \in \mathcal{A}(d,\mathbb{R})$ are represented by (Φ_i, ψ_i), $i = 1,2$, then

$$\begin{aligned} \varphi_2 \circ \varphi_1(x) &= \Phi_2 \circ \Phi_1 x + (\Phi_2\psi_1 + \psi_2) \\ \varphi^{-1}(x) &= \Phi^{-1}x + \Phi^{-1}\psi. \end{aligned} \tag{10}$$

$\mathcal{A}(d,\mathbb{R})$ is a Lie group of diffeomorphisms of $\mathbb{R}^d$, $\dim\mathcal{A}(d,\mathbb{R}) = d^2 + d$, and the Lie algebra of $\mathcal{A}(d,\mathbb{R})$ is

$$\mathrm{a}(d,R) = \mathrm{gl}(d,R)\dot{+}\mathbb{R}^d,$$

where $\dot{+}$ denotes semidirect sum (cf. Kobayashi and Nomizu [18], pp. 125–127). For a notion of affine transformations and affine vector fields on manifolds see Chapter VI of Kobayashi and Nomizu [18].

2. RDS of affine transformations

Let φ (together with the flow $\{\vartheta_t\}_{t\in T}$ on $(\Omega,\mathcal{F},P)$) be an RDS with values in $\mathcal{A}(d,\mathbb{R})$. Thus,

$$\varphi(t,\omega)(x) = \Phi(t,\omega)x + \psi(t,\omega),$$

where $\Phi(t,\omega) \in \mathrm{Gl}(d,\mathbb{R})$ and $\psi(t,\omega) \in \mathbb{R}^d$.
The linear part Φ of φ is itself an RDS of invertible matrices of $\mathbb{R}^d$. In fact, the cocycle property

$$\Phi(t+s,\omega) = \Phi(t,\vartheta_s\omega)\Phi(s,\omega)$$

for all $t,s \in T$ (P-a. s.) follows from the skew product structure of $\mathcal{A}(d,\mathbb{R})$, see (10). We shall make use of the following facts:

$$\psi(t+s,\omega) = \varphi(s,\vartheta_t\omega)(\psi(t,\omega)) = \Phi(s,\vartheta_t\omega)\psi(t,\omega) + \psi(s,\vartheta_t\omega);$$

and since $\varphi(0,\omega) = \mathrm{id}$ implies $\psi(0,\omega) = 0$ we have

$$\psi(t,\omega) = -\Phi(t,\omega)\psi(-t,\vartheta_t\omega).$$

It is convenient to introduce the transition operators

$$\begin{aligned} \varphi(s,t,\omega) &:= \varphi(t,\omega)\circ\varphi(s,\omega)^{-1}, \\ \Phi(s,t,\omega) &:= \Phi(t,\omega)\,\Phi(s,\omega)^{-1}. \end{aligned}$$

3. Particular cases

(i) *Discrete time* $T = \mathbb{Z}$

The cocycle property (4) implies for $n \geq 1$

$$\begin{aligned} \varphi(n,\omega) &= \varphi(1,\vartheta_{n-1}\omega)\circ\varphi(n-1,\omega) = \varphi(1,\vartheta_{n-1}\omega)\circ\ldots\circ\varphi(1,\omega), \\ \varphi(-n,\omega) &= \varphi(1,\vartheta_{-n}\omega)^{-1}\circ\ldots\circ\varphi(1,\vartheta_{-1}\omega)^{-1}. \end{aligned}$$

The cocycle is thus generated by the shifted iterates of $T(\omega) := \varphi(1,\omega)$, and $T(\omega)(x) = \Phi(1,\omega)x + \psi(1,\omega) = A(\omega)x + b(\omega)$, where $A(\cdot)$ generates the linear part of $\varphi(\cdot)$. The transition operators are

$$\begin{aligned} \varphi(k,n,\omega) &= T(\vartheta_{n-1}\omega)\circ\ldots\circ T(\vartheta_k\omega) \qquad \text{and} \\ \Phi(k,n,\omega) &= A(\vartheta_{n-1}\omega)\ldots A(\vartheta_k\omega), \end{aligned}$$

respectively, if $k < n$. The orbit of the affine RDS starting at a point $x \in \mathbb{R}^d$ at time $k < n$ is

$$\varphi(k,n,\omega)(x) = \Phi(k,n,\omega)x + \sum_{j=k}^{n-1} \Phi(j,n,\omega)b(\vartheta_j\omega).$$

(ii) *Continuous time $T = \mathbb{R}$, real noise case*

Given measurable functions $A : \Omega \to \mathrm{gl}(d,\mathbb{R})$ and $b : \Omega \to \mathbb{R}^d$ with $A, b \in L^1(\Omega,\mathcal{F},P)$. Then P-almost all trajectories of the stationary and ergodic processes $A(\vartheta_t\omega)$ and $b(\vartheta_t\omega)$ are locally Lebesgue integrable; measurability follows from our assumptions on $\{\vartheta_t\}_{t\in T}$.
Consider the random differential equation

$$\dot{x} = A(\vartheta_t\omega)x + b(\vartheta_t\omega).$$

The (random) vector field $x \mapsto X(\omega)(x) = A(\omega)x + b(\omega)$ takes its values in $a(d,\mathbb{R})$, hence the solution mapping $x \mapsto \varphi(t,\omega)(x)$ is an affine RDS. More specific, the transition from time s to time t is governed by

$$\varphi(s,t,\omega)(x) = \Phi(s,t,\omega)x + \int_s^t \Phi(u,t,\omega)b(\vartheta_u\omega)\,du, \tag{11}$$

where $\Phi(t,\omega)$ is the corresponding 'fundamental matrix', i.e., the linear cocycle solving

$$\dot{\Phi}(t,\omega) = A(\vartheta_t\omega)\Phi(t,\omega), \qquad \Phi(0,\omega) = \mathrm{id}.$$

(iii) *Continuous time $T = \mathbb{R}$, white noise case*

Given matrices $A_0,\ldots,A_m \in \mathrm{gl}(d,\mathbb{R})$ and vectors $b_0,\ldots,b_m \in \mathbb{R}^d$, consider the stochastic differential equation

$$dx = (A_0x + b_0)dt + \sum_{j=1}^{m}(A_jx + b_j)dW_j$$

on $\mathbb{R}^d$. The solution is an affine RDS with transitions given by

$$\varphi(s,t,\omega)(x) = \Phi(s,t,\omega)x + \int_s^t \Phi(u,t,\omega)(\bar{b}_0du + \sum_{j=1}^{m} b_jdW_j(u)),$$

where $\bar{b}_0 = b_0 - \sum_{j=1}^{m} A_jb_j$, and the linear cocycle $\Phi(t,\omega)$ is again the 'fundamental matrix' of the corresponding homogeneous equation, i.e.,

$$d\Phi = A_0\Phi dt + \sum_{j=1}^{m} A_j\Phi\, dW_j \qquad \text{with} \quad \Phi(0,\omega) = \mathrm{id}, \tag{12}$$

see Arnold [1] p. 141.

2.2 The Multiplicative Ergodic Theorem

Henceforth we will assume the following condition, which is the integrability condition needed for the Multiplicative Ergodic Theorem (MET):

$$\omega \mapsto \sup_{0<t\le t_0} \left(\log^+ \|\Phi(t,\omega)\| + \log^+ \|\Phi^{-1}(t,\omega)\|\right) \tag{13}$$

is integrable (with respect to P) for some (hence for all) $t_0 \in T$, $t_0 > 0$.
Define numbers $\lambda_1 \geq \lambda_2 \geq \ldots \geq \lambda_d$ by

$$\sum_{i=1}^{p} \lambda_i = \lim_{t\to\infty} \frac{1}{t} E\left(\log \|\textstyle\bigwedge^p \Phi(t,\cdot)\|\right), \quad 1 \leq p \leq d. \tag{14}$$

Here $\bigwedge^p$ denotes p-fold exterior products, and E refers to expectation with respect to P. The existence of the limits in (14) follows — since $t \mapsto E \log \| \bigwedge^p \Phi(t,\cdot)\|$ is subadditive — from an elementary argument, see, e. g., Dunford and Schwartz [11] Lemma VIII.1.4, pp. 618–619.
To formulate the next theorem it is convenient to renumber the λ_i's by comprising those which are equal. Then we have $\lambda_1 > \lambda_2 > \ldots > \lambda_r$ for some r, $1 \leq r \leq d$.

3. Theorem (Oseledec [21]) *Suppose Φ is a linear RDS satisfying (13). Then there exists a ϑ-invariant set $\Omega_0 \subset \Omega$ with $P(\Omega_0) = 1$ such that for all $\omega \in \Omega_0$*

(i)
$$\lambda^{\pm}(v,\omega) := \lim_{t\to\pm\infty} \frac{1}{t} \log \|\Phi(t,\omega)v\|$$

exists for all $v \in \mathbf{R}^d$ and $\lambda^{\pm}(v,\omega) \in \{\lambda_1, \lambda_2, \ldots, \lambda_r\}$ if $v \neq 0$.

(ii) *The sets*

$$E_i(\omega) = \{v \in \mathbf{R}^d \mid \lambda^+(v,\omega) = \lambda^-(v,\omega) = \lambda_i\} \cup \{0\} \tag{15}$$

are linear spaces, depending measurably on ω, with

$$E_1(\omega) \oplus E_2(\omega) \oplus \ldots \oplus E_r(\omega) = \mathbf{R}^d,$$

($\oplus$ denoting direct sum). From (15) *it follows that $\Phi(t,\omega)E_i(\omega) = E_i(\vartheta_t\omega)$ for all $t \in T$.*

Let us introduce the following notations: The subspaces

$$\begin{aligned} E^s(\omega) &= \bigoplus_{\lambda_i<0} E_i(\omega) &= \{v \in \mathbf{R}^d \mid \lim_{t\to\infty} \frac{1}{t} \log \|\Phi(t,\omega)v\| < 0\}, \\ E^u(\omega) &= \bigoplus_{\lambda_i>0} E_i(\omega) &= \{v \in \mathbf{R}^d \mid \lim_{t\to-\infty} \frac{1}{t} \log \|\Phi(t,\omega)v\| > 0\} \end{aligned}$$

are called the *stable* and the *unstable* subspace of Φ, respectively. Obviously $\Phi(t,\omega)E^{s,u}(\omega) = E^{s,u}(\vartheta_t\omega)$ for all $t \in T$ P-almost surely. Put $\lambda^s = \max_{\lambda_i<0} \lambda_i$ and $\lambda^u = \min_{\lambda_i>0} \lambda_i$.

3 The Main Result

Let $\varphi(t,\omega)$ be an RDS of affine mappings. We are looking for invariant measures for φ, i.e., for random probability measures $\omega \mapsto \mu_\omega$ satisfying

$$\varphi(t,\omega)\mu_\omega = \mu_{\vartheta_t\omega}.$$

The particular cases treated before by Brandt [8] (for $d = 1$ and $T = \mathbb{Z}$) and by Arnold and Wihstutz [2] (for $\dot{x} = Ax + b(\vartheta_t\omega)$) are contained in the following theorem. The particular cases investigated by Barnsley and Elton [4] ($T = \mathbb{Z}$, products of iid maps, and 'average contractivity'), Barnsley, Elton, and Hardin [5] ($T = \mathbb{Z}$, products of maps in Markovian dependence, and 'average contractivity'), and by Elton ($T = \mathbb{Z}$, products of stationary maps, and 'negative Lyapunov exponent') will be dealt with in Corollaries 5 and 6. Bougerol and Picard ([6],[7]) are concerned with a converse of Corollary 5. They prove that in case $T = \mathbb{Z}$ and for products of iid maps existence of an invariant Markov measure implies negativity of the top Lyapunov exponent, provided an irreducibility condition.

We use the notation $\xrightarrow{\text{e.f.}}$ to denote exponentially fast convergence.

4. Theorem *Suppose $\{\varphi(t,\omega) \mid t \in T,\ \omega \in \Omega\}$ is an RDS of affine mappings satisfying the following integrability conditions:*

$$\begin{aligned} &\log^+\|A\| + \log^+\|A^{-1}\| + \log^+\|b\| \in L^1(\Omega,\mathcal{F},P) \qquad \text{if } T = \mathbb{Z} \\ &A, b \in L^1(\Omega,\mathcal{F},P) \qquad \text{if } T = \mathbb{R} \text{ in the real noise case} \end{aligned} \tag{16}$$

(no condition is needed in the white noise case).
If the linear cocycle $\Phi(t,\omega)$ (to which the Multiplicative Ergodic Theorem applies) is hyperbolic (i.e., all Lyapunov exponents are distinct from zero) then there exists a unique invariant probability measure $\omega \mapsto \mu_\omega$ for φ. This invariant measure has the form

$$\mu_\omega = \delta_{\varkappa(\omega)},$$

where the random variable $\varkappa : \Omega \to \mathbb{R}^d$ is given as follows:
Let $\pi^{s,u}(\omega) : \mathbb{R}^d \to E^{s,u}(\omega)$ be the projection onto $E^{s,u}(\omega)$ along $E^{u,s}(\omega)$, where $E^s(\omega) \oplus E^u(\omega) = \mathbb{R}^d$ is the splitting into the stable and the unstable subspace according to the MET. Then

$$\pi^s(\omega)\varkappa(\omega) = \begin{cases} \displaystyle\sum_{n=-\infty}^{-1} \Phi(n,\omega)^{-1}\pi^s(\vartheta_n\omega)b(\vartheta_n\omega), & T = \mathbb{Z}, \\ \displaystyle\int_{-\infty}^{0} \Phi(t,\omega)^{-1}\pi^s(\vartheta_t\omega)b(\vartheta_t\omega)\,dt, & T = \mathbb{R}, \\ & \textit{real noise case}, \\ \displaystyle\lim_{t\to\infty} \pi^s(\omega)\Big(\int_{-t}^{0} \Phi(u,\omega)^{-1}(\bar{b}_0\,du + \sum_{j=1}^{m} b_j\,dW_j(u))\Big), & T = \mathbb{R}, \\ & \textit{white noise case}, \end{cases}$$

and

$$\pi^u(\omega)\varkappa(\omega) = \begin{cases} -\sum_{n=0}^{\infty} \Phi(n,\omega)^{-1}\pi^u(\vartheta_n\omega)b(\vartheta_n\omega), & T=\mathbb{Z}, \\ -\int_0^{\infty} \Phi(t,\omega)^{-1}\pi^u(\vartheta_t\omega)b(\vartheta_t\omega)\,dt, & T=\mathbb{R}, \\ & \text{real noise case,} \\ -\lim_{t\to\infty} \pi^u(\omega)\Big(\int_0^t \Phi(u,\omega)^{-1}(\bar{b}_0\,du + \sum_{j=1}^m b_j\,dW_j(u))\Big), & T=\mathbb{R}, \\ & \text{white noise case.} \end{cases}$$

PROOF (i) Existence of $\varkappa$

Since the discrete time and the continuous time real noise cases are very similar, we only write down the proofs for the two continuous time cases.
If $b \equiv 0$ we get $\varkappa \equiv 0$, so we may assume $E\,\|b\| \neq 0$.

Choose a (measurable) random basis (x_i) of $\mathbb{R}^d$ such that

$$x_1(\omega),\ldots,x_k(\omega) \in E^s(\omega),$$
$$x_{k+1}(\omega),\ldots,x_d(\omega) \in E^u(\omega),$$

where $k = \dim E^s(\omega)$. Then the adjoint basis (e_i) of $(\mathbb{R}^d)^* \cong \mathbb{R}^d$, defined by $\langle x_i, e_j\rangle = \delta_{ij}$, has the property

$$e_1(\omega),\ldots,e_k(\omega) \in (E^s(\omega))^* = (E^u(\omega))^\perp,$$
$$e_{k+1}(\omega),\ldots,e_d(\omega) \in (E^u(\omega))^* = (E^s(\omega))^\perp,$$

and for each $x \in \mathbb{R}^d$

$$\pi^s(\omega)x = \sum_{i=1}^k \langle x, e_i(\omega)\rangle x_i(\omega), \quad \pi^u(\omega)x = \sum_{i=k+1}^d \langle x, e_i(\omega)\rangle x_i(\omega) \tag{17}$$

(since $x = \sum_{i=1}^d \langle x, e_i\rangle x_i$ for all $x \in \mathbb{R}^d$ by duality of (x_i) and (e_i)).
The invariance relation $E^{s,u}(\vartheta_t\omega) = \Phi(t,\omega)E^{s,u}(\omega)$ may be rewritten as

$$\pi^{s,u}(\vartheta_t\omega)\circ\Phi(t,\omega) = \Phi(t,\omega)\circ\pi^{s,u}(\omega). \tag{18}$$

Existence of $\varkappa$ in the real noise case:

Applying the linear maps $\pi^{s,u}(\vartheta_t\omega)$ to the affine solution flow $\varphi(s,t,\omega)$ (see (11)) we obtain

$$\begin{aligned} &\pi^{s,u}(\vartheta_t\omega)\circ\varphi(s,t,\omega)(x) \\ &\quad = \pi^{s,u}(\vartheta_t\omega)\circ\Phi(s,t,\omega)x + \pi^{s,u}(\vartheta_t\omega)\int_s^t \Phi(u,t,\omega)b(\vartheta_u\omega)\,du. \end{aligned} \tag{19}$$

We first treat the stable component. Let $t > 0$ and evaluate (19) with s, t replaced by $-t, 0$ to obtain

$$\pi^s(\omega)\varphi(-t,\omega)^{-1}(x) = \pi^s(\omega)\Phi(-t,\omega)^{-1}x + \pi^s(\omega)\int_{-t}^{0}\Phi(u,\omega)^{-1}b(\vartheta_u\omega)\,du. \qquad (20)$$

We will show that (20) converges to $\pi^s(\omega)\varkappa(\omega)$ for $t \to \infty$ (independently of $x \in \mathbb{R}^d$). By (17)

$$\pi^s(\omega)\Phi(-t,\omega)^{-1}x = \sum_{i=1}^{k}\langle\Phi(-t,\omega)^{-1}x, e_i(\omega)\rangle x_i(\omega).$$

For each fixed i, $1 \le i \le d$,

$$\begin{aligned}\langle\Phi(-t,\omega)^{-1}x, e_i(\omega)\rangle &= \langle x, \Phi(-t,\omega)^{-1*}e_i(\omega)\rangle \\ &= \langle x, \Phi_{\vartheta^{-1},A^*}(t,\omega)e_i(\omega)\rangle\end{aligned}$$

by the proof of Proposition 8(R) (see Section 4). By a result of Ruelle ([22] p. 36), $(E^s(\omega))^* = (E^u(\omega))^\perp$ is the stable space for $\Phi_{\vartheta^{-1},A^*}$, so for each fixed i, $1 \le i \le k$,

$$\Phi_{\vartheta^{-1},A^*}(t,\omega)e_i(\omega) \xrightarrow{\text{e.f.}} 0 \qquad (t \to \infty).$$

Consequently,

$$\pi^s(\omega)\Phi(-t,\omega)^{-1}x \xrightarrow{\text{e.f.}} 0 \qquad (t \to \infty)$$

for arbitrary $x \in \mathbb{R}^d$. Similarly,

$$\begin{aligned}\pi^s(\omega)\int_{-t}^{0}\Phi(u,\omega)^{-1}b(\vartheta_u\omega)\,du &= \sum_{i=1}^{k}\langle\int_{-t}^{0}\Phi(u,\omega)^{-1}b(\vartheta_u\omega)\,du, e_i(\omega)\rangle x_i(\omega) \\ &= \sum_{i=1}\Big(\int_{-t}^{0}\langle\Phi(u,\omega)^{-1}b(\vartheta_u\omega), e_i(\omega)\rangle\,du\Big)x_i(\omega) \\ &= \sum_{i=1}\Big(\int_{-t}^{0}\langle b(\vartheta_u\omega), \Phi(u,\omega)^{-1*}e_i(\omega)\rangle\,du\Big)x_i(\omega).\end{aligned}$$

Again $\Phi(u,\omega)^{-1*} = \Phi_{\vartheta^{-1},A^*}(-u,\omega)$, and $e_i(\omega)$ is in the stable space of $\Phi_{\vartheta^{-1},A^*}$. Choose $\varepsilon > 0$ such that $\lambda_s + \varepsilon < 0$. The MET yields existence of $C_\varepsilon(\omega)$ such that

$$\|\Phi_{\vartheta^{-1},A^*}(t,\omega)e_i(\omega)\| \le C_\varepsilon(\omega)\,e^{t(\lambda^s+\varepsilon)} \qquad (21)$$

for all $t \ge 0$ P-a. s., and thus

$$\|\pi^s(\omega)\int_{-t}^{0}\Phi(u,\omega)^{-1}b(\vartheta_u\omega)\,du\| \le k\,C_\varepsilon(\omega)\int_{0}^{t}e^{u(\lambda^s+\varepsilon)}\|b(\vartheta_{-u}\omega)\|\,du. \qquad (22)$$

Convergence of the right hand side of (22) would follow from the existence of the Laplace transform

$$\hat{b}(\alpha,\omega) = \int_{0}^{\infty}e^{t\alpha}\|b(\vartheta_{-t}\omega)\|\,dt$$

of $\|b(\vartheta_{-t}\omega)\|$ at $\alpha = \lambda^s + \varepsilon$. This will hold if the abscissa α_0 of a. s. convergence of $\hat{b}$ satisfies $\alpha_0 = 0$. But

$$\alpha_0 = \limsup_{t\to\infty} \frac{1}{t} \log \int_0^t \|b(\vartheta_{-s}\omega)\|\, ds = \limsup_{t\to\infty} \frac{1}{t} \log(t \frac{1}{t} \int_0^t \|b(\vartheta_{-s}\omega)\|\, ds),$$

see Widder [23] p. 47. Since we had assumed $b \in L^1$, the individual ergodic theorem applies, and, since $E\,\|b\| > 0$, we obtain $\alpha_0 = 0$. (Note that in case $T = \mathbb{Z}$ one needs only the condition $\log^+ \|b\| \in L^1$, see Arnold and Wihstutz [2], p. 146.)

Altogether we have proved now that P-a. s.

$$\lim_{t\to\infty} \pi^s(\omega)\varphi(-t,\omega)^{-1}(x) = \lim_{t\to\infty} \pi^s(\omega) \int_{-t}^0 \Phi(u,\omega)^{-1} b(\vartheta_u\omega)\, du$$

exists for all $x \in \mathbb{R}^d$, and this convergence is, in fact, exponentially fast.

To handle the unstable component choose $t > 0$ and evaluate (19) with s, t replaced by $t, 0$, and proceed as above, now using the explicit form of π^u given by (17).

Existence of $\varkappa$ in the white noise case:

We only treat the stochastic integral part of the stable component, and put $m = 1$ for easier notation. We have to prove that

$$\lim_{t\to\infty} \pi^s(\omega) \int_{-t}^0 \Phi(u,\omega)^{-1} b_1\, dW(u)$$

exists (P-almost surely). By (17),

$$\begin{aligned} \pi^s(\omega) \int_{-t}^0 \Phi(u,\omega)^{-1} b_1\, dW(u) &= \sum_{i=1}^k \Big\langle \int_{-t}^0 \Phi(u,\omega)^{-1}\, dW(u)\, b_1, e_i(\omega) \Big\rangle x_i(\omega) \\ &= \sum_{i=1}^k \Big\langle b_1, (\int_{-t}^0 \Phi(u,\omega)^{-1*} dW(u)) e_i(\omega) \Big\rangle x_i(\omega) \end{aligned}$$

(we dropped the ω in the integrating Wiener process). The basic problem is that in general $\omega \mapsto e_i(\omega)$ has not the right measurability properties to be integrated against the Wiener process:

$$\int_{-t}^0 \Phi(u,\omega)^{-1*} e_i(\omega)\, dW(u)$$

does not make sense. Hence we cannot proceed directly as in the real noise case. First note that $\Phi^{-1*}(u,\omega) = \Phi_*(-u,\omega)$, where Φ_* denotes the cocycle generated by ϑ^{-1} and the linear white noise equation (12) with A_j's replaced by A_j^*'s. Thus

$$\Big(\int_{-t}^0 \Phi(u,\omega)^{-1*} dW(u)\Big) e_i(\omega) = \Big(\int_0^t \Phi_*(u,\omega)\, dW(-u)\Big) e_i(\omega),$$

where e_i is, in addition, assumed to be normalized. For $t, r \geq 0$ and $n \in \mathbb{N}$ put

$$S_n(t,t+r) = S_n(t,t+r)(\omega) = \sum_{j=[nt]}^{[n(t+r)]} \Phi_*(\frac{j}{n},\omega)\,\Delta_{-\frac{j}{n}}W$$

with $\Delta_{-\frac{j}{n}}W$ denoting increments of the Wiener process. Then

$$\int_t^{t+r} \Phi_*(u,\omega)\,dW(-u) = \lim_{n\to\infty} S_n(t,t+r) \qquad \text{in } L^2.$$

Since $(\Phi, x) \mapsto \Phi x$ is continuous and since $\|e_i(\omega)\| = 1$,

$$\begin{aligned} S_n(t,t+r)e_i(\omega) &= \sum_{j=[nt]}^{[n(t+r)]} \Phi_*(\frac{j}{n},\omega)\,e_i(\omega)\,\Delta_{-\frac{j}{n}}W \\ &\overset{n\to\infty}{\longrightarrow} \Big(\int_t^{t+r} \Phi_*(u,\omega)\,dW(-u)\Big)e_i(\omega) \qquad \text{in } L^2. \end{aligned}$$

Choose $\varepsilon > 0$ with $\lambda^s + \varepsilon < 0$. Similarly to the real noise case (see (21)) we use the MET to obtain

$$\|\Phi_*(\frac{j}{n},\omega)e_i(\omega)\| \leq C_\varepsilon(\omega)\,e^{\frac{j}{n}(\lambda^s+\varepsilon)}$$

for all $j, n \in \mathbb{N}$ (P-almost surely), hence

$$\|S_n(t,t+r)e_i(\omega)\| \leq C_\varepsilon(\omega) \sum_{j=[nt]}^{[n(t+r)]} e^{\frac{j}{n}(\lambda^s+\varepsilon)}|\Delta_{-\frac{j}{n}}W|.$$

We can now estimate

$$E\Big|\sum_{j=[nt]}^{[n(t+r)]} e^{\frac{j}{n}(\lambda^s+\varepsilon)}|\Delta_{-\frac{j}{n}}W|\Big|^2 \leq C\int_t^{t+r} e^{2u(\lambda^s+\varepsilon)}du \leq \frac{C}{-2(\lambda^s+\varepsilon)}\,e^{2t(\lambda^s+\varepsilon)},$$

where the constant C may be chosen independent of n and r. By Tchebychev's inequality we get with $h > 0$

$$P\Big(\Big|\frac{1}{C_\varepsilon(\omega)}\Big(\int_t^{t+r} \Phi_*(u,\omega)\,dW(-u)\Big)e_i(\omega)\Big| \geq h\Big) \;\leq\; \frac{C}{h^2}\cdot\frac{e^{2t(\lambda^s+\varepsilon)}}{-2(\lambda^s+\epsilon)}.$$

Putting $t = n$ and $h = \frac{1}{n}$ we obtain

$$P\Big(\Big|\frac{1}{C_\varepsilon(\omega)}\Big(\int_n^{n+r} \Phi_*(u,\omega)\,dW(-u)\Big)e_i(\omega)\Big| \geq \frac{1}{n}\Big) \leq Cn^2\frac{e^{2n(\lambda^s+\varepsilon)}}{-2(\lambda^s+\epsilon)},$$

and since $\sum_n n^2 e^{2n(\lambda^s+\varepsilon)} < \infty$, the Borel-Cantelli Lemma implies that for P-almost all ω there exists $n_0(\omega)$ such that

$$\Big|\frac{1}{C_\varepsilon(\omega)}\Big(\int_n^{n+r} \Phi_*(u,\omega)\,dW(-u)\Big)e_i(\omega)\Big| < \frac{1}{n}$$

for all $n \geq n_0(\omega)$ and for all $r \geq 0$. Consequently,

$$\Big(\int_{-t}^{0} \Phi(u,\omega)^{-1*} dW(u)\Big) e_i(\omega)$$

is a Cauchy sequence P-almost surely.

(ii) Invariance of $\mu_\omega = \delta_{\varkappa(\omega)}$ ($T = \mathbb{R}$, real noise case only)

According to (7), it suffices to show that, for each $t \in \mathbb{R}$, $\varphi(t,\omega)\varkappa(\omega) = \varkappa(\vartheta_t\omega)$ (P-a. s.). But

$$\varphi(t,\omega)\varkappa(\omega) = \Phi(t,\omega)\Big(\varkappa(\omega) + \int_0^t \Phi(s,\omega)^{-1} b(\vartheta_s\omega)\, ds\Big)$$

$$= \Phi(t,\omega)\Big((\pi^s(\omega)\varkappa(\omega) + \int_0^t \Phi(s,\omega)^{-1}\pi^s(\vartheta_s\omega) b(\vartheta_s\omega)\, ds)$$

$$\oplus\, (\pi^u(\omega)\varkappa(\omega) + \int_0^t \Phi(s,\omega)^{-1}\pi^u(\vartheta_s\omega) b(\vartheta_s\omega)\, ds)\Big)$$

$$= \Phi(t,\omega)\Big(\int_{-\infty}^{t} \Phi(s,\omega)^{-1}\pi^s(\vartheta_s\omega) b(\vartheta_s\omega)\, ds \;\oplus\; \int_{\infty}^{t} \Phi(s,\omega)^{-1}\pi^u(\vartheta_s\omega) b(\vartheta_s\omega)\, ds\Big)$$

$$= \Phi(t,\omega)\Phi(t,\omega)^{-1}\Big(\int_{-\infty}^{0} \Phi(u,\vartheta_t\omega)^{-1}\pi^s(\vartheta_u\vartheta_t\omega) b(\vartheta_u\vartheta_t\omega)\, du$$

$$\oplus \int_{\infty}^{0} \Phi(u,\vartheta_t\omega)^{-1}\pi^u(\vartheta_u\vartheta_t\omega) b(\vartheta_u\vartheta_t\omega)\, du\Big)$$

$$= \pi^s(\vartheta_t\omega)\varkappa(\vartheta_t\omega) \;\oplus\; \pi^u(\vartheta_t\omega)\varkappa(\vartheta_t\omega) \;=\; \varkappa(\vartheta_t\omega),$$

where we have used (18).

(iii) Uniqueness of invariant measures (T arbitrary, all cases)

We shall prove that if $\omega \mapsto \mu_\omega$ is a random measure satisfying $\varphi(t,\omega)\mu_\omega = \mu_{\vartheta_t\omega}$ (P-a. s.) for all $t \in T$ then $\mu_\omega = \delta_{\varkappa(\omega)}$ P-a. s. Since a measure on a product space is a Dirac measure if and only if its marginals are Dirac measures, we will be done once we have established

$$\pi^{s,u}(\omega)\mu_\omega = \delta_{\varkappa^{s,u}(\omega)}.$$

Let μ be an invariant measure. Then, for any $f \in C_b(\mathbb{R}^d)$,

$$\pi^{s,u}(\vartheta_t\omega) \circ \varphi(t,\omega)\mu_\omega(f) = \pi^{s,u}(\vartheta_t\omega)\mu_{\vartheta_t\omega}(f),$$

i. e.,

$$\int f(\pi^{s,u}(\vartheta_t\omega)\varphi(t,\omega)(x))\,d\mu_\omega(x) = \int f(x)\,d(\pi^{s,u}(\vartheta_t\omega)\mu_{\vartheta_t\omega})(x) =: \xi_f^{s,u}(\vartheta_t\omega).$$

On the other hand,

$$\pi^{s,u}(\vartheta_t\omega)\circ\varphi(t,\omega)\delta_{\varkappa(\omega)}(f) = \pi^{s,u}(\vartheta_t\omega)\delta_{\varkappa(\vartheta_t\omega)}(f),$$

i. e.,

$$f(\pi^{s,u}(\vartheta_t\omega)\varphi(t,\omega)\varkappa(\omega)) = f(\pi^{s,u}(\vartheta_t\omega)\varkappa(\vartheta_t\omega)) =: \eta_f^{s,u}(\vartheta_t\omega).$$

Consequently,

$$\begin{aligned}
&\eta_f^{s,u}(\vartheta_t\omega) - \xi_f^{s,u}(\vartheta_t\omega)\\
&\quad = f(\pi^{s,u}(\vartheta_t\omega)\varphi(t,\omega)\varkappa(\omega)) - \int f(\pi^{s,u}(\vartheta_t\omega)\varphi(t,\omega)(x))\,d\mu_\omega(x) \qquad (23)\\
&\quad = \int \Big(f(\pi^{s,u}(\vartheta_t\omega)\varphi(t,\omega)\varkappa(\omega)) - f(\pi^{s,u}(\vartheta_t\omega)\varphi(t,\omega)(x))\Big)\,d\mu_\omega(x).
\end{aligned}$$

Due to the affine structure of φ we can evaluate

$$\pi^{s,u}(\vartheta_t\omega)(\varphi(t,\omega)\varkappa(\omega) - \varphi(t,\omega)(x)) = \Phi(t,\omega)\pi^{s,u}(\omega)(\varkappa(\omega) - x).$$

For $\varepsilon > 0$ the MET yields the existence of a $B(\omega,\varepsilon)$ such that

$$\|\Phi(t,\omega)\pi^s(\omega)(\varkappa(\omega) - x)\| \le B(\omega,\varepsilon)e^{t(\lambda^s+\varepsilon)}\|\varkappa(\omega) - x\| \xrightarrow[t\to\infty]{\text{e.f.}} 0$$

as well as

$$\|\Phi(t,\omega)\pi^u(\omega)(\varkappa(\omega) - x)\| \le B(\omega,\varepsilon)e^{t(\lambda^u-\varepsilon)}\|\varkappa(\omega) - x\| \xrightarrow[t\to-\infty]{\text{e.f.}} 0$$

for all $x \in \mathbb{R}^d$.

Suppose f is uniformly continuous. Then, by virtue of the dominated convergence theorem, the last expression of (23) converges to zero as $t \to \infty$ or $t \to -\infty$, respectively, whilst the first expression is a stationary process. Thus, necessarily

$$\eta_f^{s,u}(\omega) - \xi_f^{s,u}(\omega) = 0 \qquad P\text{-a. s.}$$

for all uniformly continuous $f \in C_b(\mathbb{R}^d)$. Since uniformly continuous f suffice to identify a measure (see Gänssler und Stute [13] p. 66), we obtain $\mu_\omega = \delta_{\varkappa(\omega)}$ for P-almost all ω. □

5. Corollary *Let φ be an affine RDS satisfying* (16). *If the linear part Φ is* stable, *i. e., the top Lyapunov exponent λ_1 is negative, then there exists a unique invariant probability μ for φ given by $\mu_\omega = \delta_{\varkappa(\omega)}$,* where

$$\varkappa(\omega) = \begin{cases}
\displaystyle\sum_{n=-\infty}^{-1} \Phi(n,\omega)^{-1}b(\vartheta_n\omega), & T = \mathbb{Z},\\[2ex]
\displaystyle\int_{-\infty}^{0} \Phi(t,\omega)^{-1}b(\vartheta_t\omega)\,dt, & T = \mathbb{R},\ \textit{real noise case},\\[2ex]
\displaystyle\int_{-\infty}^{0} \Phi(t,\omega)^{-1}\Big(\bar{b}_0\,dt + \sum_{j=1}^{m} b_j\,dW_j(t)\Big), & T = \mathbb{R},\ \textit{white noise case}.
\end{cases}$$

In particular, $\mu_\omega = \delta_{\varkappa(\omega)}$ is a Markov measure.

An analogous statement holds in case all Lyapunov exponents are positive.

6. Corollary *Under the conditions of Corollary 5 — in particular, $\lambda_1 < 0$ — the following holds:*

(i)
$$\varphi(-t,\omega)^{-1}\nu \underset{t\to\infty}{\longrightarrow} \delta_{\varkappa(\omega)}$$

for any probability measure ν on $\mathbf{R}^d$.

(ii)
$$\mathcal{L}(\varphi(t,\omega)(x)) \underset{t\to\infty}{\longrightarrow} \mathcal{L}(\varkappa)$$

for all $x \in \mathbf{R}^d$, where $\mathcal{L}$ denotes the law of a random variable.

PROOF (i) For $f \in C_b(\mathbf{R}^d)$

$$\varphi(-t,\omega)^{-1}\nu(f) = \int f(\varphi(-t,\omega)^{-1}x)\, d\nu(x) \longrightarrow_{t\to\infty} \int f(\varkappa(\omega))\, d\nu(x)$$
$$= f(\varkappa(\omega)) = \int f(x)\, \delta_{\varkappa(\omega)}(x).$$

(ii) We have

$$\mathcal{L}(\varphi(-t,\omega)^{-1}x) = \mathcal{L}(\varphi(t,\vartheta_{-t}\omega)(x)) = \mathcal{L}(\varphi(t,\omega)(x)) \underset{t\to\infty}{\longrightarrow} \mathcal{L}(\varkappa) \qquad \square$$

7. Proposition *Suppose φ is an affine RDS with hyperbolic linear part Φ. The stable and unstable sets $M^{s,u}$ of φ (see (6)) satisfy*

$$\begin{aligned} M^{s,u}(x,\omega) &= x + E^{s,u}(\omega) \\ &= \{y \in \mathbf{R}^d \mid \|\varphi(t,\omega)(x) - \varphi(t,\omega)(y)\| \xrightarrow{\text{e.f.}} 0 \quad \text{for } t \to \pm\infty, \text{ resp.}\} \\ &= \{y \in \mathbf{R}^d \mid \|\Phi(t,\omega)x - \Phi(t,\omega)y\| \xrightarrow{\text{e.f.}} 0 \quad \text{for } t \to \pm\infty, \text{ resp.}\}. \end{aligned}$$

In particular, if $\lambda_1 < 0$ then P-a. s. $M^s(x,\omega) = \mathbf{R}^d$ and $M^u(x,\omega) = \{x\}$.

Note that under the above conditions the stable and unstable sets are automatically regular submanifolds of $\mathbf{R}^d$.

PROOF We have

$$\varphi(t,\omega)(y) - \varphi(t,\omega)(x) = \Phi(t,\omega)(y-x) \xrightarrow{\text{e.f.}} 0$$

as $t \to \infty$ or $t \to -\infty$, respectively, if and only if $(y-x) \in E^s(\omega) \Longleftrightarrow y \in x + E^s(\omega)$ in case $t \to \infty$ or $(y-x) \in E^u(\omega) \Longleftrightarrow y \in x + E^u(\omega)$ in case $t \to -\infty$. $\square$

4 Barnsley and Elton's 'key point'

Here we discuss relations of the results of the previous sections with those of Barnsley and Elton [4] and of Elton [12]. We continue considering affine RDS, hence our setting is more restrictive what the state space and the admissible maps are concerned. Let φ be an affine RDS with discrete time $T = \mathbb{Z}$. Thus,

$$\varphi(1,\omega)(x) = T(\omega)(x) = A(\omega)x + b(\omega),$$

where $A : \Omega \to \mathrm{Gl}(d, \mathbb{R})$ and $b : \Omega \to \mathbb{R}^d$, and $\varphi(n, \omega)$ is obtained according to (5).

First note that average contractivity implies $\lambda_1 < 0$, and average contractivity of some finite 'power of T's' is equivalent to $\lambda_1 < 0$.

The attractive (hence invariant) measure ρ for the Markov operator Q from Barnsley and Elton's Theorem is constructed by the following trick (see the proof of Theorem 1 of [4]):
Fix $x \in \mathbb{R}^d$, and consider the process

$$z_n^x(\omega) = T_0(\omega) \circ \ldots \circ T_{n-1}(\omega)x,$$

where $T_n(\omega) = T(\vartheta_n\omega)$. Then average contractivity implies P-a. s. existence of the limits

$$\lim_{n\to\infty} z_n^x(\omega) = \bar{\varkappa}(\omega),$$

where in addition $\bar{\varkappa}$ is independent of x (but the set of exceptional ω's — where the convergence fails — in general depends on x). Since an iid sequence is time reversible, we have $\mathcal{L}(T_0 \circ \ldots \circ T_{n-1}x) = \mathcal{L}(T_{n-1} \circ \ldots \circ T_0 x)$, thus $\mathcal{L}(T_{n-1} \circ \ldots \circ T_0 x) \longrightarrow \rho = \mathcal{L}(\bar{\varkappa})$. Invariance and uniqueness of ρ can now be inferred directly from Letac [20]. Letac proves (for $(T_n)_{n\in\mathbb{N}}$ an iid sequence of continuous transformations of a locally compact space X) that if $\mathcal{L}(T_{n-1} \circ \ldots \circ T_0(x))$ converges (tightly) for $n \to \infty$ to the same probability measure ρ for all $x \in X$, then ρ is the unique invariant (probability) measure for the Markov operator Q (see (8)). But note that this argument settles uniqueness only insofar Markov measures are concerned.

We will have a closer look at the relation between our $\varkappa$ and Barnsley and Elton's $\bar{\varkappa}$. (In Elton's paper [12], the iid assumption on $(T_n)_{n\in\mathbb{Z}}$ is dropped, and the invariant measure is constructed — exactly as our $\varkappa$ — going backward in time. This paper had not been accessible to us when preparing the present one.)

We first collect some facts about linear RDS. In this section we use the following notations:
$\Phi_{\vartheta,A}$ is the linear cocycle generated by the ergodic and measurable flow of P-preserving maps $\vartheta_t : \Omega \to \Omega$ and the measurable map A.
In case of discrete time $T = \mathbb{Z}$, $A : \Omega \to \mathrm{Gl}(d, \mathbb{R})$, and

$$\Phi_{\vartheta,A}(n,\omega) = \begin{cases} A(\vartheta_{n-1}\omega) \circ \ldots \circ A(\omega) & \text{for } n > 0 \\ \mathrm{id} & \text{for } n = 0 \\ A^{-1}(\vartheta_n\omega) \circ \ldots \circ A^{-1}(\vartheta_{-1}\omega) & \text{for } n < 0. \end{cases}$$

For continuous time $T = \mathbb{R}$, we restrict ourselves to the real noise system associated with $A : \Omega \to \mathrm{gl}(d, \mathbb{R})$, i. e.,

$$\dot{\Phi}_{\vartheta,A}(t,\omega) = A(\vartheta_t\omega)\Phi_{\vartheta,A}(t,\omega), \quad \Phi_{\vartheta,A}(0,\omega) = \mathrm{id}.$$

(For white noise systems cf. Remark (iv) after Theorem 9 below.)

Assume $\log^+ \|A\|, \log^+ \|A^{-1}\| \in L^1(P)$ if $T = \mathbb{Z}$ and $A \in L^1(P)$ if $T = \mathbb{R}$. Then the MET applies for $\Phi_{\vartheta,A}$, and we can define the *Lyapunov spectrum* $\Sigma(\vartheta, A)$ associated with $\Phi_{\vartheta,A}$ by

$$\Sigma(\vartheta, A) = \{(\lambda_1, d_1), \dots, (\lambda_r, d_r)\},$$

where d_i denotes the multiplicity of λ_i. Put

$$-\Sigma(\vartheta, A) := \{(-\lambda_1, d_1), \dots, (-\lambda_r, d_r)\}.$$

Then we have, in summarizing and generalizing results in Ruelle [22]:

8($\mathbb{Z}$) Proposition *For $T = \mathbb{Z}$*

(i) $\Sigma(\vartheta, A) = \Sigma(\vartheta^{-1}, A^*) =: \Sigma_1$ *and* $\Sigma(\vartheta, A^{*-1}) = \Sigma(\vartheta^{-1}, A^{-1}) = -\Sigma_1$

(ii) $\Sigma(\vartheta, A^*) = \Sigma(\vartheta^{-1}, A) =: \Sigma_2$ *and* $\Sigma(\vartheta, A^{-1}) = \Sigma(\vartheta^{-1}, A^{*-1}) = -\Sigma_2.$

8($\mathbb{R}$) Proposition *For $T = \mathbb{R}$*

(i) $\Sigma(\vartheta, A) = \Sigma(\vartheta^{-1}, A^*) =: \Sigma_1$ *and* $\Sigma(\vartheta, -A^*) = \Sigma(\vartheta^{-1}, -A) = -\Sigma_1$

(ii) $\Sigma(\vartheta, A^*) = \Sigma(\vartheta^{-1}, A) =: \Sigma_2$ *and* $\Sigma(\vartheta, -A) = \Sigma(\vartheta^{-1}, -A^*) = -\Sigma_2.$

PROOF OF PROPOSITION 8($\mathbb{Z}$) It suffices to prove (i); (ii) follows from (i) by replacing A by A^*. From the definition of Φ we get

$$\Phi_{\vartheta^{-1},A^*}(n, \omega) = \Phi_{\vartheta,A}(n, \vartheta_{-n+1}\omega)^*.$$

Now

$$\begin{aligned} E \log \|\Phi_{\vartheta^{-1},A^*}(n, \cdot)\| &= E \log \|\Phi_{\vartheta,A}(n, \vartheta_{-n+1}\cdot)^*\| \\ &= E \log \|\Phi_{\vartheta,A}(n, \vartheta_{-n+1}\cdot)\| \\ &= E \log \|\Phi_{\vartheta,A}(n, \cdot)\|, \end{aligned}$$

where we have used first $\|A\| = \|A^*\|$ for quadratic matrices A, and then ϑ-invariance of P. In fact, the same identities hold for the p-fold exterior products of Φ, $1 \le p \le d$, and, since $\bigwedge^p \Phi^* = (\bigwedge^p \Phi)^*$, we obtain

$$E \log \|\textstyle\bigwedge^p \Phi_{\vartheta^{-1},A^*}(n, \cdot)\| = E \log \|\textstyle\bigwedge^p \Phi_{\vartheta,A}(n, \cdot)\|.$$

From the definition of the Lyapunov exponents (see (14)) we conclude that

$$\Sigma(\vartheta^{-1}, A^*) = \Sigma(\vartheta, A).$$

A similar reasoning applied to

$$\Phi_{\vartheta,A^{*-1}}(n, \omega) = \Phi_{\vartheta,A}(n, \omega)^{*-1} = \Phi_{\vartheta,A}(-n, \vartheta_n\omega)^*$$

and

$$\Phi_{\vartheta^{-1},A^{-1}}(n,\omega) = \Phi_{\vartheta,A}(n,\vartheta_{-n+1}\omega)^{-1} = \Phi_{\vartheta,A}(-n,\vartheta\omega)$$

yields

$$\Sigma(\vartheta, A^{*-1}) = \Sigma(\vartheta^{-1}, A^{-1})$$
$$= \text{spectrum of the cocycle } \Phi_{\vartheta,A}(-n,\omega) = \Psi(n,\omega).$$

But

$$\lim_{n\to\pm\infty} \frac{1}{n} \log \|\Phi_{\vartheta,A}(n,\omega)v\| = \lambda_i \iff v \in E_i(\omega)$$

implies

$$\lim_{n\to\pm\infty} \frac{1}{n} \log \|\Phi_{\vartheta,A}(-n,\omega)v\| = -\lambda_i$$

for all $v \in E_i(\omega)$, hence the spectrum of $\Phi_{\vartheta,A}(-n,\omega)$ is $-\Sigma_1$ with the same splitting as $\Phi_{\vartheta,A}$. □

PROOF OF PROPOSITION 8($\mathbb{R}$) By differentiating with respect to t we obtain

$$\Phi_{\vartheta,-A^*}(t,\omega) = \Phi_{\vartheta,A}(t,\omega)^{*-1} = \Phi_{\vartheta,A}(-t,\vartheta_t\omega),$$
$$\Phi_{\vartheta^{-1},A^*}(t,\omega) = \Phi_{\vartheta,A}(-t,\omega)^{*-1} = \Phi_{\vartheta,A}(t,\vartheta_{-t}\omega),$$
$$\Phi_{\vartheta^{-1},-A}(t,\omega) = \Phi_{\vartheta,A}(-t,\omega)^{*-1}.$$

In view of this, the result follows as in the case $T = \mathbb{Z}$. □

A condition for $\Sigma_1 = \Sigma_2$

In general, there is no relation between Σ_1 and Σ_2. See, e. g., Key [16] for examples with $\Sigma_1 \neq \Sigma_2$. However, there is a convenient sufficient condition ensuring $\Sigma_1 = \Sigma_2$.

9. Theorem *If the generating process $(A(\vartheta_t\omega))_{t\in T}$ of a linear RDS $\Phi_{\vartheta,A}$ is time reversible, i. e., if*

$$\mathcal{L}\Big((A(\vartheta_t\cdot))_{t\in T}\Big) = \mathcal{L}\Big((A(\vartheta_{-t}\cdot))_{t\in T}\Big)$$

($\mathcal{L}(\chi)$ = probability law of a random variable χ), then

$$\mathcal{L}(\Phi_{\vartheta,A}) = \mathcal{L}(\Phi_{\vartheta^{-1},A}),$$

hence

$$\Sigma_1 = \Sigma(\vartheta, A) = \Sigma(\vartheta^{-1}, A) = \Sigma_2.$$

PROOF The proof goes along the same lines for discrete as for continuous time. We restrict ourselves to $T = \mathbb{Z}$. For $n > 0$

$$\Phi_{\vartheta,A}(n,\omega) = A(\vartheta_{n-1}\omega) \circ \ldots \circ A(\omega),$$
$$\Phi_{\vartheta^{-1},A}(n,\omega) = A(\vartheta_{-n+1}\omega) \circ \ldots \circ A(\omega).$$

Thus,

$$\begin{aligned}\mathcal{L}\Big(\Phi_{\vartheta,A}(n,\cdot)\Big) &= \mathcal{L}(A(\vartheta_{n-1}\cdot) \circ \ldots \circ A(\cdot)) \\ &= \mathcal{L}(A(\vartheta_{-n+1}\cdot) \circ \ldots \circ A(\cdot)) \\ &= \mathcal{L}\Big(\Phi_{\vartheta^{-1},A}(n,\cdot)\Big),\end{aligned}$$

where the second identity holds by virtue of time reversibility. It follows that for $n = 1, 2, \dots$ and $1 \le p \le d$

$$E \log \left\| \wedge^p \Big(A(\vartheta_{n-1} \cdot) \circ \dots \circ A(\cdot) \Big) \right\| = E \log \left\| \wedge^p \Big(A(\vartheta_{-n+1} \cdot) \circ \dots \circ A(\cdot) \Big) \right\|. \tag{24}$$

In view of (14), (24) implies $\Sigma_1 = \Sigma_2$. □

Remarks (i) Apparently, it suffices to assume (24) instead of time reversibility to conclude $\Sigma_1 = \Sigma_2$.

(ii) If $A(\vartheta_t \omega) = A(\xi_t(\omega))$, where $(\xi_t)_{t \in T}$ is a Markov process on some suitable state space E, then time reversibility of $(A(\vartheta_t \cdot)_{t \in T}$ follows from *reversibility* (or *detailed balance*) of $(\xi_t)_{t \in T}$, which in turn is equivalent to the generator of the Markov semigroup of $(\xi_t)_{t \in T}$ being selfadjoint in $L^2(E, \rho)$, where $\rho = \mathcal{L}(\xi_0)$ (see Ikeda and Watanabe [15] p. 280).

(iii) For $T = \mathbf{Z}$, exchangeability of $(A(\vartheta_n \cdot))_{n \in \mathbf{Z}}$ implies time reversibility, and exchangeability follows if $(A(\vartheta_n \cdot))_{n \in \mathbf{Z}}$ is iid. The latter has been assumed by Barnsley and Elton [4].

(iv) The white noise case

$$dx = A_0 x dt + \sum_{j=1}^{m} A_j x \, dW_j$$

is always time reversible due to the fact that the $\mathbf{R}^m$-valued Wiener process $W(t) = (W_1(t), \dots, W_m(t))$ satisfies $\mathcal{L}(W(\cdot)) = \mathcal{L}(W(-\cdot))$.

(v) Of course, $\Sigma_1 = \Sigma_2$ if A is symmetric, i.e., $A = A^*$ a.s., because then $\Phi_{\vartheta, A} = \Phi_{\vartheta, A^*}$ a.s.

We now consider the full affine RDS

$$\varphi(t, \omega)(x) = \Phi(t, \omega) x + \psi(t, \omega).$$

We write $\varphi_{\vartheta, A, b}$ for the affine RDS generated by ϑ, A, and b.

10. Theorem *Suppose $\Big(A(\vartheta_t \cdot), b(\vartheta_t \cdot) \Big)_{t \in T}$ is time reversible. Then*

$$\mathcal{L}\Big((\varphi_{\vartheta, A, b}(-t, \cdot)^{-1})_{t \ge 0} \Big) = \mathcal{L}\Big((\varphi_{\vartheta^{-1}, A, b}(t, \vartheta_t \cdot))_{t \ge 0} \Big). \tag{25}$$

In particular, $\varphi_{\vartheta, A, b}(-t, \cdot)^{-1}(x)$ converges P-a.s. for $t \to \infty$,

$$\lim_{t \to \infty} \varphi_{\vartheta, A, b}(-t, \cdot)^{-1}(x) =: \varkappa(\omega),$$

if and only if

$$\lim_{t \to \infty} \varphi_{\vartheta^{-1}, A, b}(t, \vartheta_t \cdot)(x) = \bar{\varkappa}(\omega)$$

exists P-a.s., in case of which $\mathcal{L}(\varkappa) = \mathcal{L}(\bar{\varkappa})$.

If $\lambda_1 < 0$ then both $\varkappa$ and $\bar{\varkappa}$ exist, and neither of them depends on $x \in \mathbf{R}^d$

Note that in general $\varkappa \ne \bar{\varkappa}$, as $\varkappa$ "depends on the past" $t \le 0$, whereas $\bar{\varkappa}$ "depends on the future" $t \ge 0$. In particular, if $T = \mathbf{Z}$ and $(T(\vartheta_n \omega))_{n \in \mathbf{Z}}$ is iid then $\varkappa$ and

$\bar{\varkappa}$ are independent. Furthermore, $\mathcal{L}(\varkappa) = \mathcal{L}(\bar{\varkappa}) = \rho$, and ρ is the unique invariant measure for the Markov chain induced by $x_n = T(\vartheta_{n-1}\omega)(x_{n-1})$, i. e., $Q\rho = \rho$, where Q denotes the Markov operator induced by $\{T(\vartheta_n\omega)\}_n$, see Lemmas 1 and 2.

PROOF Again we may — and do — restrict ourselves to the case $T = \mathbf{Z}$. Writing $\varphi = \varphi_{\vartheta,A,b}$, we have

$$\varphi(-n,\omega)^{-1} = T(\vartheta_{-1}\omega) \circ \ldots \circ T(\vartheta_{-n}\omega)$$

for $n > 0$, where $T(\omega) = \varphi(1,\omega)$. Time reversibility implies

$$\mathcal{L}\Big(\big(\varphi(-n,\cdot)^{-1}\big)_{n\geq 0}\Big) = \mathcal{L}\Big(\big(T(\vartheta_1\cdot) \circ \ldots \circ T(\vartheta_n\cdot)\big)_{n\geq 0}\Big).$$

Now

$$T(\vartheta_1\cdot) \circ \ldots \circ T(\vartheta_n\cdot) = \varphi_{\vartheta^{-1},A,b}(n,\vartheta_n\omega),$$

and (25) follows.

If all finite dimensional distributions of two stochastic processes are the same, then one of them converges almost surely if and only if the other one does so. Furthermore, the limiting random variables' distributions coincide. (Note that for this reasoning the two processes need not even be defined on the same probability space).

We give a more explicit expression for $\bar{\varkappa}$ (for $T = \mathbf{Z}$):

$$T(\vartheta_1\omega) \circ \ldots \circ T(\vartheta_n\omega)(x) = \varphi_{\vartheta^{-1},A,b}(n,\vartheta_n\omega)(x)$$

$$= \Phi_{\vartheta^{-1},A}(n,\vartheta_n\omega)x + \sum_{j=0}^{n-1} \Phi_{\vartheta^{-1},A}(j,n,\vartheta_n\omega)b(\vartheta_{j+n}\omega).$$

By Proposition 8($\mathbf{Z}$), time reversibility implies $\Sigma(\vartheta, A) = \Sigma_1 = \Sigma_2 = \Sigma(\vartheta^{-1}, A)$. Thus, if $\lambda_1 < 0$ then

$$\Phi_{\vartheta^{-1},A}(n,\vartheta_n\omega)x \xrightarrow{\text{e.f.}} 0$$

as $n \to \infty$ for all $x \in \mathbf{R}^d$. This implies

$$\bar{\varkappa}(\omega) = \lim_{n\to\infty} \Phi_{\vartheta^{-1},A}(n,\vartheta_n\omega) \sum_{j=0}^{n-1} \Phi_{\vartheta^{-1},A}(j,\vartheta_n\omega)^{-1} b(\vartheta_{n+j}\omega)$$

$$= \lim_{n\to\infty} \varphi_{\vartheta^{-1},A,b}(n,\vartheta_n\omega)(0). \qquad \square$$

For $T = \mathbf{R}$ the corresponding representation of $\bar{\varkappa}$ is

$$\bar{\varkappa}(\omega) = \lim_{t\to\infty} \varphi_{\vartheta^{-1},A,b}(t,\vartheta_t\omega)(0)$$

$$= \lim_{t\to\infty} \Phi_{\vartheta^{-1},A}(t,\vartheta_t\omega) \int_0^t \Phi_{\vartheta^{-1},A}(s,\vartheta_t\omega)^{-1} b(\vartheta_{t+s}\omega)\, ds.$$

Remark Under time reversibility we have

$$\mathcal{L}(T(\vartheta_1\omega) \circ \ldots \circ T(\vartheta_n\omega)) = \mathcal{L}(T(\vartheta_{n-1}\omega) \circ \ldots \circ T(\omega)) = \mathcal{L}(\varphi(n,\omega))$$

for each *fixed* $n \geq 1$. That's why in case $\lambda_1 < 0$

$$\mathcal{L}(\varphi(n,\omega)(x)) = \mathcal{L}(\varphi(-n,\omega)^{-1}(x)) \longrightarrow \mathcal{L}(\varkappa)$$

for any $x \in \mathbf{R}^d$.

References

[1] L. Arnold, *Stochastic Differential Equations: Theory and Applications*, Wiley, New York 1974

[2] L. Arnold and V. Wihstutz, Stationary solutions of linear systems with additive and multiplicative noise, *Stochastics* 7 (1982) 133–155

[3] M. F. Barnsley, *Fractals Everywhere*, Academic Press, San Diego 1988

[4] M. F. Barnsley and J. Elton, A new class of Markov processes for image encoding, *Adv. Appl. Probab.* 20 (1988) 14–32

[5] M. F. Barnsley, J. Elton, and D. Hardin, Recurrent iterated function systems, *Constr. Approx.* 5 (1989) 3–31

[6] P. Bougerol and N. Picard, Stationarity of GARCH processes and of some non-negative time series, preprint Nancy 1990

[7] P. Bougerol and N. Picard, Strict stationarity of generalized autoregressive processes, preprint Nancy 1990

[8] A. Brandt, The stochastic equation $Y_{n+1} = A_n Y_n + B_n$ with stationary coefficients, *Adv. Appl. Probab.* 18 (1986) 211–220

[9] H. Crauel, Markov measures for random dynamical systems, preprint Bremen 1989

[10] L. E. Dubins and D. A. Freedman, Invariant probabilities for certain Markov processes, *Ann. Math. Statist.* 37 (1966) 837–848

[11] N. Dunford and J. T. Schwartz, *Linear Operators, Part I*, Interscience, New York 1958

[12] J. Elton, A multiplicative ergodic theorem for Lipschitz maps, *Stochastic Process. Appl.* 34 (1990) 39–47

[13] P. Gänssler und W. Stute, *Wahrscheinlichkeitstheorie*, Springer, Berlin 1977

[14] J. E. Hutchinson, Fractals and Self Similarity, *Indiana Univ. Math. J.* 30 (1981) 713–747

[15] N. Ikeda and S. Watanabe, *Stochastic Differential Equations and Diffusion Processes*, North Holland, Amsterdam 1981

[16] E. Key, Lyapunov exponents for matrices with invariant subspaces, *Ann. Probab.* 16 (1988) 1721–1728

[17] Y. Kifer, *Ergodic Theory of Random Transformations*, Birkhäuser, Boston 1986

[18] S. Kobayashi and K. Nomizu, *Foundations of Differential Geometry, Volume I*, Wiley, New York 1963

[19] Y. Le Jan, Équilibre statistique pour les produits de difféomorphismes aléatoires indépendants, *Ann. Inst. Henri Poincaré* 23 (1987) 111–120

[20] G. Letac, A contraction principle for certain Markov chains and its applications, pp. 263–273 in *Random Matrices and Their Applications, Proceedings 1984*, J. E. Cohen, H. Kesten, and C. M. Newman (eds.), Contemporary Mathematics, AMS, Providence, RI, 1986

[21] V. I. Oseledec, A multiplicative ergodic theorem. Lyapunov characteristic numbers for dynamical systems. *Trans. Moscow Math. Soc.* 19 (1968) 197–231

[22] D. Ruelle, Ergodic theory of differentiable dynamical systems, *Publ. Math. I.H.E.S.* 50 (1979) 27–58

[23] D. Widder, *The Laplace Transform*, Princeton University Press, Princeton 1946

Ludwig Arnold
Institut für Dynamische Systeme
Universität Bremen
2800 Bremen 33
FR Germany

Hans Crauel
Fachbereich 9 MATHEMATIK
Universität des Saarlandes
6600 Saarbrücken 11
FR Germany

Partially supported by Volkswagen-Stiftung. This paper was written while H.C. was with the Institut für Dynamische Systeme, Universität Bremen.

Weak Convergence and Generalized Stability for Solutions to Random Dynamical Systems

John Elton and Jelel Ezzine

I. Introduction

Let (Ω, P) be a probability space. Consider the random differential equation (see Arnold and Kliemann [AK] and Crauel [C] for this terminology, and background)

$$\dot{x}(t,\omega) = F(t,\omega,x(t,\omega)), \qquad -\infty < t < \infty, \tag{1.1}$$

where $F(t,\omega,\cdot) : \mathbb{R}^d \to \mathbb{R}^d$ is a Lipschitz function for each t, ω, measurable on $\mathbb{R} \times \Omega$ and is *stationary*: that is, there exists a measure-preserving flow $\tau_t : \Omega \to \Omega$, $-\infty < t < \infty$ such that

$$F(t,\omega,\cdot) = F(0,\tau_t(\omega),\cdot).$$

The process F is *ergodic* if the flow τ_t is. Assume also that $|F(t,\omega,x)|$ is, for each x, with probability 1 locally integrable in t, i.e., $\int_s^t |F(u,\omega,x)|du < \infty$ for all s, t.

The most common example of this is of course the "inhomogeneous linear" (i.e., affine) equation

$$\dot{x} = A(t,\omega)x + B(t,\omega)$$

where $A(t,\omega)$ (resp. $B(t,\omega)$) is a stationary process of $d \times d$ (resp. $d \times 1$) matrices. We shall carry along the affine case as our motivation in what follows.

Our goal is to show that under the hypothesis of a negative Furstenberg–Kesten exponent (i.e., maximum Lyapunov exponent in the affine case), which can be defined in this general situation (in the affine case, it is the usual assumption to get stability for the associated *homogeneous* equation), that the empirical distribution of the trajectories of the solution converge to a stationary distribution which does not depend on the initial distribution, and in the ergodic case is independent of chance and is the unique, attractive stable distribution. This distribution can be realized as the distribution of a certain (not the usual one) backwards time limit. One must look a little carefully at this point: we do not require any special reversibility assumption, nor do we even require, in the affine case, that the matrices $A(t,\omega)$ are invertible. The idea is that if one pushes the *starting* time backwards toward $-\infty$ (to find the germ of the process in the infinite past, as it were) while still letting the process run forwards, then we get pointwise convergence, which implies *distributional* (definitely not pointwise) convergence of the process letting the *ending* time approach ∞ as usual. The backwards limit allows us to show that the forward trajectories are eventually close to the support of the invariant measure, never to escape again. Distributional convergence alone would only imply that we escape with ever-decreasing frequency.

Of interest from the point of view of ergodic theory is that if the initial vector $x(0,\omega)$ is made to be our backwards limit vector, then the solution $x(t,\omega)$ is not only stationary with this stable distribution, but is merely a factor of the stationary process determining the equation, i.e., $x(t,\omega) = x(0,\tau_t(\omega))$. So if the coefficient process is Markov, for example, the trajectories are a factor of a Markov process, even though they are not Markovian themselves.

Our result gives a broader view of stability: rather than only calling stable those sytems whose trajectories converge a. s. to 0, which almost never really happens because, e. g., in the affine case the inhomogeneous term B is probably not going to be exactly 0, perhaps one should call stable systems those whose trajectories behave as described above.

We were motivated by discrete-time results on iterated function systems. Note that even with *affine* maps, the attractor (support of invariant measure) is typically fractal, due to iterating randomly from a system of maps; if one iterates a *single* function, one must use a non-affine function to get a strange attractor. This is why iterated function systems are so convenient for generating fractals; see Barnsley and Sloan [BS]. Similarly, linear inhomogeneous random differential equations have interesting asymptotic behavior, even though non-random ones do not.

See [BEH] for examples of the use of non-iid (in fact Markov) map sequences to produce interesting fractal geometries.

Our proof generalizes results in Barnsley and Elton [BE] (iid map sequences), Berger and Amit [BA], Barnsley, Elton and Hardin [BEH] (Markov map sequences), and the proof is an extension of the proof in Elton [E] (stationary map sequences) to continuous time.

Applications of our result to control theory will appear in a later paper.

II. Main Results

Use any convenient norm on $\mathbb{R}^d$. For a Lipschitz function $g : \mathbb{R}^d \to \mathbb{R}^d$, define $||g|| = \sup_{x \neq y} \frac{|g(x)-g(y)|}{|x-y|}$. For g affine, this is merely the usual matrix norm of the linear part of g. In general, $|| \, ||$ is multiplicative (although not a norm).

Let $X(s,t,\omega,x)$ be the solution at time t to the equation (1.1) which starts at x at time s, i.e., $X(s,s,\omega,x) = x$. Let $H(s,t,\omega) : \mathbb{R}^d \to \mathbb{R}^d$ be given by

$$H(s,t,\omega)(x) = X(s,t,\omega,x).$$

In the affine case, let $\Phi(s,t,\omega)$ be a fundamental matrix for the associated homogeneous equation; i.e., Φ satisfies

$$\begin{aligned}\frac{d}{dt}\Phi(s,t,\omega) &= A(t,\omega)\Phi(s,t,\omega)\\ \Phi(s,s,\omega) &= I.\end{aligned}$$

Thus, $H(s,t,\omega)(x) - H(s,t,\omega)(y) = \Phi(s,t,\omega)(x-y)$ in this case.

From the differential equation (1.1),

$$\begin{aligned}&\left|\frac{d}{dt}(H(s,t,\omega)(x) - H(s,t,\omega)(y))\right|\\ &= |F(t,\omega,H(s,t,\omega)(x)) - F(t,\omega,H(s,t,\omega)(y)|\\ &\le \|F(t,\omega,\cdot)\|\,|H(s,t,\omega)(x) - H(s,t,\omega)(y)|\end{aligned}$$

so

$$\begin{aligned}&|H(s,t,\omega)(x) - H(s,t,\omega)(y)|\\ &\le |H(s,s,\omega)(x) - H(s,s,\omega)(y)| \exp\int_s^t \|F(u,\omega,\cdot)\|du\end{aligned}$$

so

$$\|H(s,t,\omega)\| \le \exp\int_s^t \|F(u,\omega,\cdot)\|du \tag{2.1}$$

so $H(s,t,\omega)$ is Lipschitz if $\|F(u,\omega,\cdot)\|$ is locally integrable in u - which will follow a. s. from another integrability assumption we will make. In the affine case, $\|H(s,t,\omega)\| = \|\Phi(s,t,\omega)\|$, the norm of the fundamental matrix.

Now H obviously satisfies (from the usual existence-uniqueness theorems)

$$\begin{aligned}H(s,t,\tau_u(\omega)) &= H(s+u,t+u,\omega)\\ H(s,t,\omega) &= H(u,t,\omega)\circ H(s,u,\omega)\\ H(s,s,\omega) &= I.\end{aligned}$$

Also log $\|H(s,t,\omega)\|$ is subadditive: $\log\|H(s,t,\omega)\| \le \log\|H(u,t,\omega)\| + \log\|H(s,u,\omega)\|$ which is true because

$$\|H(s,t,\omega)\| \le \|H(u,t,\omega)\|\,\|H(s,u,\omega)\|$$

from the above.

In the following, unless otherwise indicated, all integrals in ω are over the entire space Ω.

Lemma 1. *Assume $\int \|F(0,\omega,\cdot)\|dP(\omega) < \infty$ (by Fubini's theorem and stationarity, this implies the local integrability condition*

$$\int_s^t \|F(u,\omega,\cdot)\|du < \infty$$

a. s. for all s,t). In the affine case this reduces to $\mathbb{E}(\|A(0,\cdot)\|) < \infty$. Then

$$\lim_{t\to\infty} \frac{log\|H(s,t,\omega)\|}{t} = \lambda(\omega)$$

exists a. s. (independent of s) and is equal to, for a. a. ω,

$$\lim_{s\to-\infty} \frac{log\|H(s,t,\omega)\|}{|s|} \quad (any t).$$

λ is constant if the flow is ergodic.

Proof. First observe that log $\|H\|$ satisfies the integrable oscillation condition

$$\int \sup_{0\le s\le t\le 1} log^+\|H(s,t,\omega)\|dP(\omega)$$
$$\le \int \sup_{0\le s\le t\le 1} \int_s^t \|F(u,\omega,\cdot)\|du dP(\omega)$$
$$\le \int\int_0^1 \|F(\tau,\omega)\|du dP(\omega)$$
$$= \int \|F(0,\omega,\cdot)\|dP(\omega) \quad \text{(by stationarity)}$$
$$< \infty \qquad \text{(by hypothesis).}$$

Thus we may apply Kingman's subadditive ergodic theorem to obtain the first limit (see Krengel [K], pgs. 40 and 41). By the same argument, the second limit, call it $\tilde{\lambda}(\omega)$, exists as well, and we need to show $\lambda = \tilde{\lambda}$. To this end, assume that L^1 convergence holds as well as a. s. convergence. Then

$$\int \left| \frac{1}{|s|}\log\|H(s,t,\omega)\| - \tilde{\lambda}(\omega)\right| dP(\omega)$$
$$= \int \left| \frac{1}{|s|}\log\|H(0,t-s,\tau_s\omega)\| - \tilde{\lambda}(\omega)\right| dP(\omega)$$
$$= \int \left| \frac{1}{|s|}\log\|H(0,t-s,\omega)\| - \tilde{\lambda}(\tau_{-s}\omega)\right| dP(\omega)$$
$$\text{since } \tau_s \text{ is measure-preserving}$$
$$= \int \left| \frac{1}{|s|}\log\|H(0,t-s,\omega)\| - \tilde{\lambda}(\omega)\right| dP(\omega)$$

since $\tilde{\lambda}$ is invariant for the flow (from Kingman's theorem). Now let $s \to -\infty$; the first integral approaches 0 by definition of $\tilde{\lambda}$ and assumed L^1 convergence, so the last integral approaches 0 also, which implies $\tilde{\lambda} = \lambda$ by the uniqueness of L^1-limits and the definition of λ. Finally, a simple truncation argument reduces the general case to our assumption of L^1-convergence. □

Theorem 1. *Assume the hypothesis of Lemma 1, and assume also that* $\int |F(0,\omega,x)|dP(\omega) < \infty$ *for some (hence all)* x *(in the affine case, this amounts to assuming in addition* $\mathbb{E}(|B(0,\cdot)|) < \infty$*). Suppose*

$$\lim_{t\to\infty} \frac{log||H(s,t,\omega)||}{t} = \lambda(\omega) < 0 \quad a.s.$$

(in the affine case, this is

$$\lim_{t\to\infty} \frac{log||\Phi(s,t,\omega)||}{t} < 0 \quad a.s.$$

which is the usual assumption to get stability of solutions to a homogeneous random differential equation). Then:

(i) $\lim_{s\to-\infty} X(s,t,\omega,x) = Y(t,\omega)$ exists a. s. *and does not depend on* x*. Note that it is the initial time which* $\to -\infty$ *here.* $Y(t,\omega)$ *is a stationary process on* $\mathbb{R}$*, and is in fact a factor of the flow:*

$$Y(t,\omega) = Y(0,\tau_t\omega).$$

Thus $Y(t,\omega)$ *is ergodic if the flow is.*

(ii) $X(0,t,\omega,Y(0,\omega)) = Y(t,\omega)$ a. s.*, so if the initial condition for the random differential equation (1.1) is made to be* $Y(0,\omega)$*, the solution process is stationary, with marginal distributions* $\mu(B) = P(Y(0,\omega) \in B)$*.*

(iii) The solution process $X(s,t,\omega,x)$ *converges in distribution to* $Y(0,\omega)$ *as* $t \to \infty$ *for any initial* x *or distribution of* x*. The trajectories* $X(s,t,\omega,x)$ *do not converge pointwise as* $t \to \infty$ *(except in trivial cases), of course, but with probability one, for all* x*,*

$$\lim_{t\to\infty} \frac{1}{t}\int_0^t f(X(0,u,\omega,x))du = \mathbb{E}(f(Y(0,\cdot))|g)(\omega)$$

holds for all bounded continuous f*, where* g *is the* σ*-field of events invariant for the flow* τ_t*. In the ergodic case, the right hand side is* $\int f d\mu$*, which does not depend on* ω*. Thus the empirical distribution of almost any trajectory approaches* μ *in the ergodic case.*

(iv) Let $A =$ *support of* μ *(smallest closed set of measure 1), and let* $x \in \mathbb{R}^d$*. Then dist*$(X(0,t,\omega,x),A) \to 0$ a. s. *when* $t \to \infty$*. Thus* A *is a "black hole" for trajectories: they eventually do not escape at all, not even infrequently.*

Proof. The key is to look at the derivative of the solution with respect to the *initial* time:

$$\left|\frac{X(s+\Delta s,t,\omega,x)-X(s,t,\omega,x)}{\Delta s}\right|$$
$$=\left|\frac{H(s,t,\omega)\circ H(s+\Delta s,s,\omega)(x)-H(s,t,\omega)(x)}{\Delta s}\right|$$
$$\leq \|H(s,t,\omega)\|\left|\frac{H(s+\Delta s,s,\omega)(x)-x}{\Delta s}\right|.$$

But $I=H(s,s,\omega)=H(s+\Delta s,s,\omega)\circ H(s,s+\Delta s,\omega)$ so

$$\left|\frac{H(s+\Delta s,s,\omega)(x)-x}{\Delta s}\right|$$
$$\leq \|H(s+\Delta s,s,\omega)\|\left|\frac{x-H(s,s+\Delta s,\omega)(x)}{\Delta s}\right|,$$

and

$$\frac{x-H(s,s+\Delta s,\omega)(x)}{\Delta s}=\frac{x-X(s,s+\Delta s,\omega,x)}{\Delta s}$$

$\to -F(s,\omega,x)$ as $\Delta s\to 0$ since X satisfies (1.1) starting at x at time s. Also

$$\|H(s,s+\Delta s,\omega)\|\leq \exp\int_s^{s+\Delta s}\|F(u,\omega,\cdot)\|du \quad \text{from (2.1), so}$$

$$\|H(s,s+\Delta s,\omega)\|\to 0 \quad \text{as} \quad \Delta s\to 0. \quad \text{Thus}$$

$$\left|\frac{d}{ds}X(s,t,\omega,x)\right|\leq \|H(s,t,\omega)\|\,|F(s,\omega,x)|.$$

(In the affine case, one computes directly

$$\frac{d}{ds}X(s,t,\omega,x)=-\Phi(s,t,\omega)(A(s,\omega)x+B(s,\omega)).)$$

Strictly speaking, since we only assumed F measurable and not necessarily continuous in t, X is merely absolutely continuous in t (and in s, one may show) and not necessarily differentiable at every point, only Lebesque a. s., but that will be good enough for our purposes.

By hypothesis, for a. a. $\omega,\lambda(\omega)<0$, so using Lemma 1, $\|H(s,t,\omega)\|<\exp(|s|\lambda(\omega)/2)$ for all s sufficiently near $-\infty$.

Consider now the random variables $Z_n(\omega)=\int_{-n-1}^{-n}|F(u,\omega,x)|du$. These are identically distributed by stationarity, and they have finite expectation, since

$$\int Z_0(\omega)dP(\omega)=\int_{-1}^{0}\int|F(u,\omega,x)|dP(\omega)du=\int|F(0,\omega,x)|dP(\omega)<\infty$$

by hypothesis. Thus $\sum_{n=0}^{\infty} P(Z_n > n) = \sum_{n=0}^{\infty} P(Z_0 > n) < \infty$ (see e. g., Billingsley [Bi], pg. 282), so by the Borel–Cantelli lemma, $P(Z_n > n$ i. o.) $= 0$. Thus

$$|X(-n,t,\omega,x) - X(-n-u,t,\omega,x)| \le \int_{-n-u}^{-n} \left|\frac{d}{ds}X(s,t,\omega,x)\right| ds$$

$$\le \int_{-n-u}^{-n} \|H(s,t,\omega)\| |F(s,\omega,x)| ds \le \sum_{k=n}^{\infty} k \exp(k\lambda(\omega)/2)$$

for all n sufficiently large and $u \ge 0$. Thus

$$\lim_{n\to\infty} \sup_{u \ge 0} |X(-n,t,\omega,x) - X(-n-u,t,\omega,x)| = 0,$$

which clearly imples $\lim_{s\to-\infty} X(s,t,\omega,x) = Y(t,\omega)$ exists a. s. Since $X(s,t,\omega,x) = X(s-t,0,\tau_t\omega,x)$, we have from the above limit $Y(t,\omega) = Y(0,\tau_t\omega)$. This proves (i).

To prove (ii), $X(0,t,\omega,Y(0,\omega)) = X(0,t,\omega, \lim_{s\to-\infty} X(s,0,\omega x))$ $= \lim_{s\to-\infty}(s,t,\omega,x) = Y(t,\omega)$.

To prove (iii), note that for any x, $X(s,t,\omega,x)$ has the same *distribution* as $X(s-t,0,\omega,x)$ by stationarity; the latter converges *pointwise* (hence also in distribution) as $t\to\infty$ to $Y(0,\cdot)$, so the former converges in *distribution* (but not pointwise) to $Y(0,\cdot)$ also. Since for any x,y, $d(X(s,t,\omega,x), X(s,t,\omega,y)) \to 0$ as $t\to\infty$ by the negative Furstenberg–Kesten number (i.e., maximum Lyapunov exponent in affine case), it is easy to show the same holds if x is random with any distribution.

To get the emprirical convergence of trajectories, since $X(0,t,\omega,Y(0,\omega))$ $= Y(t,\omega)$ is stationary,

$$\lim_{t\to\infty} \frac{1}{t}\int_0^t f(X(0,u,\omega,Y(0,\omega))du = \mathbb{E}(f(Y(0,\cdot))|g)(\omega) \quad a.s.$$

by Birkhoff's pointwise ergodic theorem. But

$$|X(0,u,\omega,Y(0,\omega)) - X(0,u,\omega,x)| \le \|H(0,u,\omega)\| \, |Y(0,\omega) - x|$$

$\to 0$ a. s. as $u\to\infty$ since $\lambda(\omega) < 0$ a. s. and $\frac{log\|H(0,u,\omega)\|}{u} \to \lambda(\omega)$. Thus if f is uniformly continuous, $f(X(0,u,\omega,Y(0,\omega))$ is uniformly close to $f(X(0,u,\omega,x))$ for all sufficiently large u, so clearly

$$\lim_{t\to\infty} \frac{1}{t}\int_0^t f(X(0,u,\omega,x))du = \mathbb{E}(f(Y(0,\cdot))|g)(\omega) a.\ s.$$

also. Since the continuous functions with compact support (which are uniformly continuous) are a separable set in the sup norm topology, it is easy to see that this holds for all such f simultaneously; and then one gets this for all bounded continuous f by an approximation using Urysohn's lemma. This proves (iii).

Finally, to prove (iv), $P(Y(t,\omega) \in A) = \mu(A) = 1$ for all t, so for a. a. $\omega, Y(t,\omega) \in A$ for all $t \in Q$ (the rationals) *simultaneously* since Q is countable. Thus for $t \in Q$ and a. a. ω,

$$\begin{aligned}\text{dist}(X(0,t,\omega,x),A) &\leq \text{dist}(X(0,t,\omega,x),Y(t,\omega))\\ &= \text{dist}(X(0,t,\omega,x),X(0,t,\omega,Y(0,\omega))\\ &\leq \|H(0,t,\omega)\|\text{dist}(x,Y(0,\omega))\end{aligned}$$

$\to 0$ as $t \to \infty$, $t \in Q$, since $\lambda(\omega) < 0$. But $X(0,t,\omega,x)$ is continuous in t so the restriction to Q is not needed. □

REFERENCES

[AK] L. Arnold and W. Kliemann, *Qualitative theory of stochastic systems*, Probabilistic Analysis and Related Topics, Academic Press, **3** (1983), 1–79 pp.

[BE] M. Barnsley and J. Elton, *A new class of Markov processes for image encoding*, Adv. Appl. Probab. **20** (1988), 14–32.

[BEH] M. Barnsley, J. Elton, and D. Hardin, *Recurrent iterated function systems*, Constructive Approximations, **5** (1989), 3–31.

[BS] M. Barnsley and A. Sloan, *A better way to compress images*, Byte Magazine, January 1988 b, 215–223.

[BA] M. Berger and Y. Amit, *Products of random affine maps*, preprint.

[Bi] P. Billingsley, *Probability and Measure*. Wiley (1986).

[C] H. Crauel, *Random Dynamical Systems*, Report 175 (1988), Institüt für Dynamische Systeme, Bremen.

[E] J. Elton, *A multiplicative ergodic theorem for Lipschitz maps*, Stochastic Processes and their Applications, **34** (1990), 39–47.

[K] U. Krengel, *Ergodic Theorems*. de Gruyter (1985).

John Elton,
Iterated Systems, Inc.,
5550-A Peachtree Parkway, Suite 545,
Norcross, GA 30092, USA

Jelel Ezzine,
Systems Engineering Dept., KFUPM,
Dhahran, 31261, Saudi Arabia.

Random Affine Iterated Function Systems: Mixing and Encoding*

Marc A. Berger

Introduction

This paper is concerned with a probabilistic algorithm for image generation. The simplest form of the algorithm is illustrated in Fig. 1. The leaf is generated as follows. Pick any point $\mathbf{X}_0 \in \mathbb{R}^2$. There are four affine transformations $T : \mathbf{x} \to A\mathbf{x} + \mathbf{b}$ listed on top of this Fig., and four probabilities p_i underneath them. Choose one of these transformations at random, according to the probabilities p_i — say T_k is chosen, and apply it to $\mathbf{X}_0$, thereby obtaining $\mathbf{X}_1 = T_k\mathbf{X}_0$. Then choose a transformation again at random, independent of the previous choice, and apply it to $\mathbf{X}_1$, thereby obtaining $\mathbf{X}_2$. Continue in this fashion, and plot the orbit $\{\mathbf{X}_n\}$. The result is the leaf shown. By tabulating the frequencies with which the points $\mathbf{X}_n$ fall into the various pixels of the graphics window, one can actually plot the empirical distribution $\frac{1}{n+1}\sum_{k=0}^{n}\delta_{\mathbf{X}_k}$, using a grey scale to convert statistical frequency to color. The darker portions of the leaf correspond to high probability density.

One framework for this process is in terms of image encoding and compression. The coefficients of the transformations T_i and the probabilities p_i represent an encoding of the leaf image, and the above algorithm for generating the leaf is the decoding procedure. The most significant problem in this area is the inverse, or encoding problem of finding the parameters to encode a given target image.

In §1 the analysis of the basic algorithm is presented. The roles of the Lyapunov exponent and the "index-reversed" process are developed. In §2 the basic algorithm is extended to allow for "mixing" of such processes $\{\mathbf{X}_n\}$ as the one described above. This involves a general analysis of systems of convolution equations, and results in a broader class of images. In §3 convexity properties of the attractor are developed. These properties give information about the transformations T_i when one is given the image.

§1. Some Background

Let $G = G_m$ be the semi-group of affine transformations $g : \mathbf{x} \mapsto a\mathbf{x} + \mathbf{b}$ from $\mathbb{R}^m \to \mathbb{R}^m$. We denote by $a = a(g)$ the linear part of g, and by $\mathbf{b} = \mathbf{b}(g)$ the translational

* Research supported by Air Force Office of Scientific Research, Grant No. AFOSR-87-0137.

part of g. Let $\mu \in \mathcal{P}(G)$, the collection of Borel probabilities on G. Given $\nu \in \mathcal{P}(\mathbb{R}^m)$ we define the *convolution* $\mu * \nu$ as follows:

$$\mu * \nu(B) = \int \nu(g^{-1}B)\mu(dg) \tag{1}$$

for all Borel subsets $B \subseteq \mathbb{R}^m$. Equivalently if $g \in G$ is distributed like μ, if $\mathbf{X} \in \mathbb{R}^m$ is distributed like ν and if g and $\mathbf{X}$ are independent, then $\mu * \nu$ is the distribution of $g\mathbf{X}$. If $\nu \in \mathcal{P}(\mathbb{R}^m)$ satisfies

$$\mu * \nu = \nu \tag{2}$$

then we say that ν is μ-stationary.

Let $\nu \in \mathcal{P}(\mathbb{R}^m)$ be μ-stationary, and let C be the support $C = \mathrm{supp}(\nu)$; i.e.,

$$C = \cap\{F \subseteq \mathbb{R}^m \text{closed } : \nu(F) = 1\}$$
$$= \{\mathbf{x} \in \mathbb{R}^m : \text{every neighborhood of } \mathbf{x} \text{ has positive } \mu\text{-measure}\}.$$

Then by setting $B = C$ in (1) it follows from (2) that $\mu(G') = 1$, where $G' = \{g \in G : g^{-1}C \supseteq C\}$. In particular $G' \supseteq H$ where $H = \mathrm{supp}(\mu)$, since G' is closed in G. Thus

$$C \supseteq \overline{\bigcup_{g \in G'} gC} \supseteq \overline{\bigcup_{g \in H} gC}.$$

On the other hand by setting $B = \overline{\bigcup_{g\in H} gC}$ in (1) we find from (2) that $\nu(B) = 1$, so that $B \supseteq C$. Thus we conclude that C satisfies the *self-covering* property

$$C = \overline{\bigcup_{g \in H} gC}. \tag{3}$$

Let $H \subseteq G$. Say that a *non-empty closed* subset $S \subseteq \mathbb{R}^m$ is (i) *invariant* (under H) if $\bigcup_{g\in H} gS \subseteq S$; (ii) *self-covering* (wrt H) if $\overline{\bigcup_{g\in H} gS} = S$; and (iii) *minimal invariant* (wrt H) if it is invariant, but no proper (closed) subset of it is invariant. Observe that minimal invariant sets are self-covering, but the converse does not necessarily hold.

One case where we can be sure of the existence and uniqueness of a minimal invariant set is when H contains a strictly contractive transformation g. In this case every invariant set must contain the fixed point of g, and so the intersection of all invariant sets — being non-empty, is the only minimal invariant set. Below we shall learn of a more general condition on H under which one can assert the uniqueness of minimal invariant sets.

If the transformations in H are all uniformly strictly contractive and bounded, so that

$$\alpha = \sup_{g\in H} |a(g)| < 1, \qquad \beta = \sup_{g\in H} |\mathbf{b}(g)| < \infty, \tag{4}$$

then we can ensure the existence and uniqueness of a *compact* self-covering set. (N.B. The matrix norm is intended to be the operator norm which corresponds to the vector norm.)

Lemma I. *Under assumptions (4) there is exactly one non-empty compact subset $C \subseteq \mathbb{R}^m$ which is self-covering wrt H.*

Proof. To show the existence of C simply observe that the closed ball $B\left(0, \dfrac{\beta}{1-\alpha}\right)$ is invariant, and hence the intersection of all invariant sets must be bounded. To show the uniqueness of C we argue as follows. Suppose that the non-empty compact set $S \subseteq \mathbb{R}^m$ is also self-covering. Given $\epsilon > 0$ for any $\mathbf{y} \in S$ there exist $g \in H$ and $\mathbf{x} \in S$ such that $|\mathbf{y} - g\mathbf{x}| \leq \epsilon$. Thus

$$d(\mathbf{y}, C) \leq d(\mathbf{y}, gC) \leq d(g\mathbf{x}, gC) + \epsilon \leq \alpha d(\mathbf{x}, C) + \epsilon.$$

Thus $d(S, C) \leq \alpha d(S, C)$, and since $d(S, C) < \infty$ and $\alpha < 1$ it follows that $d(S, C) = 0$, so that $S \subseteq C$. □

Sets in $\mathbb{R}^2$ can be thought of as black and white (i.e., binary) images. In this respect self-covering sets in $\mathbb{R}^2$ are fractals, and the self-covering property (3) amounts to what Barnsley [1] refers to as the "collage," whereby C is covered by affine copies of itself. This is illustrated in Figs. 2 and 3 where, in each case, H consists of the transformations listed. Observe in Fig. 2 how C, the grey leaf, is covered by the four black leaves — each of which is an affine copy of C. This *collage property* is useful for encoding, since each of the black copies of C readily determines the corresponding affine transformation. It also shows that the support, or shape of the image only depends on the affine transformations in H, and not on the probabilities they get assigned. These probabities only affect the coloration of the image. They can be used, say, to shift around the high and low density parts of the image.

Back to our original setting suppose now that $\mu \in \mathcal{P}(G)$ has support H. If $\nu \in \mathcal{P}(\mathbb{R}^m)$ is μ-stationary then we see from (3) that C is self-covering (wrt H). Our general interest lies in the following type of problem. Given an i.i.d. sample $\{g_n\}$ from μ, construct a (simple) stochastic process $\{X_n\}$ whose empirical distributions $\frac{1}{n+1}\sum_{k=0}^{n}\delta_{\mathbf{X}_k}$ converge weakly to a μ-stationary probability, a.s. This is equivalent to requiring that with probability one the plot of the orbit of any *single* trajectory $\{\mathbf{X}_n\}$ will produce the desired image. In this case there would be no need to ever run more than one trajectory. Of course we will be working in settings where there exist μ-stationary probabilities, and we discuss this point first.

In Furstenberg and Kesten [8] it was shown that if $\{a_n\}$ is an i.i.d. sequence of $m \times m$ matrices satisfying $\mathbb{E}\log^+|a_1| < \infty$, then the limit

$$\lim_{n\to\infty}\frac{1}{n}\log|{}_na_1| = \gamma \tag{5}$$

exists and is constant a.s. ($-\infty \leq \gamma < \infty$). Here ${}_na_1$ denotes the ordered product $a_n \cdots a_1$. The number $\rho = e^\gamma$ can be thought of as the "spectral radius" of the distribution of a_1 — see Kingman [10] for a discussion of this constant. Furthermore along with (5) it was shown in [8] that

$$\lim_{n\to\infty}\frac{1}{n}\,\mathbb{E}\log|{}_na_1| = \inf_n \frac{1}{n}\,\mathbb{E}\log|{}_na_1| = \gamma. \tag{6}$$

In our setting we want the probability measure $\mu \in \mathcal{P}(G)$ to satisfy the following integrability assumptions:

(A_1) $\quad \int \left[\log^+|a(g)| + \log^+|\mathbf{b}(g)|\right]\mu(dg) < \infty.$

On account of (A_1) the constant γ defined above in (5) exists, where we take $a_n = a(g_n)$. Our next requirement is on γ:

(A_2) $\gamma < 0$.

This assumption guarantees that the iterates ${}_na_1$ converge to zero exponentially fast as $n \to \infty$, a.s. That is, for almost every ω, given $\gamma < -\kappa < 0$ there exists $N(\omega)$ such that

$$|{}_na_1(\omega)| \leq e^{-\kappa n}, \quad \forall n \geq N(\omega).$$

We refer to this assumption as the ***average contractivity*** condition. It certainly obtains whenever the transformations in H are uniformly strictly contractive — but this is not a necessary condition. On account of (6) condition (A_2) is equivalent to:

(A_2') For some n, $\mathbb{E}\log|{}_na_1| < 0$.

Given $\mu \in \mathcal{P}(G)$ there is a natural Markov chain $\{\mathbf{X}_n\}$ on $\mathbb{R}^m$ *associated with it.* Let $\{g_n\}$ be an i.i.d. sample from μ, and define

$$\mathbf{X}_{n+1} = g_{n+1}\mathbf{X}_n, \quad n \geq 0. \tag{7}$$

As the next result will show, this process has the property we seek; namely, with probability one, the empirical distributions of any (single) trajectory converge weakly to a μ-stationary probability.

Proposition II. *Under assumptions (A_1), (A_2') the distribution $\mu \in \mathcal{P}(G)$ has the following properties.*

(i) UNIQUE μ-STATIONARY DISTRIBUTION: There exists a unique $\nu \in \mathcal{P}(\mathbb{R}^m)$ which is μ-stationary.

(ii) ASYMPTOTIC STATIONARITY: For any $\nu_0 \in \mathcal{P}(\mathbb{R}^m)$, $\mu^{(n)} * \nu_0 \Rightarrow \nu$, where $\mu^{(n)}$ denotes the n-fold convolution $\mu^{(n)} = \underbrace{\mu * \cdots * \mu}_{n \text{ times}}$. (The double arrow symbol "$\Rightarrow$" denotes convergence in distribution.) *Equivalently if $\{\mathbf{X}_n\}$ is the Markov chain associated with μ, then for an arbitrary distribution on $\mathbf{X}_0$ there holds $\mathbf{X}_n \Rightarrow \mathbf{X}$ where $\mathbf{X}$ is distributed like ν.*

(iii) LAW OF LARGE NUMBERS: With probability one, for any bounded continuous $f : \mathbb{R}^m \to \mathbb{R}$

$$\frac{1}{n+1}\sum_{k=0}^{n} f(\mathbf{X}_k) \to \int f d\nu,$$

where $\mathbf{X}_0$ has an arbitrary distribution. Equivalently, with probability one the empirical distributions $\frac{1}{n+1}\sum_{k=0}^{n} \delta_{\mathbf{X}_k}$ converge weakly to ν.

The proof of this Proposition relies on the following result.

Lemma III. *Let $\{\mathbf{X}_n\}$ be a Markov chain on $\mathbb{R}^m$ obeying the "coupling condition"*

$$\lim_{n\to\infty}(\mathbf{X}_n^{\mathbf{x}} - \mathbf{X}_n^{\mathbf{y}}) = 0 \text{ a.s.} \tag{8}$$

for any $\mathbf{x}, \mathbf{y} \in \mathbb{R}^m$, where $\{\mathbf{X}_n^{\mathbf{x}}\}$ denotes the chain starting at $\mathbf{x}$. Suppose this chain has a stationary distribution $\nu \in \mathcal{P}(\mathbb{R}^m)$. Then ν is the unique stationary distribution for

$\{\mathbf{X}_n\}$; *and for an arbitrary distribution on* $\mathbf{X}_0$, *with probability one for any bounded continuous* $f : \mathbb{R}^m \to \mathbb{R}$

$$\lim_{n\to\infty} \frac{1}{n+1} \sum_{k=0}^{n} f(\mathbf{X}_k) = \int f d\nu.$$

Proof. Let $\mathbf{x} \in \mathbb{R}^m$ and let $f : \mathbb{R}^m \to \mathbb{R}$ be continuous with *compact support.* Assume without loss of generality that ν is such that the chain $\{\mathbf{X}_n^\nu\}$, with $\mathbf{X}_0$ distributed like ν, is actually stationary *ergodic.* (Such ν's are the extreme points of the set of stationary distributions for $\{\mathbf{X}_n\}$ — cf. Breiman [7, Prob. 6.11]). It follows from the Ergodic Theorem that

$$\lim_{n\to\infty} \frac{1}{n+1} \sum_{k=0}^{n} f(\mathbf{X}_k^\nu) = \int f d\nu \text{ a.s.}$$

It follows from (8) that

$$\mathbb{P}\left(\lim_{n\to\infty}(\mathbf{X}_n^\mathbf{x} - \mathbf{X}_n^\nu) = 0\right) = \int \mathbb{P}\left(\lim_{n\to\infty}(\mathbf{X}_n^\mathbf{x} - \mathbf{X}_n^\mathbf{y}) = 0\right) \nu(dy) = 1.$$

Thus since f is *uniformly* continuous it follows that

$$\lim_{n\to\infty} \frac{1}{n+1} \sum_{k=0}^{n} f(\mathbf{X}_k^\mathbf{x}) = \int f d\nu \text{ a.s.}$$

From this we get the result for an arbitrary initial distribution on $\mathbf{X}_0$. Since the space of continuous functions with compact support is both separable and convergence-determining for $\mathcal{P}(\mathbb{R}^m)$ (Billingsley [6, Prob. 6.7]), we conclude that indeed

$$\frac{1}{n+1} \sum_{k=0}^{n} \delta_{\mathbf{X}_k} \Rightarrow \nu \text{ a.s.} \qquad \square$$

Proof of Prop. II. Since

$$\mathbf{X}_n^\mathbf{x} - \mathbf{X}_n^\mathbf{y} = {}_n a_1(\mathbf{x} - \mathbf{y})$$

assumption (A_2) guarantees that our chain $\{\mathbf{X}_n\}$ from (7) obeys the coupling condition (8), and so in order to prove the Proposition one needs to construct some stationary distribution ν. To this end let $\mathbf{x} \in \mathbb{R}^m$ and consider a new process $\{\widehat{\mathbf{X}}_n^\mathbf{x}\}$ evolving as

$$\widehat{\mathbf{X}}_n^\mathbf{x} = {}_1 g_n \mathbf{x}. \tag{9}$$

Recall that ${}_1 g_n$ denotes the product $g_1 \cdots g_n$. Observe that in distinction to (7) the successive transformations g_n are applied from the inside here, rather than from the outside. This process $\{\widehat{\mathbf{X}}_n\}$ is called the *index-reversed process*, since one can think of it as

$$\widehat{\mathbf{X}}_n^\mathbf{x} = {}_{-1} g_{-n} \mathbf{x},$$

where $\{g_n : n \in \mathbb{Z}\}$ is now a two-sided sequence. (This becomes more significant in the mixed setting.) The process $\{\widehat{\mathbf{X}}_n^\mathbf{x}\}$ is no longer Markov, in general — but it has two important properties.

(a) *For each fixed n, $\mathbf{X}_n^{\mathbf{x}}$ and $\widehat{\mathbf{X}}_n^{\mathbf{x}}$ have the same distribution.*
(b) $\lim_{n\to\infty} \widehat{\mathbf{X}}_n^{\mathbf{x}} = \mathbf{X}^{\mathbf{x}}$ *exists* a.s.
Property (a) is immediate. To establish (b) one argues as follows. On account of the order in which the g_i's are applied,

$$\widehat{\mathbf{X}}_{n+1}^{\mathbf{x}} - \widehat{\mathbf{X}}_n^{\mathbf{x}} = {}_1a_n(g_{n+1}\mathbf{x} - \mathbf{x}).$$

Since $\mathbb{E}\,\log^+|g_1\mathbf{x} - \mathbf{x}| < \infty$ it follows from the Law of Large Numbers that

$$\lim_{n\to\infty} \frac{1}{n} \log^+|g_{n+1}\mathbf{x} - \mathbf{x}| = 0 \text{ a.s.} \tag{10}$$

Furthermore by (A_2)

$$\lim_{n\to\infty} \frac{1}{n}|{}_1a_n| < 0 \text{ a.s.} \tag{11}$$

Together (10) and (11) imply that

$$\limsup_{n\to\infty} \frac{1}{n} \log\left|\widehat{\mathbf{X}}_{n+1}^{\mathbf{x}} - \widehat{\mathbf{X}}_n^{\mathbf{x}}\right| < 0 \text{ a.s.}$$

so that $\sum_n \left|\widehat{\mathbf{X}}_{n+1}^{\mathbf{x}} - \widehat{\mathbf{X}}_n^{\mathbf{x}}\right| < \infty$ a.s. This establishes (b). Observe further that for any $\mathbf{y} \in \mathbb{R}^m$,

$$\widehat{\mathbf{X}}_n^{\mathbf{x}} - \widehat{\mathbf{X}}_n^{\mathbf{y}} = {}_1a_n(\mathbf{x} - \mathbf{y}) \to 0 \text{ as } n \to \infty \text{ a.s.}$$

Thus for any $\mathbf{y} \in \mathbb{R}^m$, $\lim_{n\to\infty} \widehat{\mathbf{X}}_n^{\mathbf{y}} = \mathbf{X}$ a.s., where $\mathbf{X} = \widehat{\mathbf{X}}^{\mathbf{x}}$.

The distribution ν, of $\mathbf{X}$ is μ-invariant. Thus on account of property (a) above $\mathbf{X}_n^{\mathbf{y}} \Rightarrow \mathbf{X}$ for any $\mathbf{y} \in \mathbb{R}^m$. This establishes (ii) and the rest of the Proposition follows from the Lemma above. □

Corollary IV. *Under assumptions (A_1), (A_2') there exists a unique minimal invariant set wrt $H = \text{supp}(\mu)$; namely $C = \text{supp}(\nu)$.*

Proof. Simply observe that on account of Proposition II, every invariant set under H must contain C. □

Given $H \subseteq G$ we do not know of necessary and sufficient conditions for the existence of $\mu \in \mathcal{P}(G)$ supported on H and satisfying (A_2'), even in the case when H is a finite set. Clearly, though, a sufficient condition is that H contain a strictly contractive transformation g, since then we can arrange for (A_2') to be satisfied with $n = 1$ by choosing μ close to the point mass δ_g. We move on now to describe the *mixing*.

Consider the image in Fig. 4. This image is not self-covering, but rather it satisfies

$$C = \bigcup_{i=1}^{3} T_i C \cup C', \tag{12}$$

where the T_i's are the affine transformations listed in Fig. 3 and C' is the image of the leaf in the center. Thus mixing can be achieved by allowing other "shapes" into the collage. The following Lemma explains how a condition like (12) can be arrived at.

Lemma V. *Let $\{\mathbf{X}_n\}$ be a Markov chain with initial distribution ν' and transition probabilities $P(\mathbf{x}, d\mathbf{y})$. Let $0 < p < 1$ and let σ be a random variable independent of $\{\mathbf{X}_n\}$ with distribution $\mathbb{P}(\sigma = k) = pq^k$, $k \geq 0$. Then the distribution ν satisfies*

$$\nu(d\mathbf{y}) = q \int P(\mathbf{x}, d\mathbf{y})\nu(d\mathbf{x}) + p\nu'(d\mathbf{y}) \tag{13}$$

if and only if ν is the distribution of $\mathbf{X}_\sigma$.

Proof. Sufficiency. Simply observe that

$$\mathbb{P}(\mathbf{X}_\sigma \in B) = q\mathbb{P}(\mathbf{X}_{\sigma+1} \in B) + p\mathbb{P}(\mathbf{X}_0 \in B).$$

Necessity. Iterate, obtaining

$$\nu(B) = \sum_{k=0}^{n-1} pq^k \mathbb{P}(\mathbf{X}_k \in B) + q^n \int P^{(n)}(\mathbf{x}, B)\nu(d\mathbf{x}),$$

where $P^{(n)}$ is the n-step transition probability. □

To see how this result applies to our setting, suppose that $P(\mathbf{x}, d\mathbf{y})$ is generated from a probability measure $\mu \in \mathcal{P}(G)$; i.e.,

$$P(\mathbf{x}, d\mathbf{y}) = \mu(g\mathbf{x} \in d\mathbf{y}).$$

Then the distribution of $\mathbf{X}_\sigma$ is $\nu = \mathbb{E}\mu^{(\sigma)} * \nu'$ and (13) becomes

$$\nu = q\mu * \nu + p\nu'. \tag{14}$$

Consider now the supports $C = \text{supp}(\nu)$ and $C' = \text{supp}(\nu')$. Applying (14) to the sets C and $\overline{\bigcup_{g \in H} gC} \cup C'$, where $H = \text{supp}(\mu)$, we arrive at

$$C = \overline{\bigcup_{g \in H} gC} \cup C',$$

which is what we were looking for; namely, a collage of C made out of pieces shaped like C and C'.

The key question, however, is how to generate ν empirically. The answer to this question, in the case of interest here where ν' is μ'-stationary for some $\mu' \in \mathcal{P}(G)$, was described in [2]. There the following "two-screen" setup was constructed for carrying out the mixing. Let $\{g_n\}$ and $\{g'_n\}$ be i.i.d. samples from μ and μ', respectively — these samples being independent of one another. Let $\{\mathbf{X}'_n\}$ be the Markov chain associated with μ', and let $\{\mathbf{X}_n\}$ be constructed as follows. Take $\{I_n\}$ to be a sequence of i.i.d. Bernoulli's with

$$\mathbb{P}(I_1 = 0) = q, \quad \mathbb{P}(I_1 = 1) = p,$$

the I's being independent of all the g's. Then

$$\mathbf{X}_{n+1} = \begin{cases} g_{n+1}\mathbf{X}_n & \text{if } I_{n+1} = 0, \\ \mathbf{X}'_n & \text{if } I_{n+1} = 1. \end{cases}$$

The pair process $\{(\mathbf{X}_n, \mathbf{X}'_n)\}$ is a Markov chain on $\mathbb{R}^m \times \mathbb{R}^m$, but $\{\mathbf{X}_n\}$ by itself is not. In fact as far as the $\{\mathbf{X}_n\}$ "world" is concerned, every time $I_{n+1} = 1$ so that $\mathbf{X}_{n+1}$ gets set to $\mathbf{X}'_n$, this point in the trajectory seems to "fall out of the sky." We envision this as a two-screen setup, whereby each of the chains $\{\mathbf{X}_n\}$ and $\{\mathbf{X}'_n\}$ lives on a separate "screen" (i.e., copy of $\mathbb{R}^m$), with screen 2 feeding into screen 1 at geometrically distributed times. This is illustrated in Fig. 5, and Fig. 6 is the flow-chart for this algorithm.

Proposition VI. *Assume that μ' satisfies (A_1), (A_2') and that μ satisfies (A_1). Let ν' be the unique μ'-stationary distribution, as guaranteed by Proposition II.*

(i) *UNIQUE μ-STATIONARY DISTRIBUTION: There exists a unique $\nu \in \mathcal{P}(\mathbb{R}^m)$ satisfying (14).*

(ii) *ASYMPTOTIC STATIONARITY: For an arbitrary distribution on $\mathbf{X}_0$ there holds $\mathbf{X}_n \Rightarrow \mathbf{X}$, where $\mathbf{X}$ is distributed like ν.*

(iii) *LAW OF LARGE NUMBERS: With probability one the empirical distributions $\frac{1}{n+1}\sum_{k=0}^{n} \delta(\mathbf{X}_k)$ converge weakly to ν.*

Of significance here is that μ need not satisfy the average contractivity condition (A_2). Thus in Fig. 7 μ is supported on a single 90° rotation. Something like this would not have worked without the mixing, since the Markov chain associated with μ alone is not ergodic.

Proof. The existence and uniqueness of ν follows from Lemma V. To establish (ii) we construct an index-reversed process, as in the proof of Proposition II above. Let $\{\tilde{g}'_n\}$ be another i.i.d. sample from μ', independent of the I's and the other g's. Let $\{\tilde{\mathbf{X}}'_n\}$ be the index-reversed process constructed above

$$\tilde{\mathbf{X}}'_n = {}_1\tilde{g}'_n\mathbf{x}'.$$

Define

$$\hat{\mathbf{X}}_n = \begin{cases} {}_1g_n\mathbf{x}, & \text{if } \sigma \geq n, \\ {}_1g_\sigma\tilde{\mathbf{X}}'_{n-\sigma-1} & \text{if } \sigma < n, \end{cases}$$

$$\hat{\mathbf{X}}'_n = \begin{cases} {}_1g'_n\mathbf{x}', & \text{if } \sigma \geq n, \\ {}_1g'_\sigma\tilde{\mathbf{X}}'_{n-\sigma-1} & \text{if } \sigma < n, \end{cases}$$

where, as above, σ is independent of the g's with $\mathbb{P}(\sigma = k) = pq^k$, $k \geq 0$. It can be seen that for any fixed n, $(\mathbf{X}_n, \mathbf{X}'_n)$ and $(\hat{\mathbf{X}}_n, \hat{\mathbf{X}}'_n)$ have the same distribution. Since

$$\lim_{n\to\infty}(\hat{\mathbf{X}}_n, \hat{\mathbf{X}}'_n) = ({}_1g_\sigma\mathbf{X}',\ {}_1g'_{\sigma+1}\mathbf{X}') \text{ a.s.}$$

where $\mathbf{X}' = \lim_{n\to\infty}\tilde{\mathbf{X}}'_n$, we establish (ii) as in the proof of Proposition II above. Observe that we have also exhibited a stationary distribution for the chain $\{(\mathbf{X}_n, \mathbf{X}'_n)\}$; namely, that of $({}_1g_\sigma\mathbf{X}',\ {}_1g'_{\sigma+1}\mathbf{X}')$.

Next set $I_0 = 1$ and set $S_n = \max(k \leq n : I_k = 1)$. Observe that

$$\mathbf{X}_n = \begin{cases} {}_ng_1\mathbf{X}_0, & \text{if } S_n = 0, \\ {}_ng_{S_n+1}\mathbf{X}'_{S_n-1}, & \text{if } S_n > 0. \end{cases}$$

It follows from the Borel-Cantelli Lemma that $\frac{S_n}{n} \to 1$ a.s., and thus

$$\frac{1}{n}\log^+|{}_na_{S_n+1}| \leq \frac{1}{n}\sum_{k=S_n+1}^{n}\log^+|a_k| \to 0 \text{ a.s.}$$

From this we conclude that for any $\mathbf{x}, \mathbf{y} \in \mathbb{R}^m$ there holds $\lim_{n\to\infty}(\mathbf{X}_n^{\mathbf{x}} - \mathbf{X}_n^{\mathbf{y}}) = 0$ a.s. Thus our chain $\{(\mathbf{X}_n, \mathbf{X}'_n)\}$ satisfies the coupling condition (8), and so we conclude as we did above that (iii) holds. □

Of course we could have set things up on $\mathbb{R}^{2m}$ by introducing a compound probability $\mu^* \in \mathcal{P}(G_{2m})$ corresponding to the random transformation

$$g^* : (\mathbf{x}, \mathbf{x}') \mapsto \begin{cases} (g\mathbf{x}, g'\mathbf{x}'), & \text{if } I = 0, \\ (\mathbf{x}', g'\mathbf{x}'), & \text{if } I = 1; \end{cases}$$

where g, g' are independent and distributed like μ, μ', respectively, and I is independent of them with $\mathbb{P}(I = 0) = q$, $\mathbb{P}(I = 1) = p$. But then the appropriate assumptions to make on μ, μ' would not have been as apparent. In any event it is clear that ν is just the marginal of a μ^*-stationary probability $\nu^* \in \mathcal{P}(\mathbb{R}^{2m})$.

It's clear that the mixing described above extends naturally to any number of screens with any connections between them. This is illustrated in Fig. 8, and can be described in general as follows. To construct an N-screen setup let $\mu_{ij} \in \mathcal{P}(G)$ be given, $1 \le i, j \le N$, and let (p_{ij}) be a transition probability matrix. Let $\{g_n(i,j)\}$ be i.i.d. samples from μ_{ij}, the N^2 samples being mutually independent. Let $\{I_n(i)\}$ be i.i.d. sequences, to be used for the "switching," with

$$\mathbb{P}(I_n(i) = j) = p_{ij}, \quad 1 \le j \le N.$$

The N sequences $\{I_n(i)\}$ are to be mutually independent, and independent of all the g's. Consider the Markov chain $\{(\mathbf{X}_n(1), \ldots, \mathbf{X}_n(N)\}$ on $\underbrace{\mathbb{R}^m \times \ldots \times \mathbb{R}^m}_{N \text{ times}}$ evolving as

$$X_{n+1}(i) = g_{n+1}(i,j)\mathbf{X}_n(j), \quad \text{where } j = I_{n+1}(i).$$

That is, on screen i we update $\mathbf{X}_n(i)$ by applying $g_{n+1}(i,j)$ to the point on screen j, where $j = I_{n+1}(i)$. If $\nu \in \mathcal{P}(\mathbb{R}^{Nm})$ is to be stationary for this chain, then its marginals $\nu_i \in \mathcal{P}(\mathbb{R}^m)$ must satisfy the *system of convolution equations*

$$\nu_i = \sum_{j=1}^{N} p_{ij}\mu_{ij} * \nu_j, \quad 1 \le i \le N. \tag{15}$$

From this follows that if $C_i = \operatorname{supp}(\nu_i)$, then

$$C_i = \bigcup_{j:p_{ij}>0} \overline{\bigcup_{g \in H_{ij}} gC_j}$$

where $H_{ij} = \operatorname{supp}(\mu_{ij})$. This is the *generalized collage property*, whereby each of the C_i's is covered by pieces whose shapes are allowed to resemble some subset of the shapes for $C_1, \ldots, C_N$. (This subset depends on i.) Furthermore the "index-reversed" process is easy to describe. Let $\{\Gamma_n = \Gamma_n^i\}$ be a Markov chain $\{1, \ldots, N\}$ starting at i with transition matrix (p_{ij}). Then $\mathbf{X}_n(i)$ has the same distribution as

$$\widehat{\mathbf{X}}_n(i) = g_1(\Gamma_0, \Gamma_1)g_2(\Gamma_1, \Gamma_2)\cdots g_n(\Gamma_{n-1}, \Gamma_n)\mathbf{X}_0(\Gamma_n).$$

For $j_1, \ldots, j_N \in \{1, \ldots, N\}$ let $g^*(j_1, \ldots, j_N)$ be the random affine transformation $\mathbb{R}^{Nm} \to \mathbb{R}^{Nm}$ defined by

$$g^*(j_1, \ldots, j_N) : (\mathbf{x}_1, \ldots, \mathbf{x}_N) \mapsto (g(1, j_1)\mathbf{x}_{j_1}, \ldots, g(N, j_N)\mathbf{x}_{j_N})$$

where $g(i,j)$ is distributed like μ_{ij}, and the $g(i,j)$'s are independent. Then $\{\mathbf{X}_n(1), \ldots, \mathbf{X}_n(N)\}$ is the Markov chain associated with $\mu^* \in \mathcal{P}(\mathrm{IR}^{Nm})$, which is the (compound) distribution corresponding to $g^* \in G_{Nm}$ defined by

$$g^* = g(j_1, \ldots, j_N) \text{ with probability } p_{1j_1} \cdots p_{Nj_N}.$$

In [4] a (seemingly) different phenomena, *recurrent* IFS was studied. This is motivated by the symbolic dynamics on code space corresponding to the random products. The recurrent structure enables one to specify rules as to which transformations are allowed to follow which in the IFS evolution. The model goes as follows. Suppose $\mu_i \in \mathcal{P}(G)$ are given, $1 \leq i \leq N$, and let $\{\Gamma_n\}$ be a Markov chain with transition matrix (p_{ij}). Let $\{g_n(i)\}$ be i.i.d. samples from μ_i, these N samples being independent, and consider the process

$$\mathbf{X}_{n+1} = g_{n+1}(\Gamma_{n+1})\mathbf{X}_n.$$

Observe how the transition probabilities p_{ij} determine which transformations can follow which, since transformations $g(j)$ from μ_j are allowed to follow transformations $g(i)$ from μ_i only if $p_{ij} > 0$. Again this process is not in general Markovian, but the pair process $\{(\mathbf{X}_n, \Gamma_n)\}$ is. A stationary distribution $\nu \in \mathcal{P}(\mathrm{IR}^m \times \{1, \ldots, N\})$, corresponding to joint random variables $(\mathbf{X}, \Gamma)$, would have to satisfy

$$\nu_i = \sum_{j=1}^{N} \widehat{p}_{ij}\mu_i * \nu_j, \quad 1 \leq i \leq N \tag{16}$$

where the ν_i is the conditional distribution of $\mathbf{X}$, given $\Gamma = i$. Precisely

$$\nu_i(B) = \frac{\nu(B \times \{i\})}{\pi_i},$$

and $\pi_i = \nu(\mathrm{IR}^m \times \{i\})$ are stationary probabilities for $\{\Gamma_n\}$, satsisfying

$$\pi_i = \sum_{j=1}^{N} p_{ji}\pi_j, \quad 1 \leq i \leq N.$$

In order for there to exist positive π_i's satisfying this condition, the chain $\{\Gamma_n\}$ should be recurrent. Otherwise we have to restrict the state space $\{1, \ldots, N\}$ for $\{\Gamma_n\}$ to get rid of transient states. Finally the transition probabilities $\widehat{p}_{ij}$ in (16) are the *reversed transition probabilities*

$$\widehat{p}_{ij} = p_{ji}\frac{\pi_j}{\pi_i}.$$

(see Kelly [9].)

Under suitable conditions ensuring convergence (described below) we deduce from the Ergodic Theorem that with probability one, for $f : \mathrm{IR}^m \to \mathrm{IR}$ continuous and bounded

$$\frac{1}{n+1}\sum_{k=0}^{n} f(\mathbf{X}_k)\delta_{\{i\}}(\Gamma_k) \to \int f(\mathbf{x})\nu(d\mathbf{x} \times \{i\}).$$

Equivalently, with probability one

$$\frac{1}{\pi_i}\frac{1}{n+1}\sum_{k=0}^{n}\delta_{\mathbf{X}_k}\delta_{\{i\}}(\Gamma_k) \Rightarrow \nu_i.$$

Algorithmically this amounts to *sampling the trajectory* $\{\mathbf{X}_n\}$ *only at those times when* $\Gamma_n = i$.

These same considerations show that if we consider instead the chain $\{(\mathbf{X}_n, \Gamma_{n+1})\}$ then we arrive at the convolution system

$$\lambda_i = \sum_{j=1}^{N}\widehat{p}_{ij}\mu_j * \lambda_j, \quad 1 \le i \le N. \tag{17}$$

Here, in order to generate λ_i, the sampling of $\{\mathbf{X}_n\}$ needs to be done at those times when $\Gamma_{n+1} = i$.

We have now seen algorithms for generating solutions for the different systems (15), (16), (17). It is easily seen that (16) and (17) are special cases of (15), and that (16) and (17) are equivalent to one another in the sense that a solution of one system can be used to provide a solution to the other. Indeed if the ν_i's satisfy (16) then the λ_i's defined by

$$\lambda_i = \sum_{j=1}^{N}\widehat{p}_{ij}\nu_j$$

satisfy (17). Moreover $\sum_{i=1}^{N}\pi_i\lambda_i = \sum_{i=1}^{N}\pi_i\nu_i$. Similarly if the λ_i's satisfy (17) then the ν_i's defined by

$$\nu_i = \mu_i * \lambda_i$$

satisfy (16).

It can also be shown that if the chain with transition matrix (p_{ij}) is recurrent, then (17) (or (16)) is equivalent to (15). We have to enlarge the state space from $\{1, \ldots, N\}$ to $\{1, \ldots, N\} \times \{1, \ldots, N\}$ and set up a Markov chain on pairs (i, j) corresponding to the chain on $\{1, \ldots, N\}$. Precisely, define

$$\widehat{p}_{(i,j)(i',j')} = \begin{cases} p_{jj'}, & \text{if } i' = j, \\ 0, & \text{otherwise.} \end{cases}$$

Then (17) written out leads to (15). This amounts to *setting up the states for the recurrent IFS to correspond to the screen transitions of the mixed IFS.* On the other hand if the recurrence condition is violated then (15) can no longer be obtained through (16) or (17). Indeed for the mixing in Proposition VI the transition matrix is given by

$$\begin{bmatrix} 1 & 0 \\ p & q \end{bmatrix}$$

and so $\{\Gamma_n\}$ is always absorbed at state one.

Finally we briefly mention the important extension of *place dependent probabilities.* The motivation is to allow the probabilities for choosing the various affine transformations to depend on the current position $\mathbf{x}$ of the orbit. Thus in different regions of the image one could weight the transformations differently. Of course the place dependent functions used for the probabilities would themselves have to be specified in terms of a small number of parameters, or else the data compression would be lost. The important case of piecewise constant functions for the probabilities can be used to control the symbolic dynamics, similar to the recurrent IFS. Thus by selectively zeroing out probabilities we can set different rules in different regions of the image for which transformations are to be used. An analysis of the place dependent IFS appears in [5].

§2 Mixing Theory

As above let $\{\Gamma_n\}$ be a Markov chain on $\{1, \ldots, N\}$ with transition matrix (p_{ij}), and let $\mu_{ij} \in \mathcal{P}(G)$ be given, $1 \leq i,j \leq N$. Let $\{g_n(i,j)\}$ be i.i.d. samples from μ_{ij}, the N^2 samples being mutually independent. We are concerned with existence, uniqueness and generation of solutions for the system of convolution equations

$$\sum_{j=1}^{N} p_{ij}\mu_{ij} * \nu_j = \nu_i, \quad 1 \leq i \leq N. \tag{18}$$

Initialize $\Gamma_0 = i$ and let $T = T(i)$ be the return time of the chain $\{\Gamma_n\}$ to state i. Let $\mu = \mu(i)$ be the distribution of the product $\prod_{k=1}^{T} g_k(\Gamma_{k-1}, \Gamma_k)$. We refer to μ as the distribution of *the product of the g's taken over an excursion from state i*. The product symbol $\prod$ denotes an ordered left-to-right product. That is, $\prod_{k=1}^{n} g_k = g_1 \cdots g_n$ and $\prod_{k=n}^{1} g_k = g_n \cdots g_1$.

To analyze existence and uniqueness for (18) we do not need to introduce the process $\{\mathbf{X}_n(1), \ldots, \mathbf{X}_n(N))\}$ described above in §1. Instead there is a reduction technique for the chain $\{\Gamma_n\}$ which can be used. Suppose state 1 is not absorbing, so that $p_{11} < 1$. The reduction amounts to eliminating state 1 from the state space by simply stopping the time clock whenever the chain $\{\Gamma_n\}$ is at state 1. Thus if the original chain is at state $i \neq 1$ and jumps to state 1, possibly remaining there for a while, and then jumps from 1 to $j \neq 1$ — we will say that the reduced chain makes a transition from i to j. Effectively this is just filtering out the 1's from the trajectory of $\{\Gamma_n\}$. Accordingly the transition probabilities (p'_{ij}) for this reduced chain are given by

$$\begin{aligned} p'_{ij} &= \mathbb{P}(i \to j|\text{not via state } 1)\mathbb{P}(\text{not via state } 1) + \mathbb{P}(i \to j|\text{via state } 1)\mathbb{P}(\text{via state } 1) \\ &= p_{ij} + \frac{p_{i1}p_{1j}}{1 - p_{11}} \end{aligned} \tag{19}$$

Observe that the posterior probability is given by

$$\theta_{ij} = \mathbb{P}(\text{not via state } 1|i \to j) = \frac{p_{ij}(1 - p_{11})}{p_{ij}(1 - p_{11}) + p_{i1}p_{1j}} \tag{20}$$

Theorem VII. *Assume that the μ_{ij}'s satisfy (A_1) for all i,j with $p_{ij} > 0$. Assume in addition that the excursion product distributions $\mu(i)$ obey (A_2) for all recurrent states i. Then the system (1) has a unique solution.*

Proof. If $p_{11} = 1$ then ν_1 must be the unique μ_{11}-stationary distribution. Otherwise if $p_{11} < 1$ then use Lemma V to iterate the first of eqns. (18), obtaining

$$\nu_1 = \sum_{j=2}^{N} \frac{p_{1j}}{1 - p_{11}} \lambda_{ij} * \nu_j, \tag{21}$$

where

$$\lambda_{ij} = \left[\sum_{k=0}^{\infty}(1-p_{11})p_{11}^k\mu_{11}^{(k)}\right] * \mu_{1j}.$$

Substituting (21) back into (18) we obtain

$$\nu_i = \sum_{j=2}^{N} p'_{ij}\mu'_{ij} * \nu_j, \quad 2 \le i \le N, \tag{22}$$

where p'_{ij} is the reduced probability from (19) above, and

$$\mu'_{ij} = \theta_{ij}\mu_{ij} + (1-\theta_{ij})\mu_{i1} * \lambda_{1j},$$

θ_{ij} being given by (20) above. The reduced system (22) corresponds to the reduced chain described above, and the distributions μ'_{ij} correspond to products $g_1(i,1)g_2(1,1)\cdots g_{k+1}(1,1)g_{k+2}(1,j)$ from i to j, possibly through state 1. As long as $p_{11} < 1$, so that 1 is not absorbing, this modification does not alter the classification of the remaining states, and the hypotheses of the Theorem remain valid. The proof is completed now by induction on N. □

Let $\{I_n(i)\}$ be i.i.d. sequences, $1 \le i \le N$, with

$$\mathbb{P}(I_n(i) = j) = p_{ij}, \quad 1 \le j \le N.$$

The N sequences $\{I_n(i)\}$ are to be mutually independent, and independent of all the g's. Our next result concerns the Markov chain evolving as

$$\mathbf{X}_{n+1}(i) = g_{n+1}(i,j)\mathbf{X}_n(j), \quad \text{where } j = I_{n+1}(i).$$

Theorem VIII. *Assume that the μ_{ij}'s satisfy (A_1) for all i,j with $p_{ij} > 0$. Suppose that state 1 is recurrent and that the excursion product distribution $\mu(1)$ obeys (A_2). Then $\mathbf{X}_n(1)$ converges in distribution to a random variable $\mathbf{X}$ with a $\mu(1)$-stationary distribution. Furthermore if $\rho_{i1} = \mathbb{P}_i(T(1) < \infty) = 1$ then $\mathbf{X}_n(i)$ converges in distribution to a random variable distributed like $\left[\prod_{k=1}^{T(1)} g'_k(\Gamma_{k-1},\Gamma_k)\right]\mathbf{X}$, where $\Gamma_0 = i$ and the g' sequences are independent of the g sequences and of the chain $\{\Gamma_n\}$.*

Proof. Take $\Gamma_0 = 1$ and set

$$\widehat{\mathbf{X}}_n = \widehat{\mathbf{X}}_n(1) = \left[\prod_{k=1}^{n} g_k(\Gamma_{k-1},\Gamma_k)\right]\mathbf{X}_0(\Gamma_n).$$

For each fixed n, $\mathbf{X}_n(1)$ and $\widehat{\mathbf{X}}_n$ have the same distribution. Let $T_\ell = T_\ell(1)$ be the time of the ℓth return of $\{\Gamma_n\}$ to state 1 ($T_0 = 0$), and set $h_\ell = \prod_{k=T_{\ell-1}+1}^{T_\ell} g_k(\Gamma_{k-1},\Gamma_k)$ for $1 \le \ell \le N_n$, where $N_n = N_n(1)$ is the number of visits to state 1 up to time n. Then $\widehat{\mathbf{X}}_n = \left[\prod_{\ell=1}^{N_n} h_\ell\right]\mathbf{Y}_n$ where

$$\mathbf{Y}_n = \left[\prod_{k=S_n+1}^{n} g_k(\Gamma_{k-1},\Gamma_k)\right]\mathbf{X}_0(\Gamma_n),$$

and $S_n = T_{N_n}$ is the last visit to state 1 before time n. (If $\Gamma_n = 1$ so that $S_n = n$, then this is simply $\mathbf{Y}_n = \mathbf{X}_0(1)$.) The transformations $\{h_\ell\}$ are i.i.d., each distributed like $\mu(1)$.

If state 1 is recurrent then $\lim_{n\to\infty} \frac{N_n}{n} = \pi(1)$ a.s., where $\pi(1) > 0$ is the stationary probability for state 1, π being supported on the component to which state 1 belongs. Thus if $a_\ell = a(h_\ell)$ then

$$\lim_{n\to\infty} \frac{1}{n} \log |{}_1a_{N_n}| = \pi(1) \lim_{n\to\infty} \frac{1}{n} \log |{}_1a_n| < 0 \text{ a.s.,} \tag{23}$$

the last step by virtue of our assumption (A_2) on $\mu(1)$.

Consider the differences

$$\widehat{\mathbf{X}}_{n+1} - \widehat{\mathbf{X}}_n = \left[\prod_{\ell=1}^{N_n} h_\ell\right] \mathbf{Z}_n \tag{24}$$

where

$$\mathbf{Z}_n = \left[\prod_{k=S_n+1}^{n} g_k(\Gamma_{k-1},\Gamma_k)\right] [g_{n+1}(\Gamma_n,\Gamma_{n+1})\mathbf{X}_0(\Gamma_{n+1}) - \mathbf{X}_0(\Gamma_n)] .$$

Set

$$\alpha_k = \max_{i,j:p_{ij}>0} \log^+ |a(g_k(i,j))|$$

$$\beta_n = \max_{i,j:p_{ij}>0} \log |g_{n+1}(i,j)\mathbf{X}_0(j) - \mathbf{X}_0(i)|.$$

Then

$$\log |\mathbf{Z}_n| \le \sum_{k=S_n+1}^{n} \alpha_k + \beta_n .$$

On account of assumptions (A_1), $\mathbb{E}\,\alpha_k < \infty$ and if we take $\mathbf{X}_0(i) = \mathbf{x}_i$, $1 \le i \le N$, then $\mathbb{E}\,\beta_n < \infty$. Since state 1 is recurrent $\mathbb{E}\,T(1) < \infty$ and so $\frac{S_n}{n} \to 1$ a.s. We conclude therefore that

$$\lim_{n\to\infty} \frac{1}{n} \log^+ |\mathbf{Z}_n| = 0 \text{ a.s.}$$

Using this together with (23) and (24) we see that $\widehat{\mathbf{X}}_n \to \mathbf{X}$ a.s., where $\mathbf{X}$ has the $\mu(1)$-stationary distribution. Thus $\mathbf{X}_n(1) \Rightarrow \mathbf{X}$.

If $\Gamma_0 = i$ then $\mathbf{X}_n(i)$ has the same distribution as

$$\widehat{\mathbf{X}}_n(1) = \begin{cases} \prod_{k=1}^{T(1)} g'_k(\Gamma_{k-1},\Gamma_k)\widehat{\mathbf{X}}_{n-T(1)}, & T(1) \le n, \\ \prod_{k=1}^{n} g'_k(\Gamma_{k-1},\Gamma_k)\widehat{\mathbf{X}}_0, & T(1) > n. \end{cases}$$

Since $\rho_{i1} = 1$ we know that $T(1) < \infty$ a.s. Thus from $\widehat{\mathbf{X}}_n \to \mathbf{X}$ a.s. there follows

$$\widehat{\mathbf{X}}_n(i) \to \left[\prod_{k=1}^{T(1)} g'_k(\Gamma_{k-1},\Gamma_k)\right] \mathbf{X} \text{ a.s.}$$

□

The results in §§1,2 about the convolution equation (2) and the system (18) extend to a general setting where G is a general topological semi-group acting on a complete separable metric space $(\mathcal{X}, \rho)$. The key assumptions sufficient to guarantee existence of a unique solution (2) are:

FINITE LOGARITHMIC MOMENT:

$$\mathbb{E} \log^{+} \rho(g\mathbf{x}, \mathbf{x}) < \infty, \quad \forall \mathbf{x} \in \mathcal{X}$$

COUPLING CONDITION:

$$\lim_{n\to\infty} \frac{1}{n} \log \sup_{\mathbf{x} \neq \mathbf{y}} \frac{\rho({}_n g_1 \mathbf{x}, {}_n g_1 \mathbf{y})}{\rho(\mathbf{x}, \mathbf{y})} < 0, \quad \forall \mathbf{x}, \mathbf{y} \in \mathcal{X}$$

Under these assumptions the arguments in the Proof of Prop. II carry through (see the discussion in [3]), as does the discussion of systems (18) in this section.

We close with 1-D examples.

EXAMPLE I: μ is atomic with

$$\mu(\{x \mapsto a_i x + b_i\}) = p_i \quad (a_i > 0)$$

Then if F is the d.f. for ν

$$F(x) = \sum_i p_i F\left(\frac{x - b_i}{a_i}\right).$$

The coupling condition becomes

$$\sum_i p_i \log a_i < 0.$$

For instance if

i	a_i	b_i	p_i
1	1/2	0	1/4
2	1/2	1/2	1/2
3	1/2	1	1/4

then

$$F(x) = \frac{1}{4} F(2x) + \frac{1}{2} F(2x - 1) + \frac{1}{4} F(2x - 2)$$

and the unique d.f. F is that for the sum of two independent $U(0,1)$'s. Or if

i	a_i	b_i	p_i
1	1/3	0	1/2
2	1/3	2/3	1/2

then

$$F(x) = \frac{1}{2} F(3x) + \frac{1}{2} F(3x - 2)$$

and F is the classical Cantor function, which is continuous, but only grows on the Cantor set.

EXAMPLE II: $g : x \mapsto ax + b$, where $a = a(g) > 0$ is fixed and $b = b(g)$ has d.f. Φ. Then

$$F(y) = \int_{-\infty}^{\infty} F\left(\frac{y - x}{a}\right) d\Phi(x).$$

The coupling condition is simply $a < 1$. For instance if $\Phi \sim \mathcal{N}(0, \sigma^2)$ then $F \sim \mathcal{N}\left(0, \dfrac{\sigma^2}{1 - a^2}\right)$.

§3. Encoding: Convexity Considerations

For the encoding problem one is given a target digital image, say the leaf in Fig. 1, and asked to construct affine transformations T_i, $1 \le i \le N$, and weights p_i which generate it. In this setting the measure $\mu \in \mathcal{P}(G)$ from §1 is the atomic measure

$$\mu(\{T_i\}) = p_i \quad 1 \le i \le N.$$

The stationarity condition (2) becomes

$$\nu(B) = \sum_{i=1}^{N} p_i \nu(T_i^{-1} B), \tag{25}$$

and the collage property (3) is

$$C = \overline{\bigcup_{i=1}^{N} T_i C} \tag{26}$$

When C is compact (as in Lemma I) then we can drop the closure sign from the right-hand side of (26).

For subsets $C \subseteq \mathbb{R}^m$ we denote by conv C the convex hull of C, and by ext C the set of extreme points of conv C. Observe that ext $C \subseteq \overline{C}$.

Lemma IX. *(i) For subsets $C_1, \ldots, C_N \subseteq \mathbb{R}^m$*

$$\operatorname{conv} \bigcup_{i=1}^{N} C_i \supseteq \bigcup_{i=1}^{N} \operatorname{conv} C_i.$$

(ii) For subsets $C_1, \ldots, C_N \subseteq \mathbb{R}^m$

$$\operatorname{ext} \bigcup_{i=1}^{N} C_i \subseteq \bigcup_{i=1}^{N} \operatorname{ext} C_i.$$

(iii) Let $T : \mathbb{R}^m \to \mathbb{R}^m$ be an affine transformation, and let $C \subseteq \mathbb{R}^m$. Then

$$\operatorname{conv} TC = T(\operatorname{conv} C).$$

(iv) Let $T : \mathbb{R}^m \to \mathbb{R}^m$ be an affine transformation, and let $C \subseteq \mathbb{R}^m$ be compact. Then

$$\operatorname{ext} TC \subseteq T(\operatorname{ext} C).$$

Proof. (i) Use the fact that if $A \subseteq B$ then conv $A \subseteq$ conv B.
(ii) Use the facts that
(a) if $A \subseteq B$ then ext $B \cap \overline{\operatorname{conv} A} \subseteq$ ext A; and
(b) $\operatorname{conv} \bigcup_{i=1}^{N} C_i = \operatorname{conv} \bigcup_{i=1}^{N} \operatorname{conv} C_i$.
Thus $\operatorname{ext} \bigcup_{i=1}^{N} C_i = \operatorname{ext} \bigcup_{i=1}^{N} \operatorname{conv} C_i \subseteq \bigcup_{i=1}^{N} \overline{\operatorname{conv} C_i}$; so that

$$\operatorname{ext} \bigcup_{i=1}^{N} C_i = \bigcup_{i=1}^{N} \left(\operatorname{ext} \bigcup_{i=1}^{N} C_j \cap \overline{\operatorname{conv} C_i} \right) \subseteq \bigcup_{i=1}^{N} \operatorname{ext} C_i.$$

(iii) Observe that if $\mathbf{x}_1, \ldots, \mathbf{x}_n \in \mathbb{R}^m$ and if $\sum_{j=1}^n \alpha_j = 1$ then

$$T\left(\sum_{j=1}^n \alpha_j \mathbf{x}_j\right) = \sum_{j=1}^n \alpha_j T\mathbf{x}_j.$$

(iv) Let $\mathbf{x} \in \text{ext}\, TC$ and define the compact convex set

$$S = \{\mathbf{y} \in \text{conv}\, C : T\mathbf{y} = \mathbf{x}\}.$$

Since ext $S \neq \emptyset$ it suffices to show that ext $S \subseteq$ ext C. To argue this let $\mathbf{z} \in$ ext S, and suppose we could write $\mathbf{z} = \alpha\mathbf{y}_1 + (1-\alpha)\mathbf{y}_2$ where $\mathbf{y}_1, \mathbf{y}_2 \in$ conv C, $\mathbf{y}_1 \neq \mathbf{z}$, $0 < \alpha < 1$. Then applying T to both sides of this equation, and using the fact that $\mathbf{x} \in$ ext TC we conclude that $\mathbf{y}_1, \mathbf{y}_2 \in S$. But this contradicts the fact that $\mathbf{z} \in$ ext S. □

Corollary X. *Suppose $C \subseteq \mathbb{R}^m$ satisfies*

$$C = \bigcup_{i=1}^N T_i C, \tag{27}$$

where $T_1, \ldots, T_N : \mathbb{R}^m \to \mathbb{R}^m$ are affine transformations. Then

(i)
$$\text{conv}\, C \supseteq \bigcup_{i=1}^N T_i(\text{conv}\, C);$$

(ii) if C is compact

$$\text{ext}\, C \subseteq \bigcup_{i=1}^N T_i(\text{ext}\, C).$$

Proof. (i) Use Lemma IX(i) and IX(iii).

(ii) Use Lemma IX(ii) and IX(iv). □

It follows from Corollary X that there exists an injective mapping $\psi = (\psi', \psi'') : E \to E \times \{1, \ldots, N\}$, where $E = \text{ext}\, C$, with the following property.

(P) There exist *affine* transformations $T_1, \ldots, T_N : \mathbb{R}^m \to \mathbb{R}^m$ such that $\psi(\mathbf{e}) = (\mathbf{e}', i) \Rightarrow T_i\mathbf{e}' = \mathbf{e}$.

The affine-ness in (P) is what makes it restrictive. In fact the condition (P) can be stated in a form independent of the transformations T_i. Namely

(P′) If $\psi''(\mathbf{e}_1) = \cdots = \psi''(\mathbf{e}_n)$ then

$$\text{rank}\left[\begin{array}{c|c|c} \psi'(\mathbf{e}_2) - \psi'(\mathbf{e}_1) & \cdots & \psi'(\mathbf{e}_n) - \psi'(\mathbf{e}_1) \\ \mathbf{e}_2 - \mathbf{e}_1 & & \mathbf{e}_n - \mathbf{e}_1 \end{array}\right]$$
$$= \text{rank}\left[\psi'(\mathbf{e}_2) - \psi'(\mathbf{e}_1) | \cdots | \psi'(\mathbf{e}_n) - \psi'(\mathbf{e}_1)\right].$$

(On the left we have a $2m \times (n-1)$ matrix and on the right we have an $m \times (n-1)$ matrix.) We shall say that an injective $\psi = (\psi', \psi'') : E \to E \times \{1, \ldots, N\}$ satisfying (P′) is *E-compatible.*

Next define the set qext C of *quasi-extreme* points of C: qext $C = \overline{C} \cap \partial\, \text{conv}\, C$.

Proposition XI. *Let $C \subseteq \mathbb{R}^m$ be closed and satisfy (27), where $T_1, \ldots, T_N : \mathbb{R}^m \to \mathbb{R}^m$ are affine transformations. Assume that for each i, range T_i is not supporting for C, so that*

$$\text{range } T_i \cap \overline{\text{conv } C} \not\subseteq \partial \text{ conv } C, \quad 1 \le i \le N. \tag{28}$$

Then

$$\text{qext } C \subseteq \bigcup_{i=1}^{N} T_i(\text{qext } C).$$

Proof. Let $\mathbf{y} \in$ qext C. Then $\mathbf{y} = T_i\mathbf{x}$ for some $\mathbf{x} \in C$ and some $1 \le i \le N$. For any neighborhood U of $\mathbf{x}$, T_iU contains points exterior to conv C. It thus follows from Lemma IX(iii) that $\mathbf{x} \in \partial$ conv C. □

The significance of this result is that under the additional assumption (28) we find a Q-compatible $\psi : Q \to Q \times \{1, \ldots, N\}$, where $Q =$ qext C. Typically qext C is much larger than ext C, so that the existence of a Q-compatible ψ provides much more information about the underlying transformations T_i than does the existence of an E-compatible ψ. For example in Figures 1 and 9 one can see, by applying a straight-edge to the images, the additional points that qext C contains which do not belong to ext C. The assumption (28) is essential, as the examples in Figures 10 and 11 indicate.

Lemma XII. *For any transformations $T_1, \ldots, T_N : \mathbb{R}^m \to \mathbb{R}^m$ and any subset $C \subseteq \mathbb{R}^m$, the following are equivalent:*
(a) $C = \bigcup_{i=1}^{N} T_iC$;
(b) For any subset $H \subseteq \mathbb{R}^m$

$$H \cap C = \phi \Leftrightarrow T_i^{-1}H \cap C = \phi, \quad 1 \le i \le N.$$

(c) For any subset $H \subseteq \mathbb{R}^m$

$$H \supseteq C \Leftrightarrow T_i^{-1}H \supseteq C, \quad 1 \le i \le N.$$

Proof. (a) ⇒ (b) Use the fact that

$$H \cap TC = T(T^{-1}H \cap C),$$

for any transformation $T : \mathbb{R}^m \to \mathbb{R}^m$.
(b) ⇒ (c) Apply (b) to the complement H^c.
(c) ⇒ (a) Apply (c) to $H = C$ and $H = \bigcup_{i=1}^{N} T_iC$. □

Proposition XIII. *Let $C \subseteq \mathbb{R}^m$ be a closed set satisfying (27), where $T_1, \ldots, T_N : \mathbb{R}^m \to \mathbb{R}^m$ are affine transformations. If H is a supporting hyper-plane for C then for some i either range $T_i \subseteq H$, or else $T_i^{-1}H$ is also a supporting hyper-plane for C.*

Proof. Let S be the open half-space in $\mathbb{R}^m$ disjoint from C, with $\partial S = H$. Then $S \cap C = \phi$ but $\overline{S} \cap C \ne \phi$. It follows from Lemma XII(b) that for some i, $T_i^{-1}S \cap C = \phi$, but $T_i^{-1}\overline{S} \cap C \ne \phi$. If range $T_i \not\subseteq H$ then $T_i^{-1}S$ is also a half-space, and $T_i^{-1}\overline{S} = \overline{T_i^{-1}S}$. □

Proposition XIV. *Let $C \subseteq \mathbb{R}^m$ be a closed set satisfying (27), where $T_1, \ldots, T_N : \mathbb{R}^m \to \mathbb{R}^m$ are affine transformations. Let $L, L' \subseteq \partial$ conv C and suppose $L' = T_i L$ for some i. Assume that range T_i is not supporting for C, so that*

$$\text{range } T_i \cap \overline{\text{conv } C} \not\subseteq \partial \text{ conv } C.$$

Then

$$\text{conv } L' \subseteq \partial \text{ conv } C \Rightarrow \text{ conv } L \subseteq \partial \text{ conv } C.$$

Proof. Suppose conv $L' \subseteq \partial$ conv C but conv $L \not\subseteq \partial$ conv C. Then conv L contains a point $\mathbf{x}$ in the interior of conv C. Let H be a supporting hyperplane for C containing conv L'. Then (using Lemma IX(iii))

$$T_i^{-1} H \supseteq T_i^{-1} \text{ conv } L' \subseteq \text{ conv } L.$$

Thus $\mathbf{x} \in T_i^{-1} H$, and this contradicts Proposition XIII. □

We shall say that points $\mathbf{e}_1, \ldots, \mathbf{e}_n \in \partial$ conv C are *mutually adjacent* if conv$\{\mathbf{e}_1, \ldots, \mathbf{e}_n\} \subseteq \partial$ conv C. For example if $n = 2$ then $\mathbf{e}_1, \mathbf{e}_2 \in \partial$ conv C are adjacent if the interval $[\mathbf{e}_1, \mathbf{e}_2] \subseteq \partial$ conv C. From Proposition XIV we have additional information about possible E-compatible or Q-compatible mappings $\psi = (\psi', \psi'')$. Namely under assumption (28)

(P_2) If $\psi''(\mathbf{e}_2) = \cdots = \psi''(\mathbf{e}_n)$ and if $\mathbf{e}_1, \ldots, \mathbf{e}_n$ are mutually adjacent, then $\psi'(\mathbf{e}_1), \ldots, \psi'(\mathbf{e}_n)$ must also be mutually adjacent.

We refer to this property as *adjacency-preserving.* Thus under assumption (28) any E-compatible or Q-compatible ψ must also be adjacency-preserving. Once again assumption (28) is essential, as the example in Figure 12 indicates.

REFERENCES

[1] Barnsley, M. F., *Fractals Everywhere*, Academic Press, New York, 1988.

[2] Barnsley, M. F., Berger, M. A. and Soner, H. M., Mixing Markov chains and their images, *Prob. Eng. Inf. Sci.* 2 (1988), 387–414.

[3] Barnsley, M. F., Demko, S. G., Elton, J. and Geronimo, J. S., Invariant measures for Markov processes arising from function iteration with place-dependent probabilities, *Ann. Inst. H. Poincaré*, 24 (1988), 367–394.

[4] Barnsley, M. F., Elton, J. H. and Hardin, D. P., Recurrent iterated function systems, *Const. Approx.* 5 (1989), 3–31.

[5] Berger, M. A. and Soner, H. M., Random walks generated by affine mappings, *J. Theor. Prob.* 1 (1988), 239–254.

[6] Billingsley, P., *Convergence of Probability Measures*, John Wiley & Sons, Inc., New York, 1965.

[7] Breiman, L., *Probability*, Addison-Wesley, Reading, Maassachusetts, 1968.

[8] Furstenberg, H. and Kesten, H., Products of random matrices, *Ann. Math. Stat.* 31 (1960) 457–469.

[9] Kelly, F. P., *Reversibility and Stochastic Networks*, John Wiley & Sons, Inc., New York, 1979.

[10] Kingman, J. F. C., Subadditive ergodic theory, *Ann. Prob.* 1 (1973), 883–909.

Marc A. Berger
School of Mathematics
Georgia Institute of Technology
Atlanta, GA 30332

$$T_1(x) = \begin{pmatrix} 0.8 & 0 \\ 0 & 0.8 \end{pmatrix} x + \begin{pmatrix} 0.1 \\ 0.04 \end{pmatrix} \qquad T_2(x) = \begin{pmatrix} 0.5 & 0 \\ 0 & 0.5 \end{pmatrix} x + \begin{pmatrix} 0.25 \\ 0.4 \end{pmatrix}$$

$$T_3(x) = \begin{pmatrix} 0.355 & -0.355 \\ 0.355 & 0.355 \end{pmatrix} x + \begin{pmatrix} 0.266 \\ 0.078 \end{pmatrix} \qquad T_4(x) = \begin{pmatrix} 0.355 & 0.355 \\ -0.355 & 0.355 \end{pmatrix} x + \begin{pmatrix} 0.378 \\ 0.434 \end{pmatrix}$$

$$p_1 = 0.5 \quad p_2 = 0.168 \quad p_3 = 0.166 \quad p_4 = 0.166$$

Figure 1: Maple Leaf

This image corresponds to the stationary distribution of a 2-D Markov chain. (The window here is $0 \leq x, y \leq 1$.)

Image under w_1

Image under w_2

Image under w_3

Image under w_4

Figure 2: Self-Covering of the Maple Leaf

The black leaves form a collage of the grey background leaf.

$$T_1(x) = \begin{pmatrix} 0.5 & 0 \\ 0 & 0.5 \end{pmatrix} x + \begin{pmatrix} 0.25 \\ 0.5 \end{pmatrix} \qquad T_2(x) = \begin{pmatrix} 0.5 & 0 \\ 0 & 0.5 \end{pmatrix} x + \begin{pmatrix} 0.5 \\ 0 \end{pmatrix}$$

$$T_3(x) = \begin{pmatrix} 0.5 & 0 \\ 0 & 0.5 \end{pmatrix} x + \begin{pmatrix} 0 \\ 0 \end{pmatrix}$$

$$p_1 = 0.33 \quad p_2 = 0.33 \quad p_3 = 0.34$$

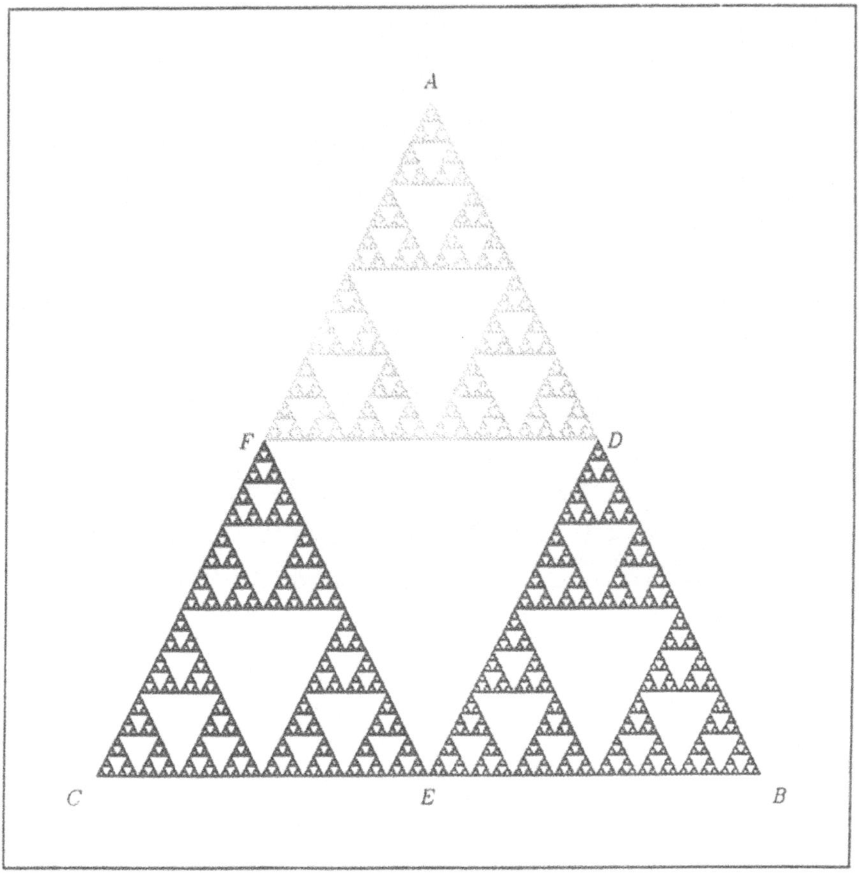

Figure 3: Self-Covering of the Sierpinski Triangle

The three triangular images ADF, FEC and DBE form a collage of the full image ABC. (The indow here is $0 \le x, y \le 1$.)

Figure 4

This image is a mixture of the leaf and triangle from Figures 1 and 3. It is collaged by three copies of itself and one copy of the leaf.

Schematic of the Simple Mixing Algorithm

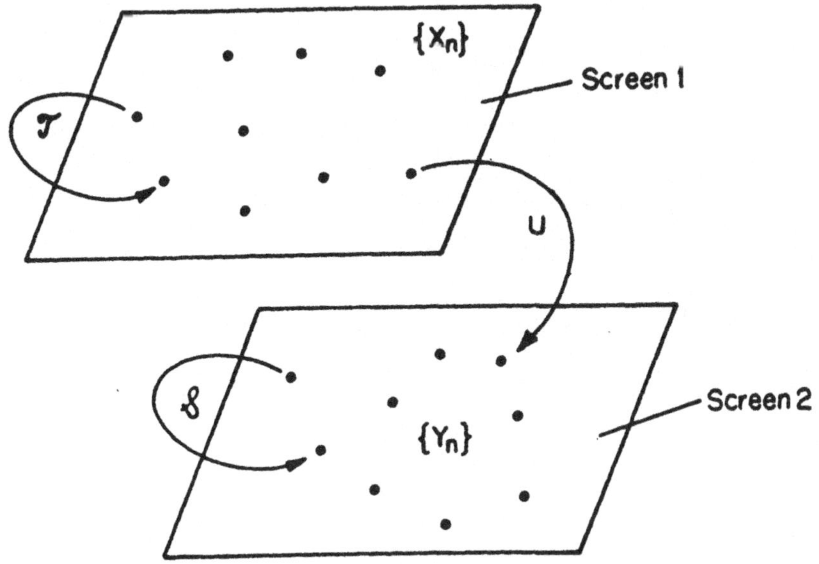

Figure 5

Only screen 2 gets plotted.

Simple Mixing Agorithm

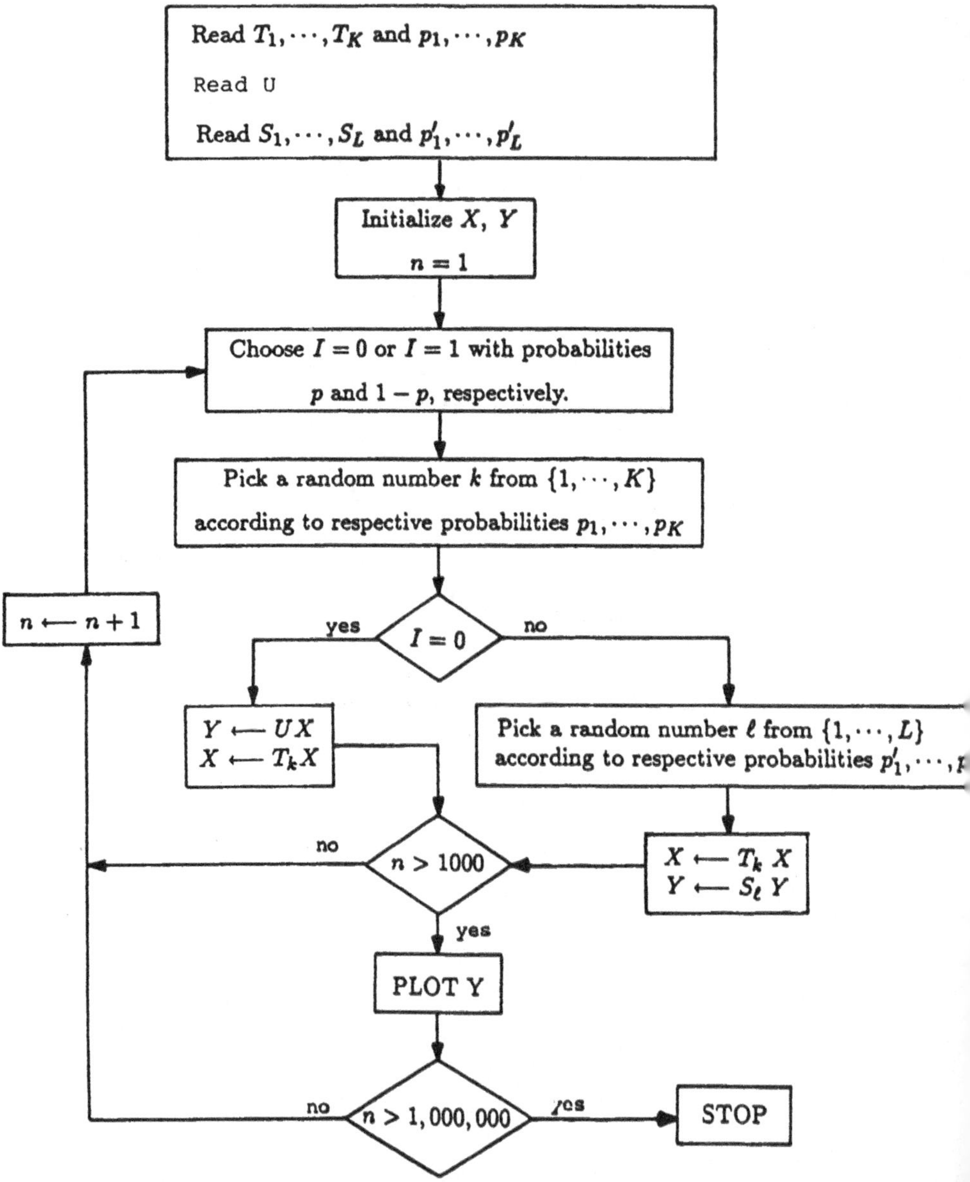

Figure 6

This algorithm mixes the individual images which are generated from the mappings T_i and

Figure 7

By generating this image as a mixture, one is able to introduce a 90° rotation in Screen 2 — even though it is not a strict contraction.

Figure 8
4-screen mixing schematic.
The switching is governed by

$$P = \begin{pmatrix} \frac{1}{3} & \frac{1}{3} & 0 & \frac{1}{3} \\ 0 & 0 & 1 & 0 \\ \frac{1}{2} & 0 & \frac{1}{2} & 0 \\ 0 & 0 & \frac{1}{2} & \frac{1}{2} \end{pmatrix}$$

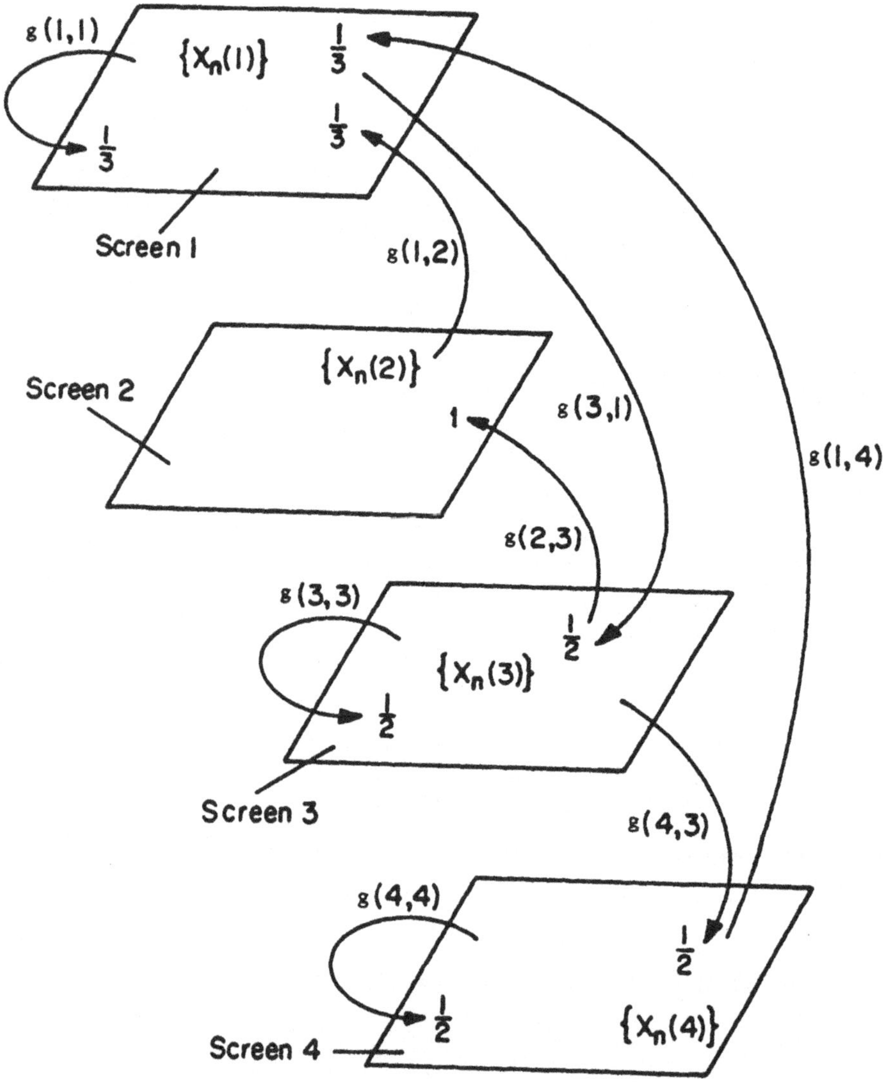

$$T_1(x) = \begin{pmatrix} 0.5 & 0.5 \\ -0.5 & 0.5 \end{pmatrix} x + \begin{pmatrix} 0.125 \\ 0.625 \end{pmatrix} \qquad T_2(x) = \begin{pmatrix} 0.5 & 0.5 \\ -0.5 & 0.5 \end{pmatrix} x + \begin{pmatrix} -0.125 \\ 0.375 \end{pmatrix}$$

$$p_1 = 0.5 \quad p_2 = 0.5$$

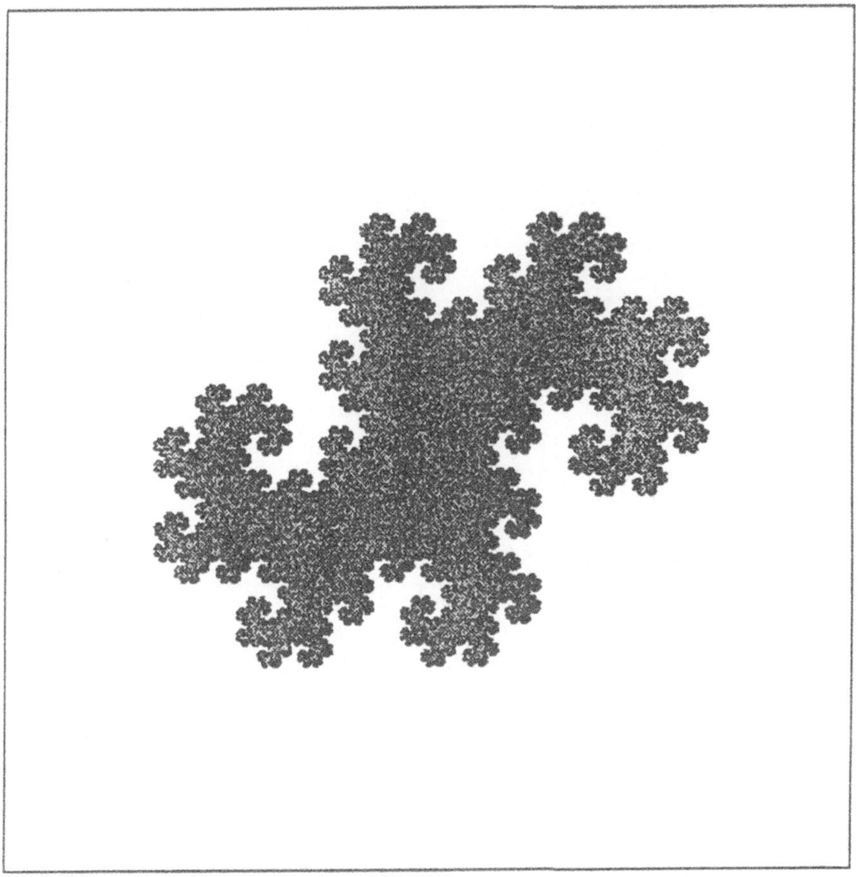

Figure 9

The interior angles of this convex hull are all 135°. (The window here is $-1 \leq x, y \leq 1$.)

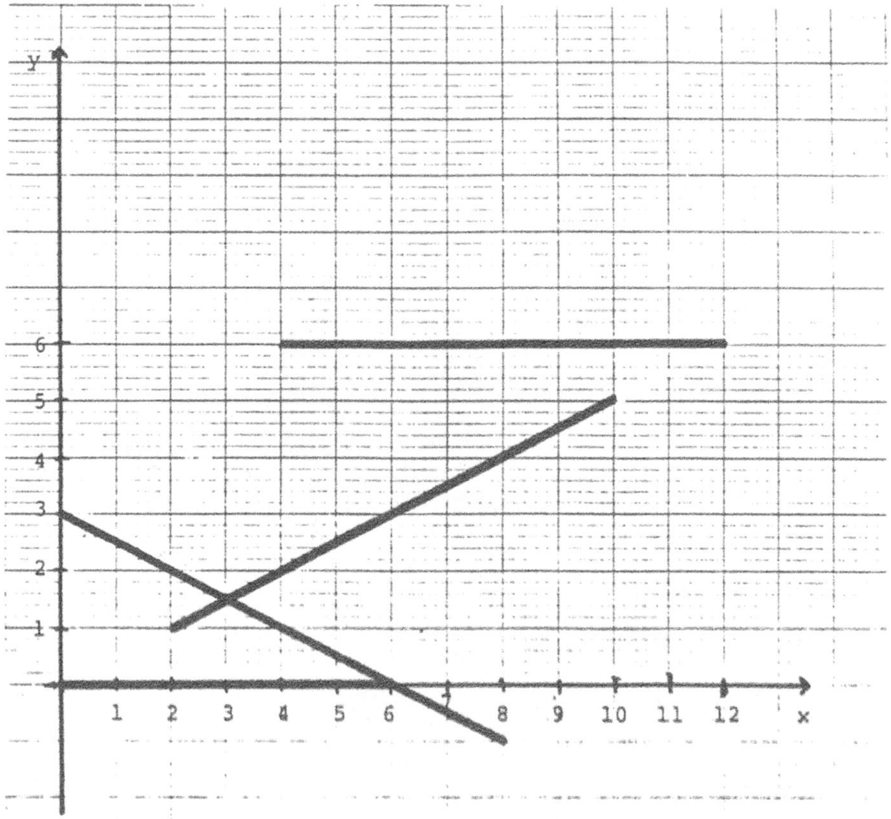

Figure 10

$$T_1(x,y) = \left(\frac{x}{2}, 0\right) \qquad T_3(x,y) = \left(\frac{2}{3}\,x + 2, 1 + \frac{1}{3}\,x\right)$$

$$T_2(x,y) = \left(\frac{2}{3}\,x, 3 - \frac{1}{3}\,x\right) \qquad T_4(x,y) = \left(\frac{2}{3}\,x + 4, 6\right)$$

The attractor C has the form $C = L_1 \cup L_2 \cup L_3 \cup L_4$ where each L_i is a line segment,

$$L_1 = [(0,0),(6,0)] \qquad L_3 = [(2,1),(10,5)]$$

$$L_2 = [(0,3),(8,-1)] \qquad L_4 = [(4,6),(12,6)]$$

The open segment $\left((4,6),(\frac{20}{3},6)\right)$ of L_4 belongs to $q\text{ext}\ C$, but its points are not images of any points from $q\text{ext}\ C$.

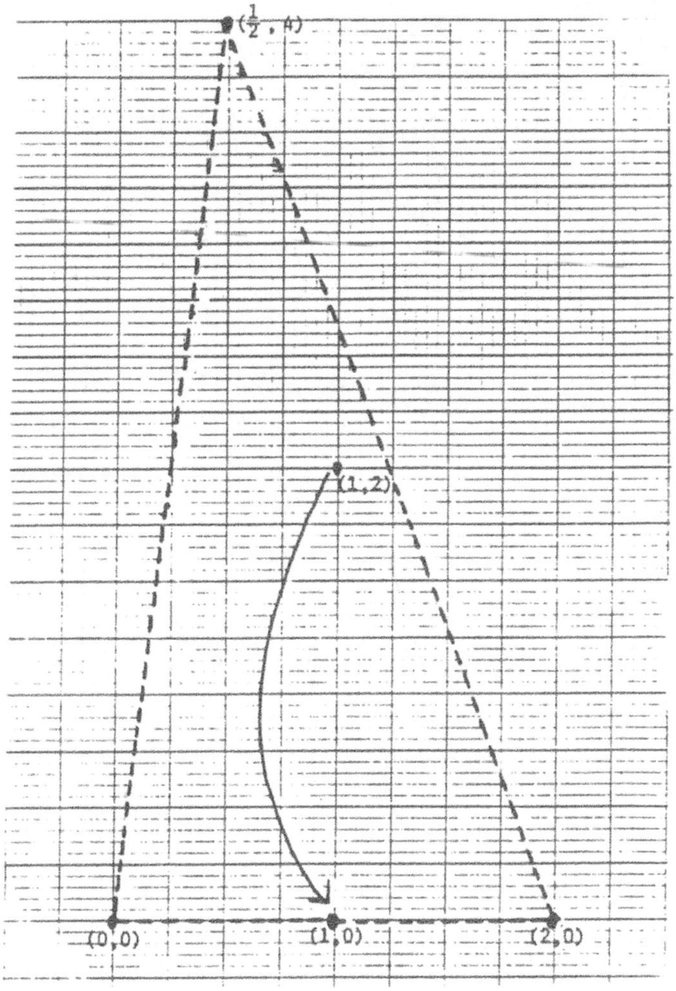

Figure 11

$$T_1(x,y) = \left(\frac{y}{2}, 0\right) \qquad T_2(x,y) = (1,2) \qquad T_3(x,y) = \left(\frac{1}{2}, 4\right)$$

Q-compatibility does not obtain in the case of degenerate maps. $(1,0) \in q\text{ext } C$, but its only antecedent is (1,2).

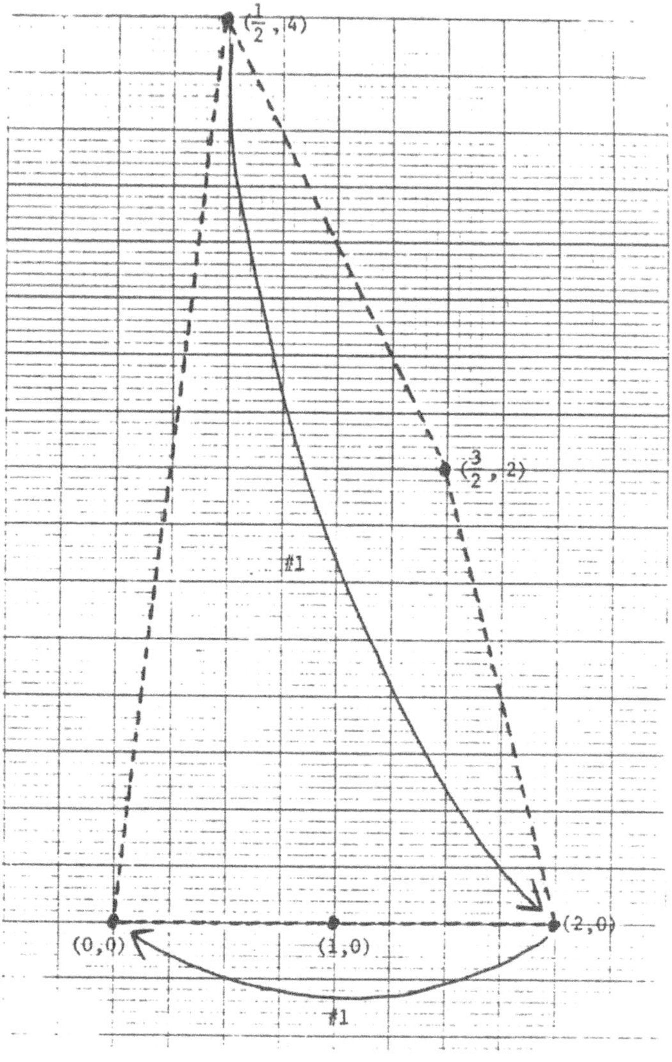

Figure 12

$$T_1(x,y) = \left(\frac{y}{2}, 0\right) \qquad T_3(x,y) = \left(\frac{1}{2}, 4\right)$$

$$T_2(x,y) = (0,3) \qquad T_4(x,y) = \left(\frac{3}{2}, 2\right)$$

Adjacency is not preserved for extreme points in the case of degenerate maps. Both (0,0) and (2,0) are assigned map # 1, but their antecedents are not adjacent extreme points.

Progress in Probability

Progress in Probability is designed for the publication of workshops, seminars and conference proceedings on all aspects of probability theory and stochastic processes, as well as their connections with and applications to other areas such as mathematical statistics and statistical physics. It acts as a companion series to *Probability and Its Applications*, a context for research level monographs and advanced graduate texts.

We encourage preparation of manuscripts in some form of TeX for delivery in camera-ready copy, which leads to rapid publication, or in electronic form for interfacing with laser printers or typesetters.

Proposals should be sent directly to the editors or to:
Birkhäuser Boston, 675 Massachusetts Avenue, Cambridge, MA 02139, U.S.A.

1 ÇINLAR/CHUNG/GETOOR. Seminar on Stochastic Processes, 1981
2 KESTEN. Percolation Theory for Mathematicians
3 ASMUSSEN/HERING. Branching Processes
4 CHUNG/WILLIAMS. Introduction to Stochastic Integration
5 ÇINLAR/CHUNG/GETOOR. Seminar on Stochastic Processes, 1982
6 BLOOMFIELD/STEIGER. Least Absolute Deviation
7 ÇINLAR/CHUNG/GETOOR. Seminar on Stochastic Processes, 1983
8 BOUGEROL/LACROIX. Products of Random Matrices with Application to Schrödinger Operator
9 ÇINLAR/CHUNG/GETOOR. Seminar on Stochastic Processes, 1984
10 KIFER. Ergodic Theory of Random Transformations
11 EBERLEIN/TAQQU. Dependence in Probability and Statistics
12 ÇINLAR/CHUNG/GETOOR. Seminar on Stochastic Processes, 1985
13 ÇINLAR/CHUNG/GETOOR/GLOVER. Seminar on Stochastic Processes, 1986
14 DEVROYE. A Course in Density Estimation
15 ÇINLAR/CHUNG/GETOOR/GLOVER. Seminar on Stochastic Processes, 1987
16 KIFER. Random Perturbations of Dynamical Systems
17 ÇINLAR/CHUNG/GETOOR/GLOVER. Seminar on Stochastic Processes, 1988
18 ÇINLAR/CHUNG/GETOOR/FITZSIMMONS/WILLIAMS. Seminar on Stochastic Processes, 1989
19 ALEXANDER/WATKINS. Spatial Stochastic Processes: A Festschrift in Honor of Ted Harris on His 70th Birthday
20 HAAGERUP/HOFFMANN-JØRGENSEN/NIELSEN. Probability in Banach Spaces 6: Proceedings of the Sixth International Conference, Sandbjerg, Denmark 1986

21 EBERLEIN/KUELBS/MARCUS. Probability in Banach Spaces 7: Proceedings of the Seventh International Conference
22 PINSKY. Diffusion Processes and Related Problems in Analysis, Volume I: Diffusions in Analysis and Geometry
23 HAHN, MASON, WEINER. Sums, Trimmed Sums and Extremes
24 ÇINLAR ET AL. Seminar on Stochastic Processes, 1990
25 CAMBANIS, SAMORODNITSKY AND TAQQU. Stable Processes and Related Topics
26 CRUZEIRO/ZAMBRINI. Stochastic Analysis and Applications: Proceedings of the 1989 Lisbon Conference
27 PINSKY/VIHSTUTZ. Diffusion Processes and Related Problems in Analysis, Vol. II. Stochastic Flows
28 DURRETT/KESTEN. Random Walks, Brownian Motion and Interacting Particle Systems

The manufacturer's authorised representative in the EU is Springer Nature Customer Service Centre GmbH, Europaplatz 3, 69115 Heidelberg, Germany. If you have any concerns regarding our products, please contact ProductSafety@springernature.com

Printed and bound by CPI Group (UK) Ltd, Croydon, CR0 4YY
23/04/2026
02095624-0005